TODAY'S TECHNICIAN™

CLASSROOM MANUAL

For Automotive Suspension & Steering Systems

SEVENTH EDITION

TODAY'S TECHNICIAN™

CLASSROOM MANUAL

For Automotive Suspension & Steering Systems

SEVENTH EDITION

Mark Schnubel

✳ Cengage

Australia • Brazil • Canada • Mexico • Singapore • United Kingdom • United States

Today's Technician: Automotive Suspension and Steering Systems, **Seventh Edition**
Mark Schnubel

SVP, GM Skills & Global Product Management: Jonathan Lau

Product Director: Matthew Seeley

Senior Product Manager: Katie McGuire

Product Assistant: Kimberly Klotz

Executive Director, Content Design: Marah Bellegarde

Learning Design Director: Juliet Steiner

Learning Designer: Mary Clyne

Vice President, Strategic Marketing Services: Jennifer Ann Baker

Marketing Director: Shawn Chamberland

Marketing Manager: Andrew Ouimet

Director, Content Delivery: Wendy Troeger

Manager, Content Deliver: Alexis Ferraro

Senior Content Manager: Meaghan Tomaso

Senior Digital Delivery Lead: Amanda Ryan

Senior Designer: Angela Sheehan

Service Provider/Compositor: SPi Global

Cover image(s): Photographicss/ShutterStock.com

For product information and technology assistance, contact us at
**Cengage Customer & Sales Support, 1-800-354-9706
or support.cengage.com.**

For permission to use material from this text or product, submit all requests online at **www.copyright.com.**

Library of Congress Control Number: 2018960546

Book Only ISBN: 978-1-337-56734-3
Package ISBN: 978-1-337-56733-6

Cengage
200 Pier 4 Boulevard
Boston, MA 02210
USA

Cengage is a leading provider of customized learning solutions with employees residing in nearly 40 different countries and sales in more than 125 countries around the world. Find your local representative at: **www.cengage.com.**

To learn more about Cengage platforms and services, register or access your online learning solution, or purchase materials for your course, visit **www.cengage.com.**

Printed at CLDPC, USA, 05-24

CONTENTS

PREFACE

Thanks to the support the *Today's Technician* series has received from those who teach automotive technology, Cengage Learning, the leader in automotive-related textbooks, is able to live up to its promise to regularly provide new editions of texts of this series. We have listened and responded to our critics and our fans and present this new updated and revised seventh edition. By revising this series on a regular basis, we can respond to changes in the industry, changes in technology, changes in the accreditation process, and to the ever-changing needs of those who teach automotive technology.

The *Today's Technician* series features textbooks that cover all mechanical and electrical systems of automobiles and light trucks. Principally, the individual titles correspond to the areas for National Institute for Automotive Service Excellence (ASE) certification.

All titles in the *Today's Technician* series include remedial skills and theories common to all of the certification areas and advanced or specific subject areas that reflect the latest technological trends.

Today's Technician: Automotive Suspension & Steering Systems, 7e is designed to give students a chance to develop the same skills and gain the same knowledge that today's successful technician has. This edition also reflects the changes in the guidelines established by the ASE Education Foundation in 2017.

The purpose of the ASE Education Foundation is to evaluate technician training programs against standards developed by the automotive industry and recommend qualifying programs for accreditation by ASE. Programs can earn ASE accreditation upon the recommendation of the ASE Education Foundation. ASE Education Foundation's national standards reflect the skills that students must master. ASE Education Foundation accreditation ensures that the training programs meet or exceed industry-recognized, uniform standards of excellence.

The technician of today and for the future must know the underlying theory of all automotive systems and be able to service and maintain those systems. Dividing the material into two volumes, a Classroom Manual and a Shop Manual, provides the reader with the information needed to begin a successful career as an automotive technician without interrupting the learning process by mixing cognitive and performance learning objectives into one volume.

The design of Cengage Learning's *Today's Technician* series was based on features that are known to promote improved student learning. The design was further enhanced by a careful study of survey results, in which the respondents were asked to value particular features. Some of these features can be found in other textbooks, while others are unique to this series.

Each Classroom Manual contains the principles of operation for each system and subsystem. The Classroom Manual also discusses design variations in key components used by the different vehicle manufacturers, and considers emerging technologies that will be standard or optional features in the near future. This volume is organized to build upon basic facts and theories. Its primary objective is to help the reader gain an understanding of how each system and subsystem operates. This understanding is necessary to diagnose the complex automobiles of today and tomorrow. Although the basics contained in the Classroom Manual provide the knowledge needed for diagnostics, diagnostic procedures appear only in the Shop Manual. An understanding of the underlying theories is also a requirement for competence in the skill areas covered in the Shop Manual.

A spiral-bound Shop Manual delivers hands-on learning experiences with step-by-step instructions for diagnostic and repair procedures. Photo Sequences are used to illustrate some of the common service procedures. Other common procedures are listed and are accompanied with fine-line drawings and photos that let the reader visualize and conceptualize the finest details of the procedure. This volume explains the reasons for performing the procedures, as well as the circumstances when each particular service is appropriate.

The two volumes are designed to be used together and are arranged in corresponding chapters. Not only are the chapters in the volumes linked together, the contents of the chapters are also linked. The linked content is indicated by marginal callouts that refer the reader to the chapter and page where the same topic is addressed in the companion volume. This valuable feature saves users the time and trouble of searching the index or table of contents to locate supporting information in the other volume. Instructors will find this feature especially helpful when planning the presentation of material and when making reading assignments.

Both volumes contain clear and thoughtfully selected illustrations, many of which are original drawings or photos specially prepared for inclusion in this series. This means that the art is a vital part of each textbook and not merely inserted to increase the number of illustrations.

The page design of this series uses available margin space to deliver helpful information efficiently without interrupting the pedagogical lesson material. This information includes examples of concepts just introduced in the text, explanations or definitions of terms that are not defined in the text, examples of common trade jargon used to describe a part or an operation, and unique applications of the system or service described in the text. Many textbooks also include this information but insert it in the main body of text; this tends to interrupt the reader's thought process. By placing this information to the side of the main text, students can read through the text uninterrupted and refer to the additional information when it is best for them.

Jack Erjavec
Series Editor

HIGHLIGHTS OF THIS EDITION—CLASSROOM MANUAL

The text was updated to include the latest technology in suspension and steering systems. Updated and expanded coverage of basic electrical theory and hybrid electric vehicle safety, hybrid vehicle steering systems, driver assist systems, active steering systems, rear active steering (RAS), CAN bus networking, computer-controlled suspension systems, and adaptive cruise control systems. Expanded coverage of tires and wheels including tire plus sizing and rim offset considerations as well as expanded alignment theory coverage and sequencing of information. The text also includes the latest technology in driver assist systems including but not limited to vehicle stability control systems, traction control systems, active roll control, lane departure warning (LDW) systems, active cruise control, collision mitigation systems, telematics, and tire pressure monitoring systems (TPMS).

PREFACE

The first chapter explains the design and purpose of basic suspension and steering systems. This chapter provides students with the necessary basic understanding of suspension and steering systems. The other chapters in the book allow the student to build upon his or her understanding of these basic systems.

The second chapter explains all the basic theories required to understand the latest suspension and steering systems described in the other chapters. Students must understand these basic theories to comprehend the complex systems explained later in the text.

The other chapters in the book are designed to be stand alone to allow reordering of topics covered to fit individual program needs and explain all the current model systems and components such as wheel bearings, tires and wheels, shock absorbers and struts, front and rear suspension systems, computer-controlled suspension systems, steering columns and linkages, power steering pumps, steering gears and systems, updated information on four-wheel steering systems currently in production, frames, and four-wheel alignment. All of the art pieces have been replaced with color photos and color diagrams throughout the text to improve visual concepts of suspension and steering systems and components.

HIGHLIGHTS OF THIS EDITION—SHOP MANUAL

The chapters in the Shop Manual have been updated to explain the diagnostic and service procedures for the latest systems and components described in the Classroom Manual. Diagnostics is a very important part of an automotive technician's job. Therefore, proper diagnostic procedures are emphasized in the Shop Manual.

All Photo Sequences are in color with several new sequences added and existing sequences updated. These Photo Sequences illustrate the correct diagnostic or service procedure for a specific system or component. These Photo Sequences allow the students to visualize the diagnostic or service procedure. Visualization of these diagnostic and service procedures helps students to remember the procedures, and perform them more accurately and efficiently. The text covers the information required for the ASE test in Suspension and Steering Systems. New and updated Job Sheets have been created to meet current ASE Education Foundation tasks.

Chapter 1 explains the necessary safety precautions and procedures in an automotive repair shop. General shop safety and the required shop safety equipment are explained in the text. The text describes safety procedures when operating vehicles and various types of automotive service equipment. Correct procedures for handling hazardous waste materials are detailed in the text.

Chapter 2 describes suspension and steering diagnostic and service equipment and the use of service manuals. This chapter also explains employer and employee obligations and ASE certification requirements.

The other chapters in the text have been updated to explain the diagnostic and service procedures for the latest suspension and steering systems explained in the Classroom Manual. New job sheets related to the new systems and components have been added in the text. All of the art pieces have been replaced with color photos and color diagrams throughout the text to improve the student's visualization of diagnostic and service procedures.

Mark Schnubel

CLASSROOM MANUAL

Features of the Classroom Manual include the following:

Cognitive Objectives

These objectives outline the chapter's contents and identify what students should know and be able to do upon completion of the chapter. Each topic is divided into small units to promote easier understanding and learning.

Terms To Know List

A list of key terms appears immediately after the Objectives. Students will see these terms discussed in the chapter. Definitions can also be found in the Glossary at the end of the manual.

Cross-References to the Shop Manual

References to the appropriate page in the Shop Manual appear whenever necessary. Although the chapters of the two manuals are synchronized, material covered in other chapters of the Shop Manual may be fundamental to the topic discussed in the Classroom Manual.

Margin Notes

The most important terms to know are highlighted and defined in the margin. Common trade jargon also appears in the margin and gives some of the common terms used for components. This helps students understand and speak the language of the trade, especially when conversing with an experienced technician.

CHAPTER 1
SUSPENSION AND STEERING SYSTEMS

Upon completion and review of this chapter, you should be able to understand and describe:

- How strength and rigidity are designed into a unitized body.
- The advantages of reduced vehicle weight.
- The design of a short-and-long arm (SLA) front suspension system.
- How limited independent rear wheel movement is provided in a semi-independent rear suspension system.
- The advantage of an independent rear suspension system.
- The purposes of vehicle tires.
- The terms positive and negative offset as they relate to vehicle wheel rims.
- Three different loads that are applied to wheel bearings.

- The purposes of shock absorbers.
- The difference between a shock absorber and a strut.
- Two different types of computer-controlled shock absorbers.
- The advantages of computer-controlled suspension systems.
- Two types of steering linkages.
- How the rack is moved in a rack and pinion steering gear.
- How a power steering pump develops hydraulic pressure.
- The result of incorrect rear wheel toe.
- The front wheel caster.
- The results of excessive negative camber.

Terms to Know

Angular bearing loads	Negative offset	
Carbon dioxide (CO_2)	Positive camber	Thrust bearing loads
Greenhouse gas	Positive caster	Thrust line
Jounce travel	Positive offset	Toe-out
Negative camber	Radial bearing load	Wheel alignment
Negative caster	Rebound travel	Wheel offset
		Wheel shimmy

INTRODUCTION

The suspension system must provide proper steering control and ride quality. Performing these functions is extremely important to maintaining vehicle safety and customer satisfaction. For example, if the suspension system allows excessive vertical wheel oscillations, the driver may lose control of the steering when driving on an irregular road surface. This loss of steering control can result in vehicle collisions and personal injury. These vertical wheel oscillations transfer undesirable vibration to the passenger compartment, which causes customer...

The suspension system and...
provide normal tire...

---

104   Chapter 4

rotate or move the component. Bearings are precision-machined assemblies, which provide smooth operation and long life. When bearings are properly installed and maintained, bearing failure is rare.

AUTHOR'S NOTE Only 21 percent of the power developed by a vehicle engine actually gets to the drive wheels. Much of the engine energy is lost in overcoming friction and wind resistance. The U.S. Department of Energy (DOE) is working with a major bearing manufacturer to explore the use and advantage of roller bearings to reduce friction in vehicle engines. The DOE estimates that if all vehicle engines used roller-bearing, low-friction technology, 100 million barrels of oil could be saved in a year.

Shop Manual
Chapter 4, page 148

A thrust bearing load may be referred to as an axial load.

BEARING LOADS

When a bearing load is applied in a vertical direction on a horizontal shaft, it is called a radial bearing load. If the vehicle weight is applied straight downward on a bearing, this weight is a radial load on the bearing. A thrust bearing load is applied in a horizontal force direction (**Figure 4-1**). For example, while a vehicle is turning a corner, horizontal force is applied to the front wheel bearings. When an angular bearing load is applied, the angle of the applied load is somewhere between the horizontal and vertical positions. This can also be referred to as a combination load.

Ball bearings have round steel balls between the inner and outer races.

BALL BEARINGS

Front and rear wheel bearings may be ball bearings or roller bearings. Either type of bearing contains these basic parts:

1. Inner race, or cone
2. Rolling elements, balls, or rollers
3. Separator, also called a cage or retainer
4. Outer race, or cup

The inner race is an accurately machined component. The inner surface of the race is mounted on the shaft with a precision fit. The rolling elements are mounted on a very

The inner race supports the inner side of the rolling elements in a bearing.

The rolling elements are the precision-machined balls or rollers between the inner and outer bearing races.

Figure 4-1 Types of bearing loads.

Author's Note

This feature includes simple explanations, stories, or examples of complex topics. These are included to help students understand difficult concepts.

A Bit of History

This feature gives the student a sense of the evolution of the automobile. This feature not only contains nice-to-know information, but also should spark some interest in the subject matter.

Summary

Each chapter concludes with summary statements that contain the important topics of the chapter. These are designed to help the reader review the contents.

Review Questions

Short-answer essay, fill-in-the-blank, and multiple-choice questions follow each chapter. These questions are designed to accurately assess the student's competence in the stated objectives at the beginning of the chapter.

SHOP MANUAL

To stress the importance of safe work habits, the Shop Manual also dedicates one full chapter to safety. Other important features of this manual include the following:

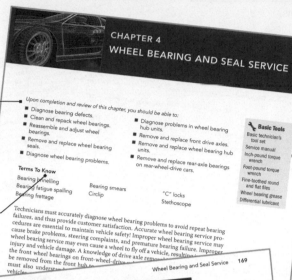

Performance-Based Objectives

These objectives define the contents of the chapter and define what the student should have learned on completion of the chapter.

Terms To Know List

Terms in this list are also defined in the Glossary at the end of the manual.

Cautions and Warnings

Cautions appear throughout the text to alert the reader to potentially hazardous materials or unsafe conditions. Warnings advise the student of things that can go wrong if instructions are not followed or if an incorrect part or tool is used.

Margin Notes

The most important terms to know are highlighted and defined in the margin. Common trade jargon also appears in the margins and gives some of the common terms used for components. This feature helps students understand and speak the language of the trade, especially when conversing with an experienced technician.

Basic Tools Lists

Each chapter begins with a list of the basic tools needed to perform the tasks included in the chapter.

References to the Classroom Manual

References to the appropriate page in the Classroom Manual appear whenever necessary. Although the chapters of the two manuals are synchronized, material covered in other chapters of the Classroom Manual may be fundamental to the topic discussed in the Shop Manual.

Special Tools Lists

Whenever a special tool is required to complete a task, it is listed in the margin next to the procedure.

Case Studies

Each chapter ends with a Case Study describing a particular vehicle problem and the logical steps a technician might use to solve the problem. These studies focus on system diagnosis skills and help students gain familiarity with the process.

Service Tips

Whenever a shortcut or special procedure is appropriate, it is described in the text. Generally, these tips describe common procedures used by experienced technicians.

ASE-Style Review Questions

Each chapter contains ASE-style review questions that reflect the performance objectives listed at the beginning of the chapter. These questions can be used to review the chapter as well as to prepare for the ASE certification exam.

Customer Care

This feature highlights those little things a technician can do or say to enhance customer relations.

Photo Sequences

Many procedures are illustrated in detailed Photo Sequences. These photographs show the students what to expect when they perform particular procedures. They also familiarize students with a system or type of equipment that the school might not have.

---

214   Chapter 5

9. Connect the ohmmeter leads from the signal return terminal in the wiring harness side of the actuator connector to a chassis ground. The ohmmeter should indicate less than 10 ohms. If the ohmmeter reading is higher than specified, check the signal return wire and the programmed ride control (PRC) module.

SERVICE TIP Wiring harness colors vary depending on the vehicle make and model year. The wiring colors in the following steps are based on Ford vehicles. If you are working on a different make of vehicle, refer to the wire colors in the vehicle manufacturer's service information.

CASE STUDY

A customer complained about a squeaking noise in the rear suspension of a 2009 Dodge Caravan. The customer said the noise occurred during normal driving at lower speeds. The technician lifted the car on a hoist and made a check of all rear suspension bushings. The spring insulators and all suspension bolts were checked visually. All of these items were in good condition, and there was no evidence of a squeaking noise as the chassis was bounced gently. Because the exact source of chassis noise can sometimes be difficult to locate, the technician performed a visual check of bushings, insulators, and fasteners in the front suspension. No problems were found in the front suspension.

One of the first requirements for successful automotive diagnosis is to obtain as much information as possible from the customer. The customer with the squeaking rear suspension in a Dodge Caravan lived in a part of the country where cold temperatures occur in the winter. This complaint occurred in January. The technician questioned the customer further about the conditions when the squeaking suspension noise was heard, and the customer revealed that the noise occurred when the temperature was severely cold. The customer also indicated that the noise disappeared at warmer temperatures.

The technician informed the customer that the only way to find the exact cause of this annoying squeak was to leave the car on the lot at the shop all night and check it first thing in the morning when it was colder. The customer complied with this suggestion, and the technician drove the car into the shop immediately the next morning. The squeaking noise occurred as the car was driven across the parking lot. The technician lifted the car on a hoist and listened with a stethoscope at each rear suspension bushing as a coworker gently bounced the rear suspension. No squeaking noise was heard at any of the rear suspension bushings. However, when the stethoscope pickup was placed on the left rear shock absorber, the squeaking noise was loud and clear. The shock absorber was quickly removed, and the squeaking noise was gone when the rear chassis was bounced gently. Replacement of the left rear shock absorber corrected this complaint and made the customer happy.

ASE-STYLE REVIEW QUESTIONS

1. A slight oil film appears on the lower shock absorber oil chamber, and the shock absorber performs satisfactorily on a road test:
Technician A says the shock absorber is satisfactory.
Technician B says the shock absorber may contain excessive pressure.
Who is correct?
A. A only
B. B only
C. Both A and B
D. Neither A nor B

2. While discussing a shock absorber and strut bounce test:
Technician A says the shock absorber is satisfactory if the bumper makes two free upward bounces.
Technician B says the bumper must be pushed downward with considerable force.
Who is correct?
A. A only
B. B only
C. Both A and B
D. Neither A nor B

8.
The rear coil springs
cedure for spring removal from the fron

INSTALLING STRUT CARTRIDGE, OFF-CAR

CUSTOMER CARE Check the cost of the strut cartridges versus the price of new struts. Give customers the best value for their repair dollar!

Many struts are a sealed unit, and thus rebuilding is impossible. However, some manufacturers supply a replacement cartridge that may be installed in the strut housing after the strut has been removed from the vehicle. Always follow the strut cartridge manufacturer's recommended installation procedure.

The following is a typical off-car strut cartridge installation procedure:

1. Install a bolt and two nuts in the upper strut-to-knuckle mounting bolt hole. Place a nut on the inside and outside of the strut flange.
2. Clamp this bolt in a vise to hold the strut.
3. Locate the line groove near the top of the strut body, and use a pipe cutter installed in this groove to cut the top of the strut body.
4. After the cutting procedure, remove the strut piston assembly from the strut **(Figure 5-26)**.
5. Remove the strut from the vise and dump the oil from the strut.
6. Place the special tool supplied by the vehicle manufacturer or cartridge manufacturer on top of the strut body. Strike the tool with a plastic hammer until the tool shoulder contacts the top of the strut body. This action removes burrs from the strut body and places a slight flare on the body.
7. Remove the tool from the strut body.
8. Install the required amount of oil in the strut, place the new cartridge in the strut body, and turn the cartridge until it settles into indentations in the bottom of the strut body.
9. Place the new nut over the cartridge.
10. Using a special tool supplied by the vehicle or cartridge manufacturer, tighten the nut to the specified torque.
11. Move the strut piston rod in and out several times to check for proper strut operation.

A strut cartridge contains the inner working part of the strut, which may be installed in the outer housing of the old strut.

Special Tool
Pipe cutter

INSTALLING STRUT CARTRIDGE, ON-CAR

WARNING If a vehicle is hoisted or lifted in any way during an on-car strut cartridge replacement, the coil spring may fly off the strut, causing vehicle damage and personal injury.

On some vehicles, the front strut cartridge may be removed and replaced with the strut

---

Figure 7-2 Electronic stethoscope.

Riding Height Measurement

Regular inspection and proper main
tant to maintain vehicle safety. The
tion. Other suspension components,
they are worn. *Since incorrect riding
measurement is critical.* Reduced re
causes rapid steering wheel return a
riding height is less than specified. Th
manufacturer's specified location, v
suspension system. On some vehicle
sured from the floor to the center of
level floor or an alignment rack (**Fig**
cedure for measuring front and rear curb riding height.

PHOTO SEQUENCE 11
Typical Procedure for Measuring Front and Rear Riding Height

P11-1 Check the trunk for extra weight.

P11-2 Check the tires for normal inflation pressure.

P11-3 Park the car on a level shop floor or an alignment rack.

Figure 11-81 The lane departure warning system will stay engaged even while cornering if the cornering maneuver is completed in less than 100 seconds.

Some LDW systems have lane-departure prevention capabilities. If the vehicle is moving toward the lane markings on the left, the ABS and stability control computer lightly pulse the brakes on the right side of the vehicle. When the wheels on the left side slow down, the wheels on the left side turn faster than the wheels on the right side, and this action steers the car away from the highway markers on the left. Other LDW systems activate the electronic (electromechanical) steering on the vehicle to steer the vehicle away from the lane markers if a lane departure is starting to occur. The LDW system will engage the electromechanical power steering system to first cause a slight vibration of the steering wheel to warn driver to take corrective action to maintain lane position and then take corrective action to maintain lane position. If the driver exerts more than 2.21 ft.-lbs (3 Nm) of pressure on the steering wheel, the LDW system will disengage to allow a lane change without using turn signal. The LDW system will continue to function even while corning as long as the corning maneuver is completed in less than 100 seconds (**Figure 11-81**). The lane departure system can be adversely affect by weather and road surfaces. While the system is effective at detecting lane lines on asphalt, it can have issues detecting lane lines on concrete roads on bright sunny days. The system can also be affected by some wet surfaces due to reflection or snow-covered surfaces that cover lane lines. A dirty windshield that affects camera image will also cause lane departure system issues and will set a diagnostic trouble code "Lane Assist – no sensor visibility present".

AUTHOR'S NOTE Statistics compiled by one of the major car manufacturers indicate that 30 percent of all vehicle accidents in the United States involve rear-end collisions. In half of these collisions, the driver did not apply the brakes before collision occurred. The statistics also indicate that 75 percent of these rear-end collisions occur below 19 mph (30 km/h).

COLLISION MITIGATION SYSTEMS

Some vehicles are equipped with a collision mitigation system that uses radar and camera information to detect a vehicle in front. The collision mitigation computer is in constant communication with the PCM and ABS computers via the vehicle network. If the radar and camera signals indicate a vehicle in front is too close and a collision is imminent, the

Author's Note

This feature includes simple explanations, stories, or examples of complex topics. These are included to help students understand difficult concepts.

---

10. While diagnosing improper rear wheel tracking:
    A. Improper rear wheel tracking may be caused when both rear springs are sagged the same amount.
    B. A bent rear suspension tie rod does not affect rear wheel tracking.
    C. Improper rear wheel tracking may result in steering pull when driving straight ahead.
    D. Improper rear wheel tracking may cause front wheel shimmy.

ASE CHALLENGE QUESTIONS

1. A squeak in the rear suspension could be caused by all of the following EXCEPT:
   A. The suspension bushing.
   B. Weak spring leaves.
   C. Worn spring antifriction pads.
   D. A defective shock absorber.

2. A vehicle with a live axle coil spring rear suspension has become hard to steer with harsh ride quality. Which of the following could be the cause of this problem?
   A. Worn lateral link bushings
   B. Weak rear coil springs
   C. Bent rear shock rod
   D. Worn stabilizer bar bushings

3. A truck with a solid axle and acts erratic during turns. Which of the following is the most likely cause?
   A. Worn rear sway bar bushings
   B. Incorrect driveline angle
   C. Incorrect ride height
   D. Loose rear axle U-bolts

4. A car with independent rear suspension has excessive rear tire wear. An inspection of the rear tires shows they are worn on the inside edge and the tread is feathered.
   Technician A says the problem could be the tires are toeing out during acceleration.
   Technician B says worn control arm bushings could be the cause of the problem.
   Who is correct?
   A. A only
   B. B only
   C. Both A and B
   D. Neither A nor B

5. The steering on a front-wheel-drive car pulls to the right.
   Technician A says the strut on the right rear suspension assembly could be the problem.
   Technician B says worn bushings of the left rear lower arm assembly could be the problem.
   Who is correct?
   A. A only
   B. B only
   C. Both A and B
   D. Neither A nor B

ASE Challenge Questions

Each technical chapter ends with five ASE challenge questions. These are not more review questions; rather, they test the students' ability to apply general knowledge to the contents of the chapter.

ASE Practice Examination

A 50-question ASE practice exam, located in the Appendix, is included to test students on the content of the complete Shop Manual.

APPENDIX A
ASE PRACTICE EXAMINATION

1. After new tires and new alloy rims are installed on a sports car, the owner complains about steering wander and steering pull in either direction while braking.
   Technician A says there may be brake fluid on the front brake linings.
   Technician B says the replacement rims may have a different offset than the original rims.
   Who is correct?
   A. Technician A
   B. Technician B
   C. Both A and B
   D. Neither A nor B

2. *Technician A* says when a vehicle pulls to one side, the problem may be caused by a pinion gear binding on the rack gear of a power rack and pinion steering gear.
   Technician B says when a power steering rack and pinion steering gear spool valve binds it may cause a vehicle to pull to one side.
   Who is correct?
   A. Technician A
   B. Technician B
   C. Both A and B
   D. Neither A nor B

3. The outside edge of the left front tire on a front-wheel-drive car is badly worn.
   Technician A says the cause could be worn ball joints.
   Technician B says the cause could be that the camber setting is excessively negative on the left front suspension.
   Who is correct?
   A. Technician A
   B. Technician B
   C. Both A and B
   D. Neither A nor B

4. The owner of a 2-wheel drive pick-up says the front tires squeal loudly during low-speed turns. The most probable cause of this condition is:
   A. Excessive positive camber.
   B. Negative caster adjustment.

   C. Improper steering axis inclination (SAI).
   D. Improper turning angle.

5. A 2-wheel drive pickup has a severe shudder when the vehicle is started from a stop with a load in the bed.
   Technician A says the problem may be worn spring eyes.
   Technician B says the problem may be axle torque wrap-up.
   Who is correct?
   A. Technician A
   B. Technician B
   C. Both A and B
   D. Neither A nor B

6. Steering pull to the right occurs only when braking. The most likely cause of this problem is:
   A. Excessive positive caster on the right-front wheel.
   B. Excessive positive camber on the left-front wheel.
   C. A loose strut rod bushing on the right-front suspension.
   D. Excessive toe-out on the right-front wheel.

7. *Technician A* says hard steering may be caused by low hydraulic pressure due to a stuck flow control valve in the pump.
   Technician B says hard steering may be caused by low hydraulic pressure due to a worn steering gear spool valve piston ring or housing bore.
   Who is correct?
   A. Technician A
   B. Technician B
   C. Both A and B
   D. Neither A nor B

8. *Technician A* says incorrect wheel alignment angles will reduce tire tread life.
   Technician B says wheel alignment is necessary after ball joint replacement.
   Who is correct?
   A. Technician A
   B. Technician B
   C. Both A and B
   D. Neither A nor B

629

Job Sheets

Located at the end of each chapter, the Job Sheets provide a format for students to perform procedures covered in the chapter. A reference to the ASE Task addressed by the procedure is included on the Job Sheet.

Name _____
Date _____
Rear Suspension Service   295

REMOVE AND SERVICE REAR SUSPENSION STRUT AND COIL SPRING ASSEMBLY

JOB SHEET 33

Upon completion of this job sheet, you should be able to remove and service rear suspension strut and coil spring assemblies.

ASE Education Foundation Task Correlation

This job sheet addresses the following **MLR** task:

B.16. Inspect, remove, and/or replace strut cartridge or assembly; inspect mounts and bushings. (P-2)

This job sheet addresses the following **AST/MAST** task:

C.10. Inspect, remove, and/or replace strut cartridge or assembly, strut coil spring, insulators (silencers), and upper strut bearing mount. (P-3)

We Support | Education Foundation

Tools and Materials
Floor jack
Hoist
Safety stands
Coil spring compressing tool

Describe the vehicle being worked on:
Year _____
VIN _____ Make _____ Model _____
Engine type and size _____

Procedure

Task Completed

1. On some applications it is necessary to remove the rear seat and trim to gain access to the upper strut mount. ☐

2. Raise vehicle on hoist. If vehicle lift is not available lift the vehicle with a floor jack, and lower the chassis onto safety stands so the rear suspension is allowed to drop downward. ☐

3. Remove rear wheel assembly.

4. Place the floor jack lift pad under the rear spindle on the side where the strut and spring removal is taking place. Raise the floor jack to support some of the suspension system weight. Is the floor jack placed under the rear spindle and supporting some of the vehicle weight? ☐ Yes ☐ No ☐
   Instructor check _____

5. Disconnect the nut from the small spring in the lower arm, and remove the brake hose and antilock brake system (ABS) wire from the strut. ☐

6. Remove the stabilizer bar link from the strut, and loosen the strut-to-spindle mounting bolts. ☐

7. Remove the three upper support nuts under the package tray trim, and lower the floor jack to remove the strut from the knuckle. Remove the strut from the chassis. ☐

SUPPLEMENTS

Instructor Resources

The *Today's Technician* series offers a robust set of instructor resources, available online at Cengage's Instructor Resource Center (http://login.cengage.com) and on DVD. The following tools have been provided to meet any instructor's classroom preparation needs:

- An Instructor's Guide provides lecture outlines, teaching tips, and complete answers to end-of-chapter questions.
- PowerPoint presentations include images, videos, and animations that coincide with each chapter's content coverage.
- Cengage Learning Testing Powered by Cognero® delivers hundreds of test questions in a flexible, online system. You can choose to author, edit, and manage test bank content from multiple Cengage Learning solutions and deliver tests from your LMS, or you can simply download editable Word documents from the DVD or Instructor Resource Center.
- An Image Gallery includes photos and illustrations from the text.
- The Job Sheets from the Shop Manual are provided in Word format.
- End-of-Chapter Review Questions are also provided in Word format, with a separate set of text rejoinders available for instructors' reference.
- To complete this powerful suite of planning tools, a pair of correlation guides map this edition's content to the ASE Education Foundation tasks and to the previous edition.

MindTap for Today's Technician: Automotive Suspension & Steering Systems, 7e

MindTap is a personalized teaching experience with relevant assignments that guide students to analyze, apply, and improve thinking, allowing you to measure skills and outcomes with ease.

- Personalized Teaching: Becomes yours with a Learning Path that is built with key student objectives. Control what students see and when they see it. Use it as-is or match to your syllabus exactly—hide, rearrange, add, and create your own content.
- Guide Students: A unique learning path of relevant readings, multimedia, and activities that move students up the learning taxonomy from basic knowledge and comprehension to analysis and application.
- Promote Better Outcomes: Empower instructors and motivate students with analytics and reports that provide a snapshot of class progress, time in course, and engagement and completion rates.

REVIEWERS

The author and publisher would like to extend a special thanks to the instructors who reviewed this text and offered invaluable feedback:

Rodney Batch
University of Northwestern Ohio
Lima, OH

Tim LeVan
University of Northwestern Ohio
Lima, OH

Cory Peck
Vatterott College
St. Louis, MO

Steve Roessner
University of Northwestern Ohio
Lima, OH

Ronald Strzalkowski
Baker College Flint, MI

Claude F. Townsend III
Oakland Community College
Bloomfield Hills, MI

CHAPTER 1
SUSPENSION AND STEERING SYSTEMS

Upon completion and review of this chapter, you should be able to understand and describe:

- How strength and rigidity are designed into a unitized body.
- The advantages of reduced vehicle weight.
- The design of a short-and-long arm (SLA) front suspension system.
- How limited independent rear wheel movement is provided in a semi-independent rear suspension system.
- The advantage of an independent rear suspension system.
- The purposes of vehicle tires.
- The terms *positive* and *negative offset* as they relate to vehicle wheel rims.
- Three different loads that are applied to wheel bearings.

- The purposes of shock absorbers.
- The difference between a shock absorber and a strut.
- Two different types of computer-controlled shock absorbers.
- The advantages of computer-controlled suspension systems.
- Two types of steering linkages.
- How the rack is moved in a rack and pinion steering gear.
- How a power steering pump develops hydraulic pressure.
- The result of incorrect rear wheel toe.
- The front wheel caster.
- The results of excessive negative camber.

Terms To Know

Angular bearing loads	Negative offset	Thrust bearing loads
Carbon dioxide (CO_2)	Positive camber	Thrust line
Greenhouse gas	Positive caster	Toe-out
Jounce travel	Positive offset	Wheel alignment
Negative camber	Radial bearing load	Wheel offset
Negative caster	Rebound travel	Wheel shimmy

INTRODUCTION

The suspension system must provide proper steering control and ride quality. Performing these functions is extremely important to maintaining vehicle safety and customer satisfaction. For example, if the suspension system allows excessive vertical wheel oscillations, the driver may lose control of the steering when driving on an irregular road surface. This loss of steering control can result in vehicle collisions and personal injury. Excessive vertical wheel oscillations transfer undesirable vibrations from the wheel(s) to the passenger compartment, which causes customer dissatisfaction with the ride quality.

The suspension system and frame must also position the wheels and tires properly to provide normal tire life and proper steering control. If the suspension system does not position each wheel and tire properly, wheel alignment angles are incorrect and usually cause excessive tire tread wear. Improper wheel and tire position can also cause the steering to pull to one side. When the suspension system positions the wheels and tires properly, the steering should remain in the straight-ahead position if the car is driven straight ahead on a reasonably straight, smooth road surface. However, if the wheels and tires are not properly positioned, the steering can be erratic, and excessive steering effort is required to maintain the steering in the straight-ahead position.

The steering system is also extremely important to maintaining vehicle safety and reduce driver fatigue. For example, if a steering system component is suddenly disconnected, the driver may experience a complete loss of steering control, resulting in vehicle collision and personal injury. Loose steering system components can cause erratic steering, which causes the driver to continually turn the steering wheel in either direction to try and keep the vehicle moving straight ahead. This condition results in premature driver fatigue.

FRAMES AND UNITIZED BODIES

Some vehicles, such as rear-wheel-drive cars, sport utility vehicles (SUVs), and trucks, have a frame that is separate from the body (**Figure 1-1**). Other vehicles have a unitized body that combines the frame and body in one unit, eliminating the external frame (**Figure 1-2**). In a unitized body, the body design, rather than a heavy steel frame, provides strength and rigidity. All parts of a unitized body are load-carrying members, and these body parts are welded together to form a strong assembly.

Figure 1-1 Vehicle frame.

Figure 1-2 Unitized body design.

The frame or unitized body serves the following purposes:

1. Allows the vehicle to support its total weight, including the weight of the vehicle and cargo
2. Allows the vehicle to absorb stress when driving on rough road surfaces
3. Enables the vehicle to absorb torque from the engine and drivetrain
4. Provides attachment points for suspension and other components

Shop Manual
Chapter 10, page 443

The unitized body provides a steel box around the passenger compartment to protect passengers in a collision. In most unitized bodies, special steel panels are inserted in the doors to protect the vehicle occupants in a side collision. Some unitized body components are manufactured from high-strength or ultra high-strength steels. The unitized body design is the most popular design today. A steel cradle is mounted under the front of the unitized body to support the engine and transaxle (**Figure 1-3**). Rubber and steel mounts support the engine and transaxle on the cradle. Large rubber bushings are mounted between the cradle and the unitized body to help prevent engine vibration from reaching the passenger compartment. Some unitized bodies have a partial frame mounted under the rear of the vehicle to provide additional strength and facilitate the attachment of rear suspension components (**Figure 1-4**).

Vehicle weight plays a significant role in fuel consumption. One automotive design engineer states, "Fuel economy improvements are almost linear with weight reduction. A 30 percent reduction in vehicle weight provides approximately a 30 percent improvement in fuel economy." If a Toyota Prius weighs 3300 lb (1497 kg) and provides 50 miles per gallon (mpg), the same Prius would provide 55 mpg if it weighed 3000 lb (1360 kg).

Left side   Right side

a   A

b   B

d   D

e   E

TOP

SIDE

Figure 1-4 Rear partial frame.

Figure 1-3 Engine cradle.

Carbon dioxide is a by-product of the gasoline, diesel fuel, or ethanol combustion process.

A greenhouse gas is a gas that collects high in the earth's atmosphere and covers the earth like a blanket, resulting in more heat retained on the planet.

In the United States, it is estimated that automobiles contribute 1.5 billion tons of CO_2 to the atmosphere each year. Coal-burning power plants produce 2.5 billion tons of CO_2 each year.

It is estimated that for every gallon of gasoline burned, 28 lb (12 7 kg) of CO_2 is released into the atmosphere. Approximately 21 lb (9 53 kg) come directly from the tailpipe of a vehicle, and 7 lb (3 18 kg) come from the extraction, transportation, and refining of the gasoline.

When coil-type suspension springs are manufactured from titanium rather than steel, the weight saved per vehicle is 44-55 lbs (20-25 kg).

The average car manufactured in the United States in 2018 contains 347 lb (157 kg) of aluminum. A decade ago, the average car contained 242 lb (110 kg) of aluminum, and in 1973, the average car contained 81 lb (37 kg) of aluminum.

Carbon dioxide (CO_2) emissions are a major concern for automotive manufacturers, because CO_2 is a greenhouse gas that contributes to global warming. Vehicle manufacturers are facing increasingly stringent CO_2 emissions standards. CO_2 emissions are proportional to fuel consumption. Reduced fuel consumption results in lower CO_2 emissions. Therefore, reducing vehicle weight results in less fuel consumption and lower CO_2 emissions. Reduced weight also contributes to improved vehicle performance.

When vehicle bodies, front and rear suspension systems, and steering systems are built from lighter-weight components, these items can make an important contribution to improved fuel economy and reduced CO_2 emissions. The components in these systems may be manufactured from high-strength steels, aluminum, magnesium, titanium, or carbon composites to reduce vehicle weight. The Corvette Z06 has a hydro-formed aluminum frame to reduce vehicle weight. Lightweight aluminum construction throughout the Jaguar XK results in a vehicle weight of 3671 lb (1665 kg), which is 450 lb (204 kg) lighter than its Mercedes-Benz SL500 competitor, which primarily uses steel construction. A new-model Ford Fiesta has been available in Europe since October 2008, and more recently in the United States and other countries. In this model, 55 percent of the unitized body structure is made from high-strength steels, making the body torsionally 10 percent stiffer than its predecessor and providing a very rigid safety cell surrounding the occupants. Ford says this model is 88 lb (44 kg) lighter than the previous model, and more fuel and CO_2 efficient.

High material and production costs have prevented the use of carbon composites in production vehicles. Carbon composites have been used in some exotic ultra high-performance cars or race cars, where cost was not a factor. Carbon composites provide reduced weight and improved crash energy absorption. New carbon composites and improved manufacturing processes may soon make the use of some carbon composite components a reality in high-production vehicles.

FRONT SUSPENSION SYSTEMS

The front and rear suspension systems are extremely important for providing proper wheel position, steering control, ride quality, and tire life. The impact of the tires striking road irregularities must be absorbed by the suspension systems. The suspension systems must supply proper ride quality to maintain customer satisfaction and reduce driver fatigue, as well as provide proper wheel and tire position to maintain directional stability when driving. Proper wheel position also ensures normal tire tread life.

Typical components in a SLA front suspension system are illustrated in **Figure 1-5**. This type of front suspension system has a long lower control arm and a shorter upper control arm. The main front suspension components serve the following purposes:

1. Upper and lower control arms—control lateral (side-to-side) wheel movement.
2. Upper and lower control arm bushings—allow upward and downward control arm movement and absorb wheel impacts and vibrations.
3. Coil springs—allow proper suspension ride height and control suspension travel during driving maneuvers.
4. Ball joints—allow the knuckle and wheels to turn to the right or left.
5. Steering knuckles—provide mounting surfaces for the wheel bearings and hubs.
6. Shock absorbers—control spring action when driving on irregular road surfaces.
7. Strut rod—controls fore-and-aft wheel movement.
8. Stabilizer bar—reduces body sway when a front wheel strikes a road irregularity.

Shop Manual
Chapter 6, page 231

Figure 1-5 Typical short-and-long arm (SLA) front suspension system.

A MacPherson strut front suspension system has no upper control arm and ball joint; instead, a strut is connected from the top of the knuckle to an upper strut mount bolted to the reinforced strut tower in the unitized body (**Figure 1-6**). The strut supports the top of the knuckle and also performs the same function as the shock absorber in an SLA suspension system. The coil spring is mounted between a lower support on the strut and the upper strut mount. Insulators are mounted between the ends of the coil spring and the mounting locations. A bearing in the upper strut mount allows the strut and coil spring to rotate with the spindle when the front wheels are turned.

Figure 1-6 Typical MacPherson strut front suspension system.

REAR SUSPENSION SYSTEMS

A typical live axle rear suspension system has a one-piece rear-axle housing. Trailing arms are connected from the rear-axle housing to the chassis through rubber bushings. The coil springs are mounted between the trailing arms and the chassis (**Figure 1-7**). Because the rear-axle housing is a one-piece assembly, vertical movement of one rear wheel causes the opposite rear wheel to be tipped outward at the top. This action increases tire tread wear and reduces ride quality and traction between the tire tread and road surface.

Many front-wheel-drive cars have a semi-independent rear suspension system with an inverted steel U-section connected between the rear spindles (**Figure 1-8**). The inverted U-section usually contains a tubular stabilizer bar. When one rear wheel strikes a road irregularity, the inverted U-section and stabilizer bar twist, allowing some independent rear-wheel movement before the wheel movement affects the opposite rear wheel. Some semi-independent rear suspension systems have a track bar and brace connected from the inverted U-section to the chassis to reduce lateral rear-axle movement (**Figure 1-9**).

Many vehicles have an independent rear suspension system, wherein each rear wheel can move independently without affecting the position of the opposite rear wheel. This type of suspension system reduces rear tire wear and provides improved steering control. Independent rear suspension systems have a number of different configurations. A MacPherson strut independent rear suspension system has a strut and coil spring assembly connected from the top of the spindle through an upper strut mount to the chassis (**Figure 1-10**). No provision for strut rotation is required, because the rear wheels are not steered. Some independent rear suspension systems have a multilink design, wherein an adjustment link connected from the rear spindle to the chassis allows rear wheel position adjustment (**Figure 1-11**).

Figure 1-7 Live axle rear suspension system with coil springs.

Figure 1-8 Semi-independent rear suspension system.

Figure 1-9 Semi-independent rear suspension system with track bar and brace.

Figure 1-10 MacPherson strut independent rear suspension system.

Figure 1-11 Short-and-long arm independent rear suspension system.

TIRES, WHEELS, AND HUBS

Tire Purpose

Tires are extremely important because they play a large part in ensuring vehicle safety and ride quality. Tires are the only point of contact between the vehicle and the road surface. Vehicle tires provide these functions:

1. Tires must support the vehicle weight safely and firmly.
2. Tires must provide a comfortable ride.
3. Tires must supply adequate traction on various road surfaces to drive and steer the vehicle.
4. Tires must contribute to proper steering control and directional stability of the vehicle.
5. Tires must absorb high stresses when cornering, accelerating, and braking.
6. Tire treads must be designed to propel water off the tread and away from the tire when driving on wet highways. This action prevents water from lifting the tires off the road surface, which decreases tire traction.

Wheel Rim Purpose

Wheel rims can be manufactured from steel, cast aluminum, forged aluminum, pressure-cast chrome-plated aluminum, or magnesium alloy. Wheel rims must retain the tires safely under all operating conditions without distortion. Tires and wheels must form airtight containers at all temperatures so air does not leak from the assembly. Wheel rims must position the tires at the proper distance inward or outward from the vertical mounting surface of the wheel. The distance between the vertical wheel rim mounting surface and the centerline of the wheel rim is called **wheel offset** (**Figure 1-12**). If the wheel centerline is located inboard from the vertical wheel mounting surface, the wheel has **positive offset**. Conversely, if the wheel centerline is located outboard from the vertical wheel mounting surface, the wheel has **negative offset**. Wheel rims typically have four to six mounting openings that fit over studs in the wheel hub. When a wheel rim is installed on the hub studs, tapered nuts are then tightened to the specified torque to retain the wheel and tire assembly on the hub. On many wheel rims, the openings in the wheel rim are tapered to match the tapers on the retaining nuts. These tapered openings and matching tapered nuts center the wheel rim on the hub.

Shop Manual
Chapter 3, page 103

⚠ **WARNING** **Wheel nuts must be tightened in the proper sequence to the specified torque. Failure to follow the proper wheel nut tightening procedure and torque may cause a wheel to come off a car while driving. This action usually results in serious personal injury and extensive vehicle and/or property damage.**

Wheel Hubs

Wheel hubs must provide a secure mounting surface for the wheel rim and tire assembly. Wheel hubs also contain the wheel bearings that provide smooth wheel rotation with reduced friction. Wheel bearings must have a minimum amount of end play to greatly

Figure 1-12 Wheel rim design.

Figure 1-13 Wheel bearing and hub assembly.

reduce wheel lateral movement. The wheel hub and bearing assemblies must carry the load supplied by the vehicle weight, and these assemblies must also guide the wheel and tire assembly (**Figure 1-13**). The vehicle weight is supplied to the wheel hub and bearing assembly in a vertical direction. This type of bearing load is called a **radial bearing load**. When the vehicle turns a corner, the wheel hubs and bearings must carry **thrust bearing loads** supplied in a horizontal direction and **angular bearing loads** supplied in a direction between the horizontal and the vertical.

SHOCK ABSORBERS AND STRUTS

Each corner of the vehicle has a shock absorber or strut connecting the suspension system to the chassis. Shock absorbers control spring action and wheel oscillations to provide a comfortable ride. Controlling spring action and wheel oscillations also improve vehicle safety because the struts help to keep each tire tread in contact with the road surface. If the struts are worn out, excessive wheel oscillations when driving on irregular road surfaces can cause the driver to lose control of the vehicle. Struts also reduce body sway and lean while turning a corner. Struts reduce the tendency of the tire tread to lift off the road surface. This action improves tire tread life, traction, steering control, and directional stability.

Struts contain a sealed lower chamber filled with a special oil. Many shock absorbers have a nitrogen gas charge on top of the oil. This gas charge helps to prevent the shock absorber oil from foaming. A circular steel mount containing a rubber bushing is attached to the bottom end of the lower chamber, and this lower mounting is bolted to the suspension system. The upper strut housing is connected to a piston rod that extends into the lower chamber. A piston valve assembly is attached to the lower end of the piston rod (**Figure 1-14**). The upper strut mount is similar to the lower mounting, and the upper mount is bolted to the chassis.

When a wheel strikes a road irregularity, the wheel and suspension move upward, and the spring in the suspension system is compressed. This action forces the lower shock absorber chamber to move upward, and the oil must flow from below the shock absorber piston and valve to the area above the valve. Upward wheel movement is called **jounce travel**. The strut valves are designed to provide precise oil flow control, and thus control the speed of upward wheel movement.

Shop Manual
Chapter 5, page 196

Figure 1-14 Shock absorber design.

Figure 1-15 Shock absorber action.

When a spring is compressed, it stores energy and then immediately expands with an equal amount of energy. When the spring expands, the tire and wheel assembly is forced downward. Under this condition, the lower strut chamber is forced downward, and oil must flow from above the shock absorber piston and valve to the area below the valve (**Figure 1-15**). Downward wheel movement is called **rebound travel**. The strut valves provide precise control of the oil flow, and this action controls spring action and wheel oscillations. Shock absorbers and valves are usually designed to provide more control during the rebound travel compared to the jounce travel.

Internal strut design is similar to shock absorber design, but struts also support the top of the steering knuckle. In most suspension systems, the lower end of the strut is attached to the top of the steering knuckle, and a special mount is connected between the upper end of the strut and the chassis (**Figure 1-16**). On front suspension systems, the upper strut mount must allow strut and spring rotation when the front wheels are turned to the right or the left. The upper strut mount isolates wheel and suspension vibrations from the chassis.

Nut, strut to mount

Strut mount

Front spring upper insulator

Jounce bumper

Spring

Front spring lower insulator

Front strut

Steering knuckle

Figure 1-16 Front strut and spring assembly.

COMPUTER-CONTROLLED SUSPENSION SYSTEMS AND SHOCK ABSORBERS

Many vehicles are equipped with computer-controlled suspension systems that provide a soft, comfortable ride for normal highway driving, and then automatically and very quickly switch to a firm ride for hard cornering, braking, or fast acceleration. Computer-controlled suspension systems reduce body sway during hard cornering, and thus contribute to improved ride quality and vehicle safety. Some computer-controlled suspension systems are driver-adjustable with up to four suspension modes to allow the driver to tailor the ride quality to the driving style.

Some computer-controlled suspension systems have electronically actuated solenoids in each shock absorber or strut. These solenoids rotate the shock absorber or strut valves to adjust the valve openings and shock absorber control (**Figure 1-17**). Other shock absorbers or struts contain a magneto-rheological fluid that is a synthetic oil containing suspended iron particles. A computer-controlled electric winding is designed into the shock absorber housing. When there is no current flow through the winding, the iron particles are randomly dispersed in the oil. Under this condition, the oil consistency is thinner and the oil flows easily through the shock absorber valves to provide a softer ride. If the suspension computer supplies current flow to the shock absorber windings, the iron particles are aligned so the oil has a jelly-like consistency (**Figure 1-18**). This action instantly provides a much firmer ride. The computer can provide a large variation in current flow through the shock absorber windings and

Figure 1-17 Strut actuator.

Shop Manual
Chapter 11, page 465

a wide range of ride control. Input sensors at each corner of the vehicle inform the suspension computer the velocity of the wheel jounce and rebound, and the computer uses these input signals to operate the shock absorber windings or actuators.

AUTHOR'S NOTE One of the advantages of computer-controlled shock absorbers and suspension systems is the speed at which modern computers can perform output functions. For example, a suspension computer can change the thickness of the magneto-rheological fluid in a shock absorber in about 1 millisecond (ms). When a wheel and tire strike a road irregularity and move upward, this fast computer action adjusts the thickness of the shock absorber fluid in relation to the wheel jounce velocity before the wheel moves downward in the rebound stroke and strikes the road surface.

Figure 1-18 Magneto-rheological fluid action in a strut or shock absorber.

Figure 1-19 Air spring and strut assembly.

Some computer-controlled suspension systems have air springs in place of coil springs (**Figure 1-19**). Front and rear height sensors inform the suspension computer regarding the suspension height, and the computer operates an air compressor and air spring control valves to control the amount of air in the air springs, and thus control suspension height. Some air suspension systems also have computer-controlled shock absorbers or struts.

STEERING SYSTEMS

Steering systems are essential to vehicle safety, steering quality, and steering control. Steering system problems can cause the steering to pull to one side when driving straight ahead, as well as causing excessive steering effort, **wheel shimmy**, and/or excessive steering wheel free play. These problems all reduce vehicle safety and increase driver fatigue. Therefore, steering systems must be properly maintained.

> **Wheel shimmy** may be defined as rapid inward and outward wheel and tire oscillations.

Steering Columns and Steering Linkage Mechanisms

The steering column connects the steering wheel to the steering gear. The steering wheel is connected to the steering shaft, and this shaft extends through the center of the steering column. The lower end of the steering shaft is connected through a universal joint or

Upper steering
column section

Instrument
panel

Lower steering
main shaft

Coupling
assembly

Universal
joint

Mounting
bracket

Lower steering
column tube

Intermediate
shaft

Steering
gear

Figure 1-20 Tilt steering column.

flexible coupling to the shaft from the steering gear. The steering shaft is supported on bearings in the steering column. Some steering columns are designed to collapse or move away from the driver if the driver is thrown against the steering wheel in a collision. Some steering columns are designed so the driver can tilt the steering wheel downward or upward to provide increased driver comfort and facilitate entering and exiting the driver's seat (**Figure 1-20**). Some steering columns also provide a telescoping action so the steering wheel can be moved closer to, or farther away from, the driver. Other steering columns do not have any tilt or telescoping action. A mounting bracket retains the steering column to the instrument panel.

On most vehicles, the ignition lock cylinder and ignition switch are mounted in the steering column. Removing the key from the ignition switch locks the steering column and the gearshift on many vehicles. The steering column usually contains a combination signal light, wipe/wash, dimmer, and cruise control switch. This switch may be called a smart switch. The switch for the hazard warning lights is also mounted in the steering column. An air bag inflator module is mounted in the top of the steering wheel, and a clock-spring electrical connector under the steering wheel maintains electrical contact between the inflator module and the air bag electrical system.

Steering linkages connect the steering gear to the steering arms on the front wheels. In a parallelogram steering linkage, a pitman arm is connected from the steering gear to a center link (**Figure 1-21**). A pivoted idler arm bolted to the chassis supports the other end of the center link. Tie rods are connected from the center link to the steering arms attached to the front wheels. Pivoted ball studs are mounted in the inner ends of the tie rods, and outer tie rod ends are threaded into the tie rod adjusting sleeves. The outer tie rod ends contain pivoted ball studs, and these tapered studs fit into matching tapered openings in the outer ends of the steering arms. The pitman arm and idler arm position the center link so the tie rods are parallel to the lower control arms. This tie rod position is very important to maintain proper steering operation.

Many vehicles have a rack and pinion steering linkage. In these linkages, the tie rods are connected through inner tie rod ends to the rack in the rack and pinion steering gear. Outer tie rod ends are connected from the tie rods to the steering arms (**Figure 1-22**). In this type of steering linkage, the steering gear mounting must position the tie rods so they are parallel to the lower control arms. The rack and pinion steering gear can be mounted on the cowl or the front cross member.

⚠ **Caution**

Regular inspection and maintenance of steering linkage components is very important for normal component life and maintaining vehicle safety.

Figure 1-21 Parallelogram steering linkage.

Figure 1-22 Rack and pinion steering gear and linkage.

Recirculating Ball Steering Gears

Some vehicles are equipped with a recirculating ball steering gear, wherein the steering shaft is attached to a worm gear in the steering gear. A ball nut with internal grooves is mounted over the worm gear. Ball bearings are mounted between the worm gear and ball nut grooves to reduce friction and provide reduced steering effort. Outer grooves on the ball nut are meshed with matching teeth on the sector shaft (**Figure 1-23**). The lower end of the sector shaft is splined to the pitman arm. When the steering wheel is turned, the ball nut moves upward or downward on the worm gear, which rotates the sector shaft to provide the desired steering action. Recirculating ball steering gears can be manual type with no hydraulic assist, or power type with hydraulic assist from the power steering pump.

Shop Manual
Chapter 14, page 611

Rack and Pinion Steering Gears

Rack and pinion steering gears and linkages are more compact than recirculating ball steering gears and parallelogram steering linkages. Therefore, rack and pinion steering gears are usually installed on smaller, front-wheel-drive vehicles. Rack and pinion steering gears transfer more road shock from the front wheels to the steering gear and steering wheel, because the tie rods are connected directly to the rack in the steering gear.

In a rack and pinion steering gear, a toothed rack is mounted on bushings in the rack housing. The rack teeth are meshed with teeth on a pinion gear mounted near one end of the gear. The pinion gear is mounted on bearings in the gear housing. The steering shaft from the steering column is attached to the upper end of the pinion gear (**Figure 1-24**).

Figure 1-23 Power recirculating ball steering gear.

Figure 1-24 Manual rack and pinion steering gear.

Fluid pumped into the left chamber

Fluid pumped into the right chamber

Figure 1-25 Power rack and pinion steering gear with sealed rack piston chambers.

When the steering wheel is turned, the rotation of the pinion gear moves the rack inward or outward to provide the desired steering action. Rack and pinion steering gears can be manual type or power assisted by fluid pressure from the power steering pump. Power rack and pinion steering gears have a piston near the center of the rack, and fluid pressure is supplied from the power steering pump to sealed chambers on either side of the rack piston to provide steering assistance (**Figure 1-25**).

WHEEL ALIGNMENT

Proper **wheel alignment** is extremely important for steering control, ride quality, and normal tire tread life. Improper wheel alignment may cause steering wander, steering pull to the right or left, or improper steering wheel return after turning a corner. Incorrect wheel alignment may contribute to harsh ride quality. Wheel alignment angles that are not within specifications may cause rapid tire tread wear.

REAR WHEEL TRACKING

The rear wheels must track directly behind the front wheels to provide proper steering control. The front and rear wheels must be parallel to the vehicle centerline to provide proper rear wheel tracking. If the rear wheels are tracking directly behind the front wheels, the **thrust line** is positioned at the geometric centerline of the vehicle (**Figure 1-26**). If the left rear wheel has excessive **toe-out**, the thrust line is moved to the left of the geometric centerline (**Figure 1-27**). This improperly positioned thrust line causes the steering to pull to the right and also results in rapid tread wear on the left rear tire.

Wheel alignment is the proper positioning of the tire and wheel assemblies on the vehicle to provide normal tire tread life, precise steering control, and satisfactory ride quality.

The **thrust line** is an imaginary line positioned at a 90-degree angle to the center of the rear-axle and extending forward.

Toe-out is a condition that occurs if the distance between the front edges of the front or rear tires is greater than the distance between the rear edges of the front or rear tires.

Geometric centerline
Thrust line

Figure 1-26 Front and rear wheels are parallel to the vehicle centerline.

Shop Manual
Chapter 9, page 375

Figure 1-27 Excessive toe-out on the left rear wheel moves the thrust line to the left of the geometric centerline.

Rear Wheel Toe

A toe-out condition occurs when the distance between the front edges of the rear tires is greater than the distance between the rear edges of the rear tires. Toe-in occurs if the distance between the rear edges of the rear tires is greater than the distance between the front edges of the rear tires (**Figure 1-28**). On front-wheel-drive vehicles, driving forces tend to push the rear spindles backward. Therefore, these vehicles are usually designed with a zero toe-in or a slight toe-in. Improper rear wheel toe causes rapid tire tread wear because the wheel and tire assembly is being pushed sideways to a certain extent as the vehicle is driven. Steering pull to one side may also be a result of improper rear wheel toe.

Rear Wheel Camber

Camber is the angle between the centerline of the wheel and tire in relation to the true vertical centerline of the wheel and tire viewed from the front. **Positive camber** occurs when the vertical centerline of the wheel and tire is tilted outward in relation to the true vertical centerline of the wheel and tire. **Negative camber** occurs if the vertical centerline of the wheel and tire is tilted inward in relation to the true vertical centerline of the wheel

Positive camber is present when the vertical centerline of the tire is tilted outward in relation to the true vertical centerline of the tire.

Negative camber occurs if the vertical centerline of the tire is tilted inward in relation to the true vertical centerline of the tire.

Figure 1-28 Rear wheel toe.

Loosen both inner
control arm bolts
and move in or out
to adjust camber

Upper
control arm

Eccentric
cam
(camber)

OR

Figure 1-29 Rear wheel camber.

and tire (**Figure 1-29**). Excessive positive camber causes rapid wear on the outside edge of the tire tread, whereas excessive negative camber results in rapid wear on the inside edge of the tire tread. Because a tilted wheel and tire assembly always rolls in the direction of the tilt, improper camber angle on the front wheels may cause steering pull. Many front-wheel-drive vehicles have a slightly negative rear wheel camber that improves cornering stability.

Front Wheel Camber

Front wheel camber is the same as rear wheel camber except for the camber setting. Many front suspension systems are designed with a slightly positive camber.

Front Wheel Caster

Caster is the tilt of a line through the tire centerline (steering axis) in relation to the true vertical tire centerline viewed from the side. **Positive caster** occurs when the centerline of the tire is tilted rearward in relation to the true vertical centerline of the tire. **Negative caster** occurs when the centerline of the tire is tilted forward in relation to the centerline of the tire viewed from the side (**Figure 1-30**). Excessive negative caster causes the steering to wander when the vehicle is driven straight ahead. Excessive positive caster causes increased steering effort and rapid steering wheel return after turning a corner. Harsh ride quality is also a result of excessive positive caster because this condition causes the caster line to be aimed at road irregularities.

Positive caster is the angle of a line through the center of the tire that is tilted rearward in relation to the true vertical centerline of the tire viewed from the side.

Negative caster is the angle of a line through the center of the tire that is tilted forward in relation to the true vertical centerline of the tire viewed from the side.

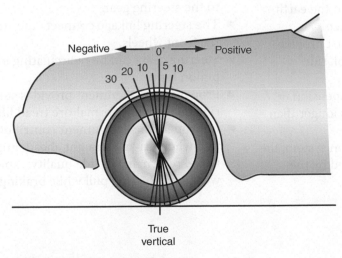

Negative ◄——— 0° ———► Positive

30 20 10 5 10

True
vertical

Figure 1-30 Positive and negative caster.

A BIT OF HISTORY

Front Wheel Toe

Front wheel toe may be defined the same as rear wheel toe. However, the front wheel toe specification usually differs from the rear wheel toe specification. On a rear-wheel-drive car, driving forces tend to move the steering knuckles to a toe-out position. Therefore, these vehicles usually have a slight toe-in. Drive axle forces on a front-wheel-drive vehicle tend to move the front steering knuckles to a toe-in position, and this condition requires a slight toe-out specification.

SUMMARY

- The suspension system must provide proper steering control and ride quality.
- The steering system must maintain vehicle safety and reduce driver fatigue.
- All parts of a unitized body are load-carrying members, and these parts are welded together to form a strong assembly.
- Carbon dioxide (CO_2) is a by-product of the gasoline, diesel fuel, or ethanol combustion process.
- Greenhouse gases collect in the earth's upper atmosphere and form a blanket around the earth, which traps heat near the earth's surface.
- Front and rear suspension systems must provide proper wheel position, steering control, ride quality, and tire life.
- A short-and-long arm (SLA) front suspension system has a lower control arm that is longer than the upper control arm.
- In a MacPherson strut front suspension system, the top of the steering knuckle is supported by the lower end of the strut.

- Rear suspension systems can be live axle, semi-independent, or independent.
- Wheel rims are manufactured from steel, cast aluminum, forged aluminum, pressure-cast chrome-plated aluminum, or magnesium alloy.
- Wheel hubs contain the wheel bearings and support the load supplied by the vehicle weight.
- Bearing loads can be radial, thrust, or angular.
- Shock absorbers control spring action and wheel oscillations.
- The steering column connects the steering wheel to the steering gear.
- The steering linkage connects the steering gear to the front wheels.
- Steering gears can be recirculating ball or rack-and-pinion type.
- Proper wheel alignment provides steering control, ride quality, and normal tire tread life.
- Improper wheel alignment contributes to steering pull when driving straight ahead, improper steering wheel return, harsh ride quality, rapid tire tread wear, and steering pull while braking.

REVIEW QUESTIONS

Short Answer Essays

1. Explain how the engine and transaxle are supported in a front-wheel-drive vehicle with a unitized body.

2. Explain the purpose of coil springs in an SLA front suspension system.

3. Describe how the top of the steering knuckle is supported in a MacPherson strut front suspension system.

4. Explain the disadvantages of a live axle rear suspension system.

5. Explain the sources of CO_2 related to gasoline production and vehicle operation.

6. Describe the design of a wheel rim with positive offset.

7. Explain a radial bearing load.

8. Describe jounce wheel travel.

9. Explain the operation of a computer-controlled shock absorber that contains magneto-rheological fluid.

10. Explain why CO_2 is a harmful gas.

Fill-in-the-Blanks

1. In a rack and pinion steering gear, the inner ends of the tie rods are attached to the ends of the _____.

2. The _____ provides a steel box around the passenger compartment to provide passenger protection in a collision.

3. Tires are extremely important because they play a large part in providing _____ and _____.

4. Proper wheel alignment is extremely important to providing _____ _____, _____ _____, and normal tire _____ _____.

5. Excessive toe-out on a rear wheel may cause _____ and rapid _____.

6. The vehicle thrust line should be positioned at the _____ of the vehicle.

7. If a wheel has negative camber, the tire centerline is tilted _____ in relation to the true vertical tire centerline.

8. On a wheel rim the tapered openings match the tapers on the _____.

9. Each corner of the vehicle has a _____ or _____ connected from the suspension system to the chassis.

10. Steering column connects the steering wheel to the _____.

Multiple Choice

1. Unitized vehicle bodies have these special features and applications:
   A. Unitized bodies are typically used in large rear-wheel-drive vehicles.
   B. Unitized bodies have special steel panels in the hood and trunk lid to protect the vehicle occupants in a collision.
   C. Some members of a unitized body are load-carrying members.
   D. Unitized bodies have a steel box around the passenger compartment.

2. The purpose of the upper and lower control arms in an SLA front suspension system is to:
   A. Allow the steering knuckle to turn to the right or left.
   B. Control spring action on irregular road surfaces.
   C. Reduce body sway.
   D. Control lateral (side-to-side) wheel movement.

3. A semi-independent rear suspension system has:
   A. A rear-axle with an inverted steel U-section containing a tubular stabilizer bar.
   B. Upper and lower control arms.
   C. A one-piece rear-axle housing.
   D. A lower ball joint.

4. All of these statements about wheel rims are true EXCEPT:

   A. Wheel rims must position the tires at the proper distance inward or outward from the vertical wheel mounting surface.

   B. A wheel rim with a positive offset has the center of the tire positioned inboard from the vertical wheel mounting surface.

   C. A wheel rim with a neutral offset has the center of the tire positioned at the vertical center of the brake rotor.

   D. A wheel rim with a negative offset has the center of the tire positioned outboard from the vertical wheel mounting surface.

5. A thrust-type bearing load is applied in a:

   A. Horizontal direction.

   B. Vertical direction.

   C. Angular direction.

   D. Radial direction.

6. Shock absorbers control spring action and help to prevent:

   A. Improper wheel and tire position.

   B. Wheel oscillations.

   C. Excessive steering effort.

   D. Slow steering wheel return after a turn.

7. In a parallelogram steering linkage, the tie rods are parallel to the:

   A. Steering arms.

   B. Center link.

   C. Upper control arms.

   D. Lower control arms.

8. All of these statements about recirculating ball steering gears are true EXCEPT:

   A. Ball bearings are mounted between the worm gear and the ball nut.

   B. The lower end of the sector shaft is splined to the idler arm.

   C. As the steering wheel is turned, the ball nut moves upward or downward on the worm gear.

   D. The steering shaft is attached to the worm gear.

9. Excessive rear wheel toe-out on the left rear wheel causes:

   A. Steering pull to the right.

   B. Harsh ride quality.

   C. Reduced steering effort.

   D. Excessive wheel oscillations.

10. Proper wheel alignment is extremely important to provide all of the following EXCEPT:

    A. Steering control.

    B. Ride quality.

    C. Reduce body sway.

    D. Normal tire tread life.

CHAPTER 2
BASIC THEORIES

Upon completion and review of this chapter, you should be able to understand and describe:

- Newton's Laws of Motion.
- Work and force.
- Power.
- The most common types of energy and energy conversions.
- Inertia and momentum.
- Friction.
- Mass, weight, and volume.
- Static unbalance.
- Dynamic unbalance.
- The compressibility of gases and the noncompressibility of liquids.
- Atmospheric pressure and vacuum.
- Venturi operation.
- Voltage.
- Current.
- Resistance.
- Ohm's law.

Terms To Know

Ampere (A)	Horsepower	Torque
Atmospheric pressure	Inertia	Vacuum
Compressible	Mass	Voltage (V)
Dynamic balance	Momentum	Volume
Energy	Ohms (Ω)	Weight
Force	Ohm's law	Work
Friction	Static balance	

INTRODUCTION

An understanding of the basics is essential before you attempt a study of complex systems and components. Basic theories such as static balance, dynamic balance, and compressibility must be understood prior to a study of the components and systems in this book. If you have studied basic theories previously, the information in this chapter may be used as a review. A thorough study of this chapter will provide all the necessary background information you need before you study the suspension and steering systems in this book.

 A BIT OF HISTORY

The automotive industry in the United States is a very large, dynamic industry. Total production of passenger cars, trucks, buses, and commercial vehicles has increased from 4192 in 1900 to 12,770,714 in 2000. Since 1920, the lowest vehicle production was 725,215 in 1945.

NEWTON'S LAWS OF MOTION

First Law

A body in motion remains in motion, and a body at rest remains at rest, unless some outside force acts on it. When a car is parked on a level street, it remains stationary unless it is driven or pushed. If the gas pedal is depressed with the engine running and the transmission in drive, the engine delivers power to the drive wheels and this force moves the car.

Second Law

A body's acceleration is directly proportional to the force applied to it, and the body moves in a straight line away from the force. For example, if the engine power supplied to the drive wheels increases, the vehicle accelerates faster.

Third Law

For every action, there is an equal and opposite reaction. A practical application of this law occurs when the wheel on a vehicle strikes a bump in the road surface. This action drives the wheel and suspension upward with a certain force, and a specific amount of energy is stored in the spring. After this action occurs, the spring forces the wheel and suspension downward with a force equal to the initial upward force.

WORK AND FORCE

Force is defined as energy applied to an object.

Work is defined as the result of applying a force.

When a **force** moves a certain mass a specific distance, **work** is done. When work is accomplished, the mass may be lifted or slid on a surface (**Figure 2-1**). Because force is measured in pounds and distance is measured in feet, the measurement for work is footpounds (ft.-lb). In the metric system, work is measured in Newton-meters (Nm). If a force moves a 3000-lb vehicle for 50 ft., 150,000 ft.-lb of work are produced. Mechanical force acts on an object to start, stop, or change the direction of the object. It is possible to apply force to an object and not move the object. For example, you may push with all your strength on a car stuck in a ditch and not move the car. Under this condition, no work is done. Work is only accomplished when an object is started, stopped, or redirected by mechanical force.

ENERGY

Energy may be defined as the ability to do work.

When **energy** is released to do work, it is called kinetic energy. This type of energy may also be referred to as energy in motion. Stored energy may be called potential energy. Energy is available in one of six forms:

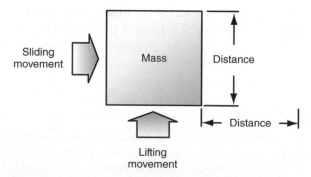

Figure 2-1 Work is accomplished when a mass is lifted or slid on a surface.

1. *Chemical energy* is contained in the molecules of different atoms. In the automobile, chemical energy is contained in the molecules of gasoline and also in the molecules of electrolyte in the battery.
2. *Electrical energy* is required to move electrons through an electric circuit. In the automobile, the battery is capable of producing electrical energy to start the vehicle, and the alternator produces electrical energy to power the electrical accessories and recharge the battery.
3. *Mechanical energy* is the ability to move objects. In the automobile, the battery supplies electrical energy to the starting motor, and this motor converts the electrical energy to mechanical energy to crank the engine. Because this energy is in motion, it may be called kinetic energy.
4. *Thermal energy* is energy produced by heat. When gasoline burns, thermal energy is released.
5. *Radiant energy* is light energy. In the automobile, radiant energy is produced by the lights.
6. *Nuclear energy* is the energy within atoms when they are split apart or combined. Nuclear power plants generate electricity with this principle. This type of energy is not used in the automobile.

ENERGY CONVERSION

Energy conversion occurs when one form of energy is changed to another form. Since energy is not always in the desired form, it must be converted to a usable form. Some of the most common automotive energy conversions are discussed in the following sections.

Chemical to Thermal Energy Conversion

Chemical energy in gasoline or diesel fuel is converted to thermal energy when the fuel burns in the engine cylinders.

Thermal to Mechanical Energy Conversion

Mechanical energy is required to rotate the drive wheels and move the vehicle. The piston and crankshaft in the engine and the drivetrain are designed to convert the thermal energy produced by the burning fuel into mechanical energy (**Figure 2-2**).

Explosion produces
pressure instantly

Figure 2-2 Thermal energy in the fuel is converted to mechanical energy in the engine cylinders. The piston, crankshaft, and drivetrain deliver this mechanical energy to the drive wheels.

Electrical to Mechanical Energy Conversion

The windshield wiper motor converts electrical energy from the battery or alternator to mechanical energy to drive the windshield wipers.

Mechanical to Electrical Energy Conversion

The alternator is driven by mechanical energy from the engine. The alternator converts this energy to electrical energy, which powers the electrical accessories on the vehicle and recharges the battery.

INERTIA

Inertia is defined as the tendency of an object at rest to remain at rest or the tendency of an object in motion to stay in motion.

The **inertia** of an object at rest is called static inertia, whereas dynamic inertia refers to the inertia of an object in motion. Inertia exists in liquids, solids, and gases. When you push and move a parked vehicle, you overcome the static inertia of the vehicle. If you catch a ball in motion, you overcome the dynamic inertia of the ball. Inertial forces are also referred to as apparent forces. Among the inertial forces are centrifugal and Coriolis forces. An example of inertial force can be observed if we place an object such as an apple on the passenger seat of a car while moving. To the driver the apple appears motionless. But when the brakes are applied suddenly with force, the apple travels forward in the direction of travel with great force and hits the dashboard, while the driver is restrained in the seat by the seat belt (**Figure 2-3**). From the driver's point of view, the apple's state of motion changes as it moves forward. It is the inertial force that causes the apple to move.

An example of centrifugal force can be observed when a driver makes a sharp and rapid turn into a tight corner. Again, if the driver has an apple sitting on the passenger seat as the vehicle is accelerated around a tight corner, the apple moves away from the center of rotation of the curve flying off the seat (**Figure 2-4**).

The Coriolis force is an inertial force caused when an object moves in relation to a rotating reference system. When the earth is the reference system the Coriolis force is caused by the rotation of the earth. This can be observed with pendulum swinging on a

Figure 2-3 As a driver slams on the brakes of a moving car, the apple that was sitting motionless on the passenger seat now flies forward due to inertial force.

Figure 2-4 As a driver makes a sharp cornering turn, the apple that was sitting motionless on the passenger seat moves away from the center of rotation of the curve, flying off the seat.

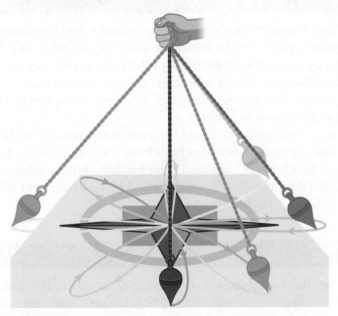

Figure 2-5 An example of the Coriolis force can be observed with a pendulum on a long wire. The pendulum will swing in overlapping elliptical rings. Over a 24-hour period of time, the plane of the oscillating pendulum will rotate 360 degrees.

long wire (**Figure 2-5**). The pendulum will swing in overlapping elliptical rings. Over a 24-hour period of time, the plane of the oscillating pendulum will rotate 360 degrees in a clockwise direction if it is located in the Northern Hemisphere. The Coriolis force is used in vehicle sensors to measure the yaw rate of the vehicle in motion and on some mass airflow sensors. Coriolis force plays an important role in ocean currents and the dynamics of weather.

MOMENTUM

When a force overcomes static inertia and moves an object, the object gains **momentum**.

Momentum is the product of an object's weight times its speed. Momentum is a type of mechanical energy. An object loses momentum if another force overcomes the dynamic inertia of the moving object.

FRICTION

Friction may be defined as the resistance to motion when the surface of one object is moved over the surface of another object.

Coefficient of drag (Cd) may also be called aerodynamic drag.

Friction may occur in solids, liquids, and gases. When a car is driven down the road, friction occurs between the air and the car's surface. This friction opposes the momentum, or mechanical energy, of the moving vehicle. Since friction creates heat, some of the mechanical energy from the vehicle's momentum is changed to heat energy in the air and body components. The mechanical energy from the engine must overcome the vehicle's inertia and the friction of the air striking the vehicle. Body design has a very dramatic effect on the amount of friction developed by the air striking the vehicle. The total resistance to motion caused by friction between a moving vehicle and the air is referred to as coefficient of drag (Cd). The study of Cd is not only very complicated but also very important. At 45 miles per hour (72 kilometers per hour), half of the engine's mechanical energy is used to overcome air friction, or resistance. Therefore, reducing a vehicle's Cd can be a very effective method of improving fuel economy and reducing CO_2 emissions.

MASS, WEIGHT, AND VOLUME

Mass is the measurement of an object's inertia.

Weight is the measurement of the earth's gravitational pull on the object.

Volume is the length, width, and height of a space occupied by an object.

A lawn mower is much easier to push than a 2500-pound vehicle because the lawn mower has very little inertia compared to the vehicle. A space ship might weigh 100 tons on earth where it is affected by the earth's gravitational pull. In outer space beyond the earth's gravity and atmosphere, the space ship is almost weightless. Here on earth, **mass** and **weight** are measured in pounds and ounces in the English system. In the metric system, mass and weight are measured in grams or kilograms.

   Volume is a measurement of size, and it is related to mass and weight. For example, a pound of gold and a pound of feathers both have the same weight, but the pound of feathers occupies a much larger volume. In the English system, volume is measured in cubic inches, cubic feet, cubic yards, or gallons. The measurement for volume in the metric system is cubic centimeters or liters.

TORQUE

Torque is a force that does work with a twisting or turning force. However, movement does not have to occur.

When you pull a wrench to tighten a bolt, you supply **torque** to the bolt. If you pull on a wrench to check the torque on a bolt, and the bolt torque is sufficient, torque is applied to the bolt, but no movement occurs. This torque, or twisting force, is calculated by multiplying the force and the radius. For example, if you supply a 10-lb force on the end of a 2-ft. wrench to tighten a bolt, the torque is $10 \times 2 = 20$ ft.-lb (**Figure 2-6**). If the bolt turns during torque application, work is done. When a bolt does not rotate during torque application, no work is accomplished.

POWER

Horsepower is a measurement for the rate, or speed, at which work is done.

James Watt calculated that a horse could move 330 lb for 100 ft. in 1 minute (**Figure 2-7**). If you multiply 330 lb by 100 ft., the answer is 33,000 ft.-lb of work. Watt determined that one horse could do 33,000 ft.-lb of work in 1 minute. Thus, 1 **horsepower** (HP) is equal

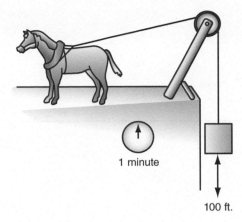

Figure 2-7 1 horsepower is produced when 330 lb are moved 100 ft. in 1 minute.

Figure 2-6 Force applied to a wrench produces torque. If the bolt turns, work is accomplished.

to 33,000 ft.-lb per minute, or 550 ft.-lb per second. 2 HP could do this same amount of work in one-half minute, or 4 HP could complete this work in one-quarter minute. If you push a 3000-lb (1360-kilogram) car for 11 ft. (3.3 meters) in one-quarter minute, you produce 4 HP. From this brief discussion about horsepower, we can understand that as power increases, speed also increases, or the time to do work decreases.

🖋📖 A BIT OF HISTORY

James Watt, of Scotland, is credited with being the first person to calculate power. He measured the amount of work that a horse could do in a specific time.

PRINCIPLES INVOLVING TIRES AND WHEELS IN MOTION

If you roll a cone-shaped piece of metal on a smooth surface, the cone does not move straight ahead. Instead, the cone moves toward the direction of the tilt on the cone. When you are riding a bicycle and want to make a left turn, if you tilt the bicycle to the left, it is much easier to complete the turn. The reason for this action is that a tilted, rolling wheel tends to move in the direction of the tilt. Similarly, if a tire and wheel on a vehicle are tilted, the tire and wheel tend to move in the direction of the tilt (**Figure 2-8**). This principle is used in front wheel alignment.

The casters on a piece of furniture are angled so the center of the caster wheel is some distance from the pivot center (**Figure 2-9**). When the furniture is moved, the casters turn on their pivots to bring the caster wheels into line with the pushing force on the furniture (**Figure 2-10**). This caster action causes the furniture to move easily in a straight line.

The weight of a bicycle rider is projected through the bicycle front fork to the road surface, and the tire pivots on the vertical centerline of the wheel when the handlebars are turned. Notice the centerline of the front fork is tilted rearward in relation to the vertical centerline of the wheel (**Figure 2-11**). Since the tire pivot point is behind the front fork centerline where the weight is projected against the road surface, the front wheel tends to return to the straight-ahead position after a turn. The wheel also tends to remain in the straight-ahead position as the bicycle is driven. This principle is applied in automotive front wheel alignment.

Shop Manual
Chapter 2, page 62

Tilted rolling tires and wheels always move in the direction of the tilt.

Figure 2-8 A tilted, rolling wheel tends to move in the direction of the tilt.

Figure 2-9 Distance between the wheel center and the pivot center on a caster.

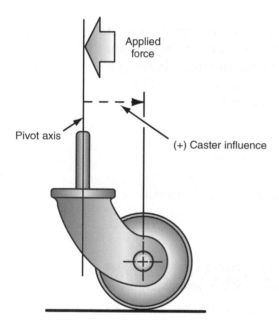

Figure 2-10 Caster aligned with the pushing force provides straight-ahead movement when furniture is pushed.

Figure 2-11 If a wheel centerline and pivot point are behind the front fork centerline pivot point, the wheel tends to return to the straight-ahead position after a turn. The wheel also tends to remain in the straight-ahead position as the bicycle is driven.

PRINCIPLES INVOLVING THE BALANCE OF WHEELS IN MOTION

Static balance refers to the equal distribution of weight around the center of a tire-and-wheel assembly viewed from the side.

When the weight of a tire and wheel assembly is distributed equally around the center of wheel rotation viewed from the side, the wheel and tire have proper **static balance**. Under this condition, the tire and wheel assembly has no tendency to rotate by itself, regardless of the wheel position. If the weight is not distributed equally around the center of wheel rotation, the wheel and tire are statically unbalanced (**Figure 2-12**). As the wheel and tire rotate, centrifugal force acts on this static unbalance and causes the wheel to "tramp" or hop. This wheel and tire action can often be seen while looking at tires on vehicles that may pass you on a highway.

When a ball is rotated on the end of a string, the ball and string form an angle with the axis of rotation. If the rotational speed is increased, the ball and string form a 90-degree

Figure 2-12 Static wheel unbalance caused by unequal weight distribution around the center of the wheel rotation.

Figure 2-13 A weight tends to rotate at a 90-degree angle in relation to the axis of rotation.

angle with the axis of rotation (**Figure 2-13**). Any weight will always tend to rotate at a 90-degree angle to the axis of rotation.

If two balls are positioned on a metal bar so their weight is equally distributed on the centerline of the rotational axis of rotation, the path of rotation remains at a 90-degree angle to the centerline of the axis when the bar is rotated. Under this condition, the metal bar and the balls are in **dynamic balance** (**Figure 2-14**).

If weights are placed on a metal bar so their weights are not equally distributed in relation to the centerline of the rotational axis of rotation, the weights still tend to rotate at a 90-degree angle in relation to the rotational axis. This action forces the pivot out of its vertical axis (**Figure 2-15**). When the bar is rotated 180 degrees, the bar is forced out of its vertical axis in the opposite direction. If this condition is present, the bar has a wobbling action as it rotates. Under this condition, the bar is said to have dynamic unbalance, but static balance is maintained.

> **Dynamic balance** refers to the equal distribution of weight on each side of a tire-and-wheel centerline viewed from the front or rear.

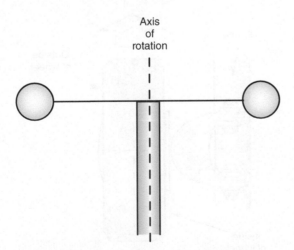

Figure 2-14 When weight on a metal bar is distributed equally on the centerline of the axis of rotation, the bar and balls remain in dynamic balance during rotation.

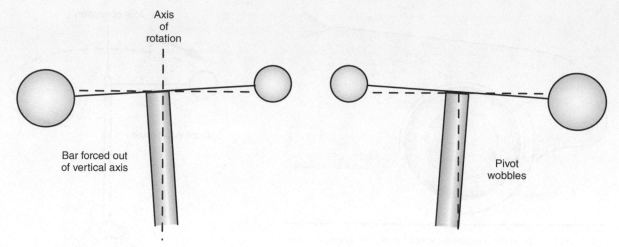

Figure 2-15 When weight is not equally distributed in relation to the centerline of rotational axis, dynamic unbalance causes a wobbling action during rotation.

Molecular energy may be defined as the kinetic energy available in atoms and molecules because of the constant electron movement within the atoms and molecules.

Similarly, when the weight on a tire and wheel is not distributed equally on both sides of the tire centerline viewed from the front, the tire and wheel are dynamically unbalanced. This condition produces a wobbling action as the tire and wheel rotate. The weight must be distributed equally in relation to the tire centerline to provide proper dynamic balance (**Figure 2-16**). These principles are used in automotive wheel balancing.

AUTHOR'S NOTE Improper static or dynamic wheel balance causes excessive tire tread wear, increased wear on suspension components, and driver fatigue.

PRINCIPLES INVOLVING LIQUIDS AND GASES

Molecular Energy

Remember that kinetic energy refers to energy in motion. Since electrons are constantly in motion around the nucleus in atoms or molecules, kinetic energy is present in all matter.

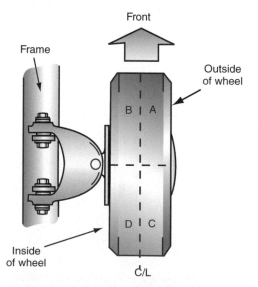

Figure 2-16 Weight must be distributed equally in relation to the tire-and-wheel centerline.

Kinetic energy in atoms and molecules increases as the temperature increases. A decrease in temperature reduces the kinetic energy. Molecules in solids move slowly compared to those in liquids or gases. Gas molecules move quickly compared to liquid molecules. Since gas molecules are in constant motion, they spread out to fill all the space available. At higher temperatures, gas molecules spread out more, whereas at lower temperatures, gas molecules move closer together.

Temperature

Temperature affects all liquids, solids, and gases. The volume of any matter increases as the temperature increases. Conversely, the volume decreases in relation to a reduction in temperature. When the gases in an engine cylinder are burned, the sudden temperature increase causes rapid gas expansion, which pushes the piston downward and causes engine rotation (**Figure 2-17**).

Pressure and Compressibility

Since liquids and gases are both substances that flow, they may be classified as fluids. If a nail punctures an automotive tire, the air escapes until the pressure in the tire is equal to atmospheric pressure outside the tire. When the tire is repaired and inflated, air pressure is forced into the tire. If the tire is inflated to 32 pounds per square inch (psi), or 220 kilopascals (kPa), this pressure is applied to every square inch on the inner tire surface. Pressure is always supplied equally to the entire surface of a container. Since air is a gas, the molecules have plenty of space between them. When the tire is inflated, the pressure in the tire increases, and the air molecules are squeezed closer together, or compressed. Under this condition, the air molecules cannot move as freely, but extra molecules of air can still be forced into the tire. Therefore, gases such as air are said to be **compressible**.

The air in the tire may be compared to a few balls on a billiard table without pockets. If a few more balls are placed on the table, the balls are closer together, but they can still move freely (**Figure 2-18**).

Pressure may be defined as a force exerted on a given surface area.

Pressure and temperature are directly related. If you increase one, you also increase the other.

Pressure and volume are inversely proportional. If volume is decreased, pressure is increased.

POWER STROKE

Steady pressure full length of stroke

Figure 2-17 Hot, expanding gases push the piston downward and rotate the crankshaft.

Balls are still free to move and bounce, but with less space and more activity.

More balls, or molecules, compressed into same space

Figure 2-18 Gases can be compressed much like more balls can be placed on a billiard table containing a few balls.

Similar to a liquid, molecules occupy all space. In a container filled with liquid, molecules occupy all the space, and more molecules cannot be added to the container.

Balls could roll and move

Balls, or molecules, fill all available space

Figure 2-19 Liquids are noncompressible, just as more balls cannot be added to a billiard table with no pockets that is completely filled with balls.

If the vehicle is driven at high speed, friction between the road surface and the tires heats the tires and the air in the tires. When air temperature increases, the pressure in the tire also increases. Conversely, a temperature decrease reduces pressure.

If 100 cubic feet (2.8 cubic meters) of air is forced into a large truck tire, and the same amount of air is forced into a much smaller car tire, the pressure in the car tire is much greater.

Molecules in a liquid may be compared to a billiard table without pockets that is completely filled with balls. These balls can roll around, but no additional balls can be placed on the table because the balls cannot be compressed. Similarly, liquid molecules cannot be compressed (**Figure 2-19**).

Liquid Flow

If a tube is filled with billiard balls and the outlet is open, more balls may be added to the inlet. When each ball is moved into the inlet, a ball is forced from the outlet. If the outlet is closed, no more balls can be forced into the inlet (**Figure 2-20**).

The billiard balls in the tube may be compared to molecules of power steering fluid in the line between the power steering pump and steering gear. Since noncompressible fluid fills the line and gear chamber, the force developed by the pump pressure is transferred through the line to the gear chamber (**Figure 2-21**).

Balls can roll in and out as long as outlet is clear

Tube

Open outlet

Figure 2-20 Billiard ball movement in a tube filled with balls compared to liquid flow.

Pump packs molecules into tube

Molecules of fluid push against rack piston like rolling billard balls with force behind

Figure 2-21 Power steering pump pressure supplied to the steering gear chamber.

Hydraulic Fluid as a Flexible Machine

Hydraulic pressure has the same effect as a mechanical lever, because both of these items can multiply the input force to do more work. If a mechanical lever has a fulcrum at the center point and a 10-lb, or 4.5-kilogram (kg), weight on one end of the lever, 10 lb (4.5 kg) of weight are required on the other end of the lever to force the lever downward and raise the weight on the opposite end. If the lever is 5 ft. long and the fulcrum is placed 1 foot from the weight to be lifted, the lever has a 4 to 1 mechanical advantage. Under this condition, 2.5 lb (1.1 kg) of weight are required on the other end of the lever to lift the 10-lb (4.5-kg) weight (**Figure 2-22**). Therefore, a mechanical lever multiplies input force and makes it easier to do work.

When the power steering pump pressure supplied to the steering gear chamber is 1000 psi (7000 kPa), this pressure is applied to every square inch in the steering gear chamber (**Figure 2-23**). This pressure applied to the rack piston in the gear chamber acts like a mechanical lever and helps move the rack piston. Since the rack is connected through steering linkages and arms to the front wheels, the force on the rack piston helps the driver move the front wheels to the left or right during a turn (**Figure 2-24**).

Atmospheric Pressure

Since air is gaseous matter with mass and weight, it exerts pressure on the earth's surface. A 1-square-inch column of air extending from the earth's surface to the outer edge of the atmosphere weighs 14.7 psi at sea level. Therefore, **atmospheric pressure** is 14.7 psi at sea level (**Figure 2-25**).

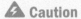

⚠ Caution

Never completely block a pump outlet unless instructed to do so in the vehicle manufacturer's service manual. This action can result in extremely high pump pressure, damaged components, and personal injury.

An equal volume of hot air weighs less and exerts less pressure on the earth's surface than cold air.

Atmospheric pressure may be defined as the total weight of the earth's atmosphere.

Figure 2-22 A mechanical lever multiplies input force and makes it easier to do work.

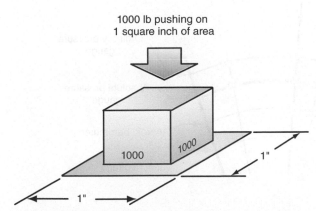

Figure 2-23 Hydraulic pressure is applied equally to every square inch in a container.

Figure 2-24 Steering gear, linkages, and arms connected to the front wheels.

Figure 2-25 A column of air 1 in. square extending from the earth's surface at sea level to the outer edge of the atmosphere weighs 14.7 lb.

Atmospheric Pressure and Temperature

When air becomes hotter it expands, and this hotter air is lighter compared to an equal volume of cooler air. This hotter, lighter air exerts less pressure on the earth's surface compared to cooler air.

If the temperature decreases, air contracts and becomes heavier. Therefore, an equal volume of cooler air exerts more pressure on the earth's surface compared to hotter air.

Atmospheric pressure decreases as altitude increases.

Atmospheric Pressure and Altitude

When you climb above sea level, atmospheric pressure decreases. The weight of a column of air is less at 5000 ft. (1524 meters) elevation compared to sea level. As altitude continues to increase, atmospheric pressure and weight continue to decrease. At an altitude of several hundred miles above sea level, the earth's atmosphere ends, and there is only vacuum beyond that point.

VACUUM

Atmospheric pressure is generally considered to be 14.7 lb per square inch (psi) at sea level. Pressures greater than atmospheric pressure may be measured in psi, whereas pressures below atmospheric pressure are measured in **vacuum**, or psi absolute (psia) (**Figure 2-26**).

Vacuum may be defined as the absence of atmospheric pressure. A vacuum may be called a low pressure because it is a pressure less than atmospheric pressure.

Figure 2-26 Pressure and vacuum scales.

A conventional pressure gauge is used to measure pressures greater than atmospheric pressure. This type of pressure gauge indicates 0 psi at atmospheric pressure, and as the pressure increases it can read up to 15 psi. A conventional vacuum gauge indicates 0 inches of mercury (in. Hg) when atmospheric pressure is applied, and as the vacuum increases this gauge reads from 0 in. Hg to 29.91 in. Hg, and this may be considered a perfect vacuum. An absolute pressure gauge indicates absolute pressure in pounds per square inch gauge (psig). An absolute pressure gauge indicates 0 at a perfect vacuum, 15 psig at atmospheric pressure, and 30 psig at 15 psi on a conventional pressure gauge. An aneroid barometer reads in in. Hg absolute pressure, and thus it reads 0 in. Hg absolute at a perfect vacuum and 29.92 in. Hg absolute when atmospheric pressure is present.

Pressures above atmospheric pressure measured in psi are found in these automotive systems:

1. Fuel
2. Oil
3. Cooling
4. Power steering
5. Air conditioning
6. Turbocharger boost
7. Air springs
8. Gas-filled shock absorbers
9. Brake system during brake application

Pressures below atmospheric pressure measured in inches of vacuum are found in these automotive systems:

1. Manifold vacuum signal
2. Ported vacuum signal above the throttle
3. Carburetor venturi
4. Air-conditioning evacuation

Vacuum could be measured in psi, but in. Hg is most commonly used for this measurement. Let us assume that a manometer is partially filled with mercury, and atmospheric pressure is allowed to enter one end of the tube. If vacuum is supplied to the other end of the manometer, the mercury is forced downward by the atmospheric pressure. When this movement occurs, the mercury also moves upward on the side where the vacuum is supplied. If the mercury moves downward 10 in., or 25.4 centimeters (cm), where the atmospheric pressure is supplied, and upward 10 in. (25.4 cm) where the vacuum is supplied, 20 in. Hg is supplied to the manometer (**Figure 2-27**). The highest possible, or perfect, vacuum is approximately 29.9 in. Hg.

Vacuum and atmospheric pressure are used in several automotive systems. For example, atmospheric pressure is present outside the compressor inlet on an electronic air suspension system. When the compressor is running, it creates a vacuum at the inlet and in the compressor cylinder. This pressure difference causes air to move from the atmosphere surrounding the inlet into the cylinder. The compressor develops high pressure at the discharge valve, and this pressure forces air into the air springs when the pressure is lower than at the compressor outlet (**Figure 2-28**).

Pumps use high and low pressure to move liquids or gases. For example, as a power steering pump rotates, it creates a high pressure at the pump outlet by pumping against a restriction, and a low pressure at the inlet by moving fluid through a restriction. This pressure difference causes fluid to flow through the power steering system.

Liquids, solids, and gases tend to move from an area of high pressure to a low-pressure area.

Figure 2-27 A vacuum of 20 in. Hg (68 kPa) supplied to mercury in a manometer.

Figure 2-28 An air suspension compressor creates vacuum in the cylinder, and atmospheric pressure at the inlet forces air into the cylinder. The compressor provides high pressure at the discharge valve, which moves air into the air springs.

A venturi may be defined as a narrow area in a pipe through which a liquid or a gas is flowing.

VENTURI PRINCIPLE

If a gas or a liquid is flowing through a pipe and the pipe diameter is narrow in one place, the flow speeds up in the narrow area. This increase in speed causes a lower pressure in the narrow area, which may be defined as a venturi (**Figure 2-29**). Power steering pumps use a venturi principle to assist in the control of pump pressure.

Technicians must understand basic principles to comprehend the complex systems on modern vehicles. When basic principles are mastered, then technicians understand both the reason for, and the result of, these specific service procedures.

Air speed increases, pressure decreases

Figure 2-29 A venturi increases airflow speed and produces a vacuum.

ELECTRICAL PRINCIPLES

We often think of suspension and steering components and systems as mechanical, but many subsystems are controlled or powered by electricity. Examples of these include the active steering system and the computer-controlled suspension system. Therefore, a basic understanding of some of the electrical principles, including voltage, amperage, and resistance, is required. The following section is meant to be a review or overview of these concepts. It is recommended that a technician receive in-depth electrical and electronic training as part of their overall education. Electronics is part of every major system on the automobile, and electrical failures have become routine complaints, though not always routine repairs.

The Basics: Voltage, Amperage, and Resistance

For many it is often easier to think of electricity in terms of water flowing in your home (hydraulic system principles) as there is a visible cause and effect. The flow of electricity is similar to the flow of water through your household plumbing. Where the water pressure is similar to electrical pressure or voltage, the water flow is similar to current flow in a conductor or amperage, and a restriction such as a kink in a hose is similar to the resistance in an electrical system. However, unlike water, electricity does not flow out the end of the wire and pour onto the ground if left open.

Voltage (V) or electromotive force (EMF) is the electrical pressure and is measured in volts (V); it may be either direct current (DC) or alternating current (AC). In the automotive industry, especially on hybrid electric vehicle (HEV) and electric vehicle (EV) platforms, we deal with both. Current is the flow or rate of flow of electrons under pressure in a conductor between two points having a difference in potential and is measured in **amperes (A)** or amperage. A complete circuit is required for current to flow. A DC circuit requires a complete circuit or loop between positive (+) and negative (−) for current to flow. Resistance is the friction in an electrical circuit, which restricts the flow of electrons under pressure and is measured in **ohms (Ω)**. Electrical resistance is a load on the moving current that must be present to do any useful work. Resistance controls the amount of current flow in an electrical circuit. Electrical devices that use electricity to operate have a greater amount of resistance than a conductor (wire) and are considered loads in a circuit. A motor, light bulb, or solenoids are examples of electrical loads in a circuit; a load in a circuit is the electrical device that consumes electricity. Poor connections and corrosion are examples of unwanted electrical loads in a circuit. Resistance in an electrical circuit is measured in ohms (Ω). If there is resistance (Ω) in the conductor, electrons will not flow as readily.

For all practical automotive purposes electricity only flows through a good conductor; in the majority of automotive wiring this is copper. It takes a high voltage (V) to flow

Voltage is a unit of measurement of electrical pressure.

Ohm is a unit of measurement of electrical resistance.

current (A) through a poor conductor. For example, air is a poor conductor, but with a high enough voltage even air can flow electricity. Lightning with millions of volts is capable of flowing current through air! High voltage electrical systems on HEV and EV platforms do not contain voltage as high as lightning, but it is not the 12 V system either. To jump an inch of air, it takes approximately 10,000 V. Have you ever seen a faulty secondary ignition wire (spark plug wire) arcing to the engine? The voltage on the HEV and EV orange wiring harness is generally between 200 and 300 V, which is too low to jump through air, but the capacitors and condensers in the system may contain much higher voltage.

To summarize:

- Voltage is the pressure that moves electrons and is measured in either AC or DC volts (V).
- Current or amperage is the flow or volume of electrons flowing in a conductor and is measured in amperes (A).
- Electrical resistance is a load or opposition on the moving electrons (current) in a circuit and is measured in ohms (Ω).

Ohm's Law

Ohm's law defines the relationship between voltage, resistance, and current.

The amount of current flow is determined by the amount of resistance in the loop of the circuit. In a fixed voltage circuit, if the amount of resistance in the loop of the circuit is high, the current flow will be low and if the amount of resistance in the loop of the circuit is low, the current flow will be high. This inverse relationship is known as **Ohm's law** and is summarized by the mathematical equation in **Figure 2-30**. Ohm's law states that it requires 1 V of electrical pressure to move 1 A through 1 Ω of resistance. Mathematically,

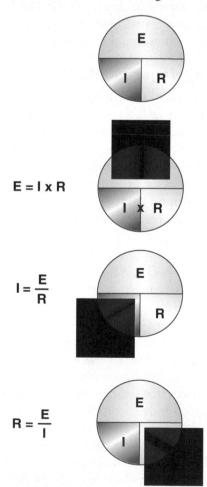

$$E = I \times R$$

$$I = \frac{E}{R}$$

$$R = \frac{E}{I}$$

Figure 2-30 With Ohm's law, if you know two of the three electrical factors, you can calculate the third.

Ohm's law is expressed by the following equations, using the symbols E for voltage, I for current, and R for resistance:

To find voltage $E = I \times R$

To find current $I = E \div R$

To find resistance $R = E \div I$

For example, if a 12.6 V circuit has 2 Ω of resistance, you can use Ohm's law to determine the current flowing in the circuit as follows:

$I = E \div R$

$I = 12.6 \div 2$

$I = 6.3 \text{ A}$

The current flowing in the circuit is 6.3 A.

These equations can be used to calculate the voltage, current, or resistance of a circuit. It is the understanding of this relationship that is most important to you as a technician. You are not generally crunching numbers; instead you are making measurements on a circuit with a digital multimeter (DMM) and trying to understand what this information means. Ohm's law can help to clear up what the meter is telling you for a given reading.

Types of Circuits and Using Digital Multimeters

There are three basic types of circuits that we will be dealing with, and a few ways to test these circuits for voltage, amperage, and resistance with a DMM. It is imperative that you understand these basic concepts and how to test a circuit with a DMM to avoid damage to the electrical circuit and components and/or damage to your expensive DMM. Remember Snap-On does not warranty misuse of equipment and service managers do not understand expensive mistakes! Take your time and think before you test. Remember the best place to begin your diagnosis is with the electrical diagram and system operation description contained in the vehicle service information database. As stated earlier this section is meant to be a review and not an in-depth training section, please refer to *Today's Technician Automotive Electricity & Electronics* for more information and training.

A series circuit is a circuit that provides a single path for current flow from the electrical source through all the circuit's components and back to the source (**Figure 2-31**).

Series Circuit laws:

- Current flow is the same at any point in the circuit (**Figure 2-32**).
- The sum of the individual voltage drops equals the source voltage (**Figure 2-33**).
- Total circuit resistance is the sum of the individual circuit resistances (**Figure 2-34**).

Figure 2-31 A simple series circuit including a switch (A), a fuse (B), and a lamp (C). For all article purposes the only load in the circuit is the lamp.

Figure 2-32 Current flow is the same at any point in the circuit.

Figure 2-33 The sum of the individual voltage drops equals the source voltage (2V + 4V + 6V = 12V or source voltage).

Figure 2-34 Total circuit resistance is the sum of the individual circuit resistances, $1\Omega + 2\Omega + 1\Omega = 4\Omega$ of total circuit resistance.

Figure 2-35 There are multiple paths for current to flow in a parallel circuit. If R_1 were to fail, R_2 would still work.

Parallel circuit is a circuit that provides two or more paths for current flow through all the circuit's components and back to the source (**Figure 2-35**). In a parallel circuit each path has separate resistances that operate independently or in conjunction with one another depending on the design of the circuit. Current can flow through more than one branch at a time, and voltage is the same across each branch of the circuit. In this type of circuit, the failure of a component in one branch does not affect the operation of the components in the other branches of the circuit.

Parallel Circuit laws:

■ Voltage is the same across each branch of the parallel circuit (**Figure 2-36**) and the voltage in each branch is used by the load(s) in that branch. The voltage dropped across each parallel branch will be the same; but, if the branch contains more than

Figure 2-36 Voltage is the same across each branch of the parallel circuit, and the voltage in each branch is used by the load(s) in that branch.

one resistor, the voltage drop across each of them will depend on the resistance of each resistor in the branch.

- Total current in a parallel circuit is equal to the sum of the individual branch currents (**Figure 2-37**).
- Total resistance in a parallel circuit is always less than the smallest resistive branch (**Figure 2-38**).

Figure 2-37 Total current in a parallel circuit is equal to the sum of the individual branch currents.

Total circuit resistance is 0.667 Ω

Figure 2-38 Total resistance in a parallel circuit is always less than the smallest resistive branch.

Figure 2-39 The series-parallel circuit has some loads that are in series and some that are in parallel with each other.

$$R_T = \frac{1}{1/R_1 + 1/R_2 + 1/R_3 + \ldots + 1/R_{10}}$$

$$R_T = \frac{1}{\dfrac{1}{1/3\,\Omega + 1/6\,\Omega + 1/1\,\Omega}}$$

$$R_T = \frac{1}{0.333 + 0.166 + 1}$$

$$R_T = \frac{1}{1.5}$$

$$R_T = 0.667\ \Omega$$

The series-parallel circuit has some loads that are in series and some that are in parallel with each other (**Figure 2-39**).

A voltmeter can be used to check for available voltage at the battery, terminals of any component or connectors. It can also be used to test voltage drops across electrical circuits, component loads, connectors, and switches. A voltmeter is connected in parallel with the circuit being tested (**Figure 2-40**). In **Figure 2-41** voltage is being tested in a closed 12 V series circuit with two loads. At test point A, the voltage should be the source voltage of 12 V. At point B, the 1 Ω resistor would have dropped half the voltage (there are two 1 Ω loads in the circuit) and the meter should read 6 V. At test point C, all

Figure 2-40 A voltmeter is connected in parallel with the circuit being tested.

Figure 2-41 A voltmeter testing for voltage at various points in the series circuit.

the voltage should have been used up by the two loads in the circuit and the meter reading should be 0 V. These readings would indicate normal circuit operation.

One of the most useful tests to perform is the voltage drop test. In a circuit all of the voltage provided by the source power is used (dropped) by the circuit with nothing left over. The voltage is used by resistance in wiring, connectors, switches, and loads. This loss or use of voltage is called a voltage drop and is the amount of electrical energy converted into another form of energy. Voltage dropped in wiring, connectors, and switches is converted into heat energy. When a circuit or branch of a circuit has only one load (resistance) source voltage is dropped across that load (**Figure 2-42**). If there is more than one load in a circuit each load will use a portion of the voltage. The total of all voltage drops in a circuit should equal source voltage. All of the voltage must be used by the circuit. You should verify that the circuit is turned on and that source voltage is available at the load and that the load drops source voltage. If the component (load) that is in the circuit drops source voltage, then the component is faulty. In **Figure 2-43**, if both Lamp 1 and Lamp 2 are the same resistance value (size) they will both share the source voltage equally. Lamp 1 will use 6 V and Lamp 2 will use 6 V. If there is unwanted resistance in a circuit the load in the circuit will not drop all source voltage (**Figure 2-44**). There is some allowable voltage drop by circuit components other than the load, but it is generally limited to a maximum of:

0.2 V (200 mV) for wires and cables
0.3 V (300 mV) for a switch
0.1 V (100 mV) for a ground
0 V for a connection or connectors

Common circuit faults:

- Short circuits
- Short to ground
- Opens in a circuit
- High resistance (may be caused by corrosion or poor connections)
- Low voltage

Figure 2-42 Only the load will drop source voltage; the wiring and the switch should not drop significant voltage.

Available voltage point A =   12.00 V
Minus available voltage point B =   6.00 V
Voltage drop across lamp 1 =   6.00 V

Available voltage point C =   6.00 V
Minus available voltage point D =   0.00 V
Voltage drop across lamp 2 =   6.00 V

Total voltage drop between points A and D = 12.00 V

Figure 2-43 There should be source voltage before the first load. If there are two loads in a series circuit each load will drop a portion of the source voltage. There should be no voltage after the last load.

Figure 2-44 Only the load will drop source voltage; the wiring and the switch should not drop significant voltage. if there is unwanted resistance in a circuit, source voltage will not be available for the light bulb and the light bulb would be dimmer than normal or may not light at all.

TABLE 2-1 Circuit Test Chart

Type of Defect	Test Unit	Expected Results
Open	Ohmmeter	∞ infinite resistance between conductor ends
	Test light	No light after open
	Voltmeter	∅ volts at end of conductor after the open
Short to ground	Ohmmeter	∅ resistance to ground
	Test light	Lights if connected across fuse
	Voltmeter	Generally not used to test for ground
Short	Ohmmeter	Lower than specified resistance through load component ∅ resistance to adjacent conductor
	Test light	Light will illuminate on both conductors
	Voltmeter	A voltage will be read on both conductors
Excessive resistance	Ohmmeter	Higher than specified resistance through circuit
	Test light	Light illuminates dimly
	Voltmeter	Voltage will be read when connected in parallel over resistance

Table 2-1 shows common testing units and expected results when diagnosing the above common circuit faults. When using an ammeter to measure amperage, be sure to first turn power off in the circuit before connecting the DMM. The circuit must be opened (disconnected) from the load being tested. The DMM is placed in series in the circuit, recompleting the disconnected circuit (**Figure 2-45**). Always verify that the DMM is capable of handling the highest expected amperage in the circuit being tested. What size fuse protects the circuit being tested? Many DMMs are only internally protected to 10 A by their internal fuse and many automotive circuits can provide 20 or more amps. In these cases, an inductive amp probe is the best choice of tools to use to avoid DMM damage (**Figure 2-46**). The inductive probe eliminates the need to connect the DMM in series and is a safe noninvasive method of measuring amperage in a circuit.

When using an ohmmeter to measure resistance, the power from the circuit must be removed and circuit or component should be isolated (**Figure 2-47**). Ohmmeter leads are placed across or parallel with the component or circuit being tested.

Figure 2-45 When testing amperage draw the circuit must be opened (disconnected) from the load being tested and the multimeter is placed in series in the circuit.

Figure 2-46 A multimeter with an inductive amp probe is a safe, noninvasive method of measuring amperage in a circuit.

Fuse removed
to de-energize
circuit

Figure 2-47 When measuring resistance with an ohmmeter, the meter is connected in parallel with power removed from the circuit and the circuit or component isolated.

SUMMARY

- Work is the result of applying a force.
- Force is measured in pounds and distance.
- Energy is the ability to do work; there are six basic types of energy.
- Inertia is the tendency of an object at rest to remain at rest, or the tendency of an object in motion to remain in motion.
- An object gains momentum when force overcomes static energy and moves the object.
- Friction is the resistance to motion when one object is moved over another object.
- Mass is a measurement of an object's inertia.
- Weight is a measurement of the earth's gravitational pull on an object.

- Volume is the length, width, and height of a space occupied by an object.
- Power is a measurement for the speed at which work is done.
- A rolling, tilted wheel tends to move in the direction of the tilt.
- If the pivot point at the tire centerline is behind the centerline where the vehicle weight is projected on the road surface, the wheel tends to remain in the straight-ahead position.
- To obtain static balance, weight must be distributed equally around the center of wheel rotation as viewed from the side.

- Weight must be distributed equally on both sides of the tire centerline viewed from the front of the tire to maintain proper dynamic balance.
- Torque is a twisting force that does work.
- Atmospheric pressure is the total weight of the earth's atmosphere.
- Vacuum is the absence of atmospheric pressure.
- Voltage (V) or electromotive force (EMF) is the electrical pressure and is measured in volts (V); it may be either direct current (DC) or alternating current (AC).

- Current is the flow or rate of flow of electrons under pressure in a conductor between two points having a difference in potential and is measured in amperes (A) or amperage.
- Resistance is defined as the opposition to current flow and is measured in ohms (Ω).
- Ohm's law defines the relationship between voltage, current, and resistance. It is the basic electrical law.

REVIEW QUESTIONS

Short Answer Essays

1. Describe Newton's first law of motion, and give an application of this law in automotive theory.

2. Explain Newton's second law of motion, and give an example of how this law is used in automotive theory.

3. Describe Newton's third law of motion, and give an example of how this law applies to an automotive suspension system.

4. Describe six different forms of energy.

5. Describe four different types of energy conversion.

6. Explain the difference between static and dynamic inertia.

7. Explain why a rotating, tilted wheel moves in the direction of the tilt.

8. Explain why the front wheel of a bicycle tends to remain in the straight-ahead position as the bicycle is driven.

9. Define static and dynamic balance.

10. Briefly define the term *current*.

Fill-in-the-Blanks

1. Improper static wheel balance causes wheel _____ when driving the vehicle.

2. If the tire pivot point is behind the wheel-and-tire centerline where the vehicle weight is projected against the road surface, the wheel tends to remain in the _____ position when driving the vehicle.

3. _____ is a unit of measurement of electrical resistance.

4. _____ is the flow or rate of flow of electrons under pressure in a conductor between two points having a difference in potential and is measured in _____.

5. When one object is moved over another object, the resistance to motion is called _____.

6. _____ is the electrical pressure and it may be either direct current (DC) or alternating current (AC).

7. Torque is a force that does work with a/an _____ action.

8. Power is a measurement for the rate at which _____ is done.

9. To obtain proper dynamic balance, the weight must be distributed equally on both sides of a wheel-and-tire centerline viewed from the _____ of the tire.

10. Vacuum is defined as the absence of _____.

Multiple Choice

1. When an engine is running, the alternator converts:
   A. Thermal energy to mechanical energy.
   B. Electrical energy to mechanical energy.
   C. Chemical energy to electrical energy.
   D. Mechanical energy to electrical energy.

2. Work is accomplished during all of these conditions EXCEPT:

   A. When a mechanical force starts an object in motion.

   B. When a mechanical force is applied to an object, but the object does not move.

   C. When a mechanical force stops an object in motion.

   D. When a mechanical force redirects an object in motion.

3. All these statements about Newton's Laws of Motion are true EXCEPT:

   A. For every action there is an equal and opposite reaction.

   B. A body in motion remains in motion unless an outside force acts on it.

   C. A body's acceleration is directly proportional to the force applied to it.

   D. A body moves in an arc away from the force acting upon the object.

4. Torque is calculated by:

   A. Multiplying the force and the radius from the force to the object.

   B. Dividing the radius by the force.

   C. Adding the force and the distance of force movement.

   D. Subtracting the radius from the force to the object and the weight of the object.

5. When working with gases and liquids:

   A. Gases are not compressible.

   B. Liquids are not compressible.

   C. Temperature does not affect gas volume.

   D. Pressure is applied unevenly to the inside of a container surface.

6. When applying the principles of work and force:

   A. Work is accomplished when force is applied to an object that does not move.

   B. In the metric system the measurement for work is cubic centimeters.

   C. No work is accomplished when an object is stopped by mechanical force.

   D. If a 50-lb object is moved 10 ft., 500 ft.-lb of work are produced.

7. All of the following are true of voltage drops EXCEPT:

   A. Voltage drop can be measured with a voltmeter.

   B. All of the source voltage in a circuit must be dropped.

   C. Corrosion in a circuit does not cause a voltage drop.

   D. Voltage drop is the conversion of electrical energy into another form of energy.

8. A lever is 10 ft. (3 m) long and the fulcrum is positioned 1 ft. (0.304 m) from the end of the lever. A 5-lb (2.26-kg) weight is placed on the end of the lever nearest the fulcrum. The weight required on the opposite end of the lever to lift the 5-lb (2.26-kg) weight is:

   A. 0.368 lb (0.166 kg).

   B. 0.555 lb (0.251 kg).

   C. 0.714 lb (0.323 kg).

   D. 0.748 lb (0.339 kg).

9. A tire-and-wheel assembly that does not have proper dynamic balance has:

   A. Weight that is not distributed equally on both sides of the tire centerline.

   B. Worn tread on the inside edge of the tire.

   C. A tire with improperly positioned steel belts.

   D. Weight that is not distributed equally around the center of the tire and wheel.

10. When a car is driven on the road above 50 mph (80 km/h), the left front tire has a tramping action. The MOST likely cause of this condition is:

    A. Improper left front dynamic wheel balance.

    B. A bent left front wheel rim.

    C. Improper left front static wheel balance.

    D. Improper rim offset.

CHAPTER 3

TIRES AND WHEELS

Upon completion and review of this Chapter, you should be able to understand and describe:

- General tire function.
- Typical tire construction, and identify the purpose of each component in a tire.
- Three types of tire ply and belt designs.
- Tire ratings, and explain the meaning of each designation in the rating.
- The purpose of the tire performance criteria (TPC) rating.
- The difference between all-season tires and conventional tires.
- Two different types of tire load ratings.
- The Uniform Tire Quality Grading (UTQG) designations.
- The precautions to be observed when selecting replacement tires.
- Tire contact area, free tire diameter, and rolling tire diameter.

- The tire motion forces while a tire-and-wheel assembly is rotating on a vehicle.
- The importance of tire design quality as it relates to the tire motion forces.
- Wheel offset.
- Plus tire sizing.
- Wheel tramp, and explain how static unbalance causes wheel tramp.
- Wheel shimmy, and describe how dynamic unbalance results in wheel shimmy.
- Various types of tire pressure monitoring systems.
- The advantage of nitrogen tire inflation.
- Noise, vibration, and harshness analysis.

Terms To Know

All-season tires
Amplitude
Bead filler
Bead wire
Belt cover
Belted bias-ply tires
Bias-ply tires
Compact spare tires
Conicity
Cord plies
Dynamic balance
Electronic vibration analyzer (EVA)
Hertz
Hydroplaning
Liner
Load rating

Magnesium alloy wheels
Mud and snow tires
Natural vibration frequencies
Nitrogen tire inflation
Puncture sealing tires
Radial-ply tires
Replacement tires
Resonance
Road force variation
Run-flat tires
Sidewalls
Speed rating
Static balance
Synthetic rubber
Temperature rating
Tire belts

Tire chains
Tire contact area
Tire free diameter
Tire performance criteria (TPC)
Tire placard
Tire pressure
Tire rolling diameter
Tire treads
Tire valve
Traction ratings
Tread wear ratings
Uniform Tire Quality Grading (UTQG) System
Wheel rims
Wheel tramp

INTRODUCTION

Although tires are often taken for granted, they contribute greatly to the ride and steering quality of a vehicle. Tires also play a significant role in vehicle safety. Improper types of tires, incorrect inflation pressure, and worn-out tires create a safety hazard. When tires and wheels are out of balance, tire wear and driver fatigue are increased, which can create a driving hazard. Tires serve the following functions:

1. Tires cushion the vehicle ride to provide a comfortable ride for the occupant.
2. Tires must firmly support the vehicle weight.
3. Tires must develop traction to drive and steer the vehicle under a wide variety of road conditions. In other words, they must transmit traction and braking forces to the road.
4. Tires contribute to directional stability of the vehicle and must absorb all the stresses of accelerating, braking, and centrifugal force in turns.

TIRE DESIGN

Tires equipped on today's passenger cars and light trucks are of the radial ply design. Tire construction varies depending on the manufacturer and the type of tire. A typical modern tire contains these components (**Figure 3-1**):

The **bead wire** is a group of wire strands that retain the bead on the wheel rim.

The tire **liner** is made from synthetic gum rubber and seals the inside of the tire.

1. **Bead wire**
2. Bead filler
3. **Liner**, which functions as the air seal membrane
4. Sidewall with hard side compound
5. Rayon carcass plies
6. Steel belts
7. Jointless belt cover
8. Hard undertread compound
9. Hard high-grip tread compound

Figure 3-1 The typical tire design and construction.

The bead wire contains several turns of bronze-coated steel wire in a continuous loop. This bead wire is molded into the tire at the inner circumference and wrapped in the cord plies. The bead wire anchors the tire to the wheel. The **bead filler** above the bead reinforces the sidewall and acts as a rim extender.

Tire **sidewalls** are made from a blend of rubbers, which absorbs shocks and impacts from road irregularities, prevents damage to the plies, and also contains antioxidants and other chemicals that are gradually released to the surface of the sidewall during the life of the tire. These antioxidants help keep the sidewall from cracking and protect it from ultraviolet radiation and ozone attack. Since the sidewall must be flexible to provide ride quality, minimum thickness of this component is essential. Tire manufacturers have reduced sidewall thickness by 40 percent in recent years to reduce weight and heat buildup, improve ride quality, reduce rolling resistance, and improve fuel economy. Some off-road and performance tires are equipped with raised rubber that extends above the rim area and is referred to as a rim guard to offer added protection. A lettering and numbering arrangement for tire identification is located on the outside of the sidewall.

Shop Manual
Chapter 3, page 109

The **cord plies** surround both beads and extend around the inner surface of the tire to enable the tire to carry its load. The plies are molded into the sidewalls. Each ply is a layer of rubber with parallel cords imbedded in its body. The load capacity of a tire may be increased by adding more cords in each ply or by installing additional plies. The most common materials in tire plies are polyester, rayon, and nylon. Passenger car tires usually have two-cord plies, whereas the tires of many trucks and recreation vehicles are rated to carry the heavier loads of these vehicles and are equipped with a higher ply-rated tire. In general, tires with more plies have stiffer sidewalls, which provide less cushioning and reduced ride quality.

Steel is the most common material in **tire belts**, although other belt materials such as polyester have been used to some extent. Many tires contain two steel belts. The tire belts restrict ply movement and provide tread stability and resistance to deformation. This belt action provides longer tread wear and reduces heat buildup in the tire. Steel belts expand as wheel speed and tire temperature increase. Centrifugal force and belt expansion tend to tear the tire apart at high speeds and temperatures. Therefore, high-speed tires usually have a nylon jointless **belt cover**. This nylon belt cover contracts as it is heated and helps hold the tire together, providing longer tire life, improved stability, and better handling. Additionally, a tire may have a layer of DuPont Kevlar for added puncture proof protection under the tread layer.

Tire belts are positioned between the tread and the cord plies.

Tire **treads** are made from a blend of rubber compounds that are very resistant to abrasion wear. Spaces between the tire treads allow tire distortion on the road without scrubbing, which accelerates wear. Modern automotive tires contain two layers of tread materials. The first tread layer is designed to provide cool operation, low rolling resistance, and durability. The outer layer, or tread, is designed for long life and maximum traction. Tread rubber is a blend of many different synthetic and natural rubbers. Tire manufacturers may use up to thirty different types of **synthetic rubber** and eight types of natural rubber in their tires. Manufacturers blend these synthetic and natural rubbers in both tread layers to provide the desired traction and durability. Tire treads must provide traction between the tire and the road surface when the vehicle is accelerating, braking, and cornering. This traction must be maintained as much as possible on a wide variety of road surfaces. For example, on wet pavement, tire treads must be designed to drain off water between the tire and the road surface. This draining action is extremely important for maintaining adequate acceleration, braking, and directional control. Lines cut across the tread provide a wiping action, which helps dry the tire-road contact area. Most tire manufacturers add a PermaBlack compound to the rubber in their tires to maintain new-tire appearance throughout the life of the tire.

Tire treads are mounted between the sidewalls.

Synthetic rubber compounds are developed in laboratories, whereas natural rubber is found in nature.

A BIT OF HISTORY

In 1834, Charles Goodyear was a hardware merchant from Philadelphia with a great interest in a new substance imported from Brazil called rubber. When rubber was first imported to the United States, many entrepreneurs were interested in manufacturing products from rubber. However, these entrepreneurs soon discovered that rubber became bone-hard in cold weather, and then turned to a glue-like substance in very hot weather, and these characteristics ended the rubber manufacturing business at that time. Charles Goodyear persisted in experimenting with rubber! He experimented with mixing magnesia powder, nitric acid, and sulfur with rubber to reduce its stickiness. These substances improved the rubber quality, but it still was not perfected. Goodyear was often financially bankrupt and he had difficulty obtaining enough money to feed his children. He was even jailed on some occasions for failure to pay his bills, and ill health often plagued him.

By accident, Goodyear discovered that extreme heat changed the rubber to a weatherproof, gum-like material. He found that pressurized steam applied to rubber for four to six hours at 270°F (132°C) provided the most uniform rubber. Goodyear was now able to obtain financial backing to produce rubber. Goodyear was plagued with patent right problems and those trying to steal his rubber patents. However, he did succeed in discovering modern rubber! Neither Charles Goodyear nor his family were ever connected with the company named in his honor, today's Goodyear Tire and Rubber Company.

Today there is a cultivated rubber tree for every two human beings on Earth, and three million people are employed collecting the rubber from the trees in various countries. The United States imports approximately half of the rubber in the world and synthesizes an equal amount from petroleum. About 300,000 people in the United States are employed in the rubber industry, and they produce approximately $6 billion worth of products each year.

The synthetic gum rubber liner is bonded to the inner surface of the tire to seal the tire. All modern passenger car and light truck tires are the tubeless type. In these tires, the tire bead must provide an airtight seal on the rim, and both the tire and the wheel rim must be completely sealed. The inner liner of the tire provides an air tight seal.

Some heavy-duty truck tires and off-road equipment have inner tubes mounted inside the tire. On tube-type tires, the air is sealed in the inner tube, and the sealing qualities of the tire and wheel rim are not important.

Designing tires is a very complex engineering operation. The average all-season tire contains the components (by weight) listed in **Table 3-1**. Tire design varies depending on the operating conditions and the load capacity of the tire. A tire designed for improved steering and handling characteristics has a nylon bead reinforcement and a hard bead filler

TABLE 3-1 Tire Components by Weight

A typical P195/75-14 all-season tire contains:	
Synthetic rubber (30 types)	2.49 kg
Carbon Black (8 types)	2.27 kg
Natural Rubber (8 types)	2.04 kg
Chemicals, waxes, oils, pigments, etc. (40 types)	1.36 kg
Steel cord for belts	0.68 kg
Polyester and nylon	0.45 kg
Bead wire	0.23 kg
Total weight	9.42 kg

Figure 3-2 Tires designed for improved steering and handling with a nylon bead reinforcement and a hard bead filler with a slim tapered profile.

with a slim tapered profile (**Figure 3-2**). This type of tire is suitable for sports car operation because the design stiffens the tire and reduces tire deflection during high-speed cornering. However, this type of tire may provide slightly firmer ride quality.

TIRE PLY AND BELT DESIGN

The three basic tire construction designs that have been used are the bias, belted bias, and belted radial. In **bias-ply** or **belted bias-ply tires**, the cords crisscross each other. These cords are usually at an angle of 25° to 45° to the tire centerline. The belt-ply cord angle is usually 5° less than the cord angle in the tire casing. Two plies and two belts are most commonly used, but four plies and four belts may be used in some tires. Compared to a bias-ply tire, a belted bias-ply tire has greater tread rigidity. The belts reduce tread motion during road contact. This action provides extended tread life compared to a bias-ply tire.

In **radial-ply tires**, the ply cords are arranged radially at a right angle to the tire centerline (**Figure 3-3**). Steel belts are most commonly used in radial tires, but other belt materials such as fiberglass, nylon, and rayon have also been used. The steel or fiberglass cords in the belts are crisscrossed at an angle of 10° to 30° in relation to the tire centerline. Many radial tires have two plies and two belts. Radial tires provide less rolling resistance, better steering characteristics, and longer tread life than bias-ply tires. As was stated earlier all modern vehicles are equipped with radial-ply tires.

> **Belted bias-ply tires** have steel or polyester belts between the cord plies and the tread.

Tread

Tread belts

Carcass plies

BELTED RADIAL   BELTED BIAS   BIAS

Figure 3-3 Three types of tire construction: bias-ply, belted bias-ply, and belted radial-ply.

Regardless of the type of tire construction, the tire must be uniform in diameter and width. Radial runout refers to variations in tire diameter. A tire with excessive radial runout causes a tire thumping problem as the car is driven. When a tire has excessive variations in width, this condition is called lateral runout. A tire with excessive lateral runout causes the chassis to "waddle" when the car is driven.

The tire plies and belts must be level across the tread area. If the plies and/or belts are not level across the tread area, the tire is cone shaped. This condition is referred to as tire **conicity**. This can occur for several reasons; if the tires' internal components are misaligned during the manufacturing curing process it is possible that unequal forces internally could develop causing conicity. With this type of manufacturing defect, the tire pull will be present immediately after tire installation on a vehicle, or following the first time the tires are rotated. When a front tire has conicity, the steering may pull to one side as the car is driven straight ahead. This is referred to as a radial pull. A rear tire with conicity will not affect the steering as much as a front tire with conicity. If the radial pull develops later in the tires service life it is likely due to vehicle misalignment or driving conditions that have caused a slight angular wear pattern on the tire. This is when the tire tread wears faster on one side than the other (inside versus outside). If a tire pull is suspected, the front tires can be cross rotated. If the pull shifts direction or goes away, suspect a radial pull; further diagnosis will be required to determine which tire is affected. Often rotating the front tires to the rear tire position will eradicate this problem and even out the tire wear pattern. Additionally, the root cause of the initial improper wear pattern should be addressed, such as setting the vehicle alignment to manufacturer's preferred specifications.

> Tire **conicity** refers to a condition where the plies and/or belts are not level across the tire tread. When a tire has conicity, the plies and/or belts are somewhat cone shaped and may cause the vehicle to pull in one direction when driven.

TIRE TREAD DESIGN

Vehicle tires have many different tire treads designed to provide the tire performance desired by the tire and vehicle manufacturers. A typical modern tire tread has these features:

1. An interlocking tread pattern for improved tread gripping quality on icy or slick roads (**Figure 3-4**).
2. Deeply carved aquachutes to propel water off the tread and away from the tire to reduce the possibility of **hydroplaning** when driving on wet pavement (**Figure 3-5**).
3. Reinforced tread shoulders to improve tread gripping quality when turning corners on dry pavement (**Figure 3-6**).

> **Hydroplaning** occurs when water on the pavement is allowed to remain between the pavement and the tire tread contact area. This action reduces friction between the tire tread and the road surface, and can contribute to a loss of steering control.

Figure 3-4 Interlocking tread pattern for improved traction on icy roads.

Figure 3-5 Aquachutes propel water off the tire tread.

Figure 3-6 Reinforced tread shoulders improve tread gripping quality while cornering.

Symmetrical tread   Asymmetrical tread   Directional tread

Figure 3-7 Tire tread pattern designs generally fall into symmetrical, asymmetrical, and directional designs.

Tire tread pattern designs generally fall into symmetrical, asymmetrical, and directional designs (**Figure 3-7**). Most tires manufactured have a nondirectional symmetrical tread design and can be mounted on a rim in either direction with no defined inside or outside preference on sidewall. Some tires feature an asymmetrical tire tread design that has a different tire tread pattern on the inside versus the outside tread surface area of the tire. An asymmetrical tire has a defined outside sidewall with markings identifying the inner side and outer side of the tire (**Figure 3-8**). If all four tires are the same size, the tires are rotated the same as symmetrical tires.

Directional tire tread designs have a unique pattern that is designed to rotate in a specific direction for optimum performance. Direction tread patterns can be either symmetrical or asymmetrical (**Figure 3-9**). A directional tire has a V-shaped tread pattern, and the tire has a defined sidewall marking indicating the direction of rotation with an arrow. When mounting four directional tires, two will be mounted with the arrow pointing in the clockwise direction and two will be mounted with the arrow pointing in the counterclockwise direction. When all four tires are the same size they can only be rotated front-rear and must remain on the original side of the vehicle (**Figure 3-10**). If the vehicle

Asymmetrical   Sidewall label distinguishes
tread pattern   inner and outer sides of tire

Figure 3-8 An asymmetrical tire has a defined sidewall marking indicating the outside or inside sidewall.

V-shaped tread pattern

Must install according to the direction of the arrow on the sidewall

Figure 3-9 A directional tire has a V-shaped tread pattern and the tire has a defined sidewall marking indicating the direction of rotation with an arrow.

Figure 3-10 With directional tires, when all four tires are the same size they can only be rotated front-rear and must remain on the original side of the vehicle.

is equipped with directional tires but the front and rear tires are different sizes, then the tires are position specific and tire rotation is prohibited.

All-season tires are designed for all road conditions that a typical vehicle could experience in North America including wet, dry, hot, cold, winter weather and they have an all-season designation of M+S which stands for mud and snow on the tire side wall (**Figure 3-11**), variations of this symbol include M&S, M/S, and MS. These tires have a 37 percent higher average snow traction compared to non-all-season tires. All-season tires may have slightly improved performance in areas such as wet traction, rolling resistance,

Figure 3-11 All-season tires are designated with the M+S symbol.

Figure 3-12 All-season tires are generalist and while they perform adequately in most conditions, they do not provide the improved traction of a specialized winter tire for harsh snow and ice conditions.

and tread life. Improvements in tread design and tread compounds provide the superior performance in all-season tires. Though they are designated as an all-season tire, they do not perform as well as a winter/ice tire in extreme cold and harsh weather conditions (**Figure 3-12**).

Winter tires are designed to perform in extreme cold conditions, and remain flexible even at 7°F (-14°C) to provide improved traction performance. Winter tires carry the triple-peak mountain snowflake symbol, which is an indication that the tire meets minimum standards for harsh conditions traction (**Figure 3-13**). The Rubber Manufacturers Association sets the standards for approved tires that may carry the all-season and winter tire designation. Some winter tires are studdable, meaning they have a tread design that allows the installation of metal studs for enhanced ice and hard-packed snow performance (**Figure 3-14**). Studded tires provide improved traction on ice, but these tires are prohibited by law in many states because their use resulted in road surface damage. The states that do allow studded tires limit their use to certain winter months (**Figure 3-15**). A winter tire also often has sipes, which are small zigzag slits cut into the tread blocks to improve traction on slippery road surfaces. Some all-season tires are also equipped with sipes.

Figure 3-13 Winter tires are designated with M+S and have a triple peak mountain and snowflake symbol.

Figure 3-14 Winter tire tread design has a deep lug pattern, sipes for improved ice and slush traction and may also accept studs.

STUDDED TIRE RESTRICTIONS BY STATE & PROVINCE

STATE/ PROVINCE	PERMISSABLE DATES	STATE/ PROVINCE	PERMISSABLE DATES	STATE/ PROVINCE	PERMISSABLE DATES
ALABAMA	PROHIBITED	ALASKA	SEPT 30 - APR 15	ARIZONA	OCT 1 - MAY 1
ARKANSAS	NOV 15 - APR 15	CALIFORNIA	NOV 1 - APR 30	COLORADO	PERMITTED NO RESTRICTIONS
CONNECTICUT	NOV 15 - APR 30	DELAWARE	OCT 15 - APR 15	D.C.	OCT 15 - APR 15
FLORIDA	PROHIBITED	GEORGIA	PEMITTED ONLY FOR SNOW & ICE	HAWAII	PROHIBITED
IDAHO	OCT 15 - APR 30	ILLINOIS	PROHIBITED	INDIANA	OCT 1 - MAY 1
IOWA	NOV 1 - APR 1	KANSAS	NOV 1 - APR 15	KENTUCKY	PERMITTED NO RESTRICTIONS
LOUISIANA	PROHIBITED	MAINE	OCT 1 - MAY 1	MARYLAND	PROHIBITED
MASSACHUSETTS	NOV 2 - APR 30	MICHIGAN	PROHIBITED	MINNESOTA	PROHIBITED
MISSISSIPPI	PROHIBITED	MISSOURI	NOV 1 - MAR 31	MONTANA	OCT 1 - MAY 31
NEBRASKA	NOV 1 - APR 1	NEVADA	OCT 1 - APR 30	NEW HAMPSHIRE	NO RESTRICTIONS
NEW JERSEY	NOV 15 - APR 1	NEW MEXICO	NO RESTRICTIONS	NEW YORK	OCT 16 - APR 30
NORTH CAROLINA	NO RESTRICTIONS	NORTH DAKOTA	OCT 15 - APR 15	OHIO	NOV 1 - APR 15
OKLAHOMA	NOV 1 - APR 1	OREGON	NOV 1 - APR 1	PENNSYLVANIA	NOV 1 - APR 15
RHODE ISLAND	NOV 15 - APR 1	SOUTH CAROLINA	ALL YEAR. 1/16 Studs	SOUTH DAKOTA	OCT 1 - APR 30
TENNESSEE	OCT 1 - APR 15	TEXAS	PROHIBITED	UTAH	OCT 15 - MAR 31
VERMONT	NO RESTRICTIONS	VIRGINIA	OCT 15 - APR 15	WASHINGTON	NOV 1 - APR 1
WEST VIRGINIA	NOV 1 - APR 15	WISCONSIN	PROHIBITED	WYOMING	NO RESTRICTIONS
ALBERTA	NO RESTRICTIONS	BRITISH COLUMBIA	OCT 1 - APR 30	MANITOBA	OCT 1 - APR 30
NEW BRUNSWICK	OCT 16 - APR 30	NEWFOUNDLAND	NOV 1 - APR 30	N.W. TERRITORIES	NO RESTRICTIONS
NOVA SCOTIA	OCT 15 - APR 30	ONTARIO	PROHIBITED	P.E.I.	OCT 1 - MAY 31
QUEBEC	OCT 1 - MAY 1	SASKATCHEWAN	NO RESTRICTIONS	YUKON TERRITORY	NO RESTRICTIONS

Figure 3-15 Many states restrict the use of studded tires to winter months; some states prohibit their use all together.

One tire manufacturer installs 2 to 3 billion microscopic hollow shells or micro-bubbles in the tread material on one brand of their tires. These hollow shells are installed to 60 percent of the tread depth, and add rigidity to the tread material. When these hollow shells contact the road surface they break open, and the shell edges provide a gripping effect to improve traction. When driving on wet pavement, each time a hollow shell contacts the road surface and breaks open, a small amount of water is pulled into the hollow shell (**Figure 3-16**). Because this action is occurring at the millions of hollow shells in contact with the road surface, water is removed from between the tread and the road surface. This action reduces the possibility of hydroplaning. As the tire continues to rotate, water is expelled from the hollow shells when they move out of contact with the road surface.

HOW SHELLED MICRO-BUBBLES WORK

CONTACT	SUCTION	DRAINAGE
At contact with road surface, the shelled bubble breaks to create empty space	Space sucks out water and breaks water film between tread rubber and road surface	Water in the space is drained by rotation of the tire

Figure 3-16 Advantages of micro-bubbles in the tire tread.

TIRE DEFECTS

Tire defects may be the result of manufacturing anomalies, road impact, or misuse. Manufacturing defects may be covered by the tire manufacturer's warranty. Possible defects include the following:

1. Separations and bulges, including tread or sidewall separation, or an open tread splice
2. Tread chunking, tearing, groove cracking, and shoulder cracking
3. Sidewall circumference fatigue, cracking, and weather cracking
4. Bead chafing, broken wires, or a pulled bead
5. Breaks in the sidewall or tread areas
6. Excessive radial or lateral runout
7. Conicity

If this inner liner is pinched or damaged, air can leak between the tire plies, forming a bubble on the sidewall or under the tread of the tire. This is often the result of tire impact damage such as hitting a pothole or curb, and is not a manufacturing defect. Inner liner damage can also result from an extreme underinflation condition that causes excessive heat buildup and flexing of the sidewall of the tire. It is critical that a tire suspected of being damaged or that has been driven underinflated be removed and inspected thoroughly before attempting to inflate or repair the assembly. A tire that is driven on while flat is almost always irreparable and must be replaced.

TIRE RATINGS AND SIDEWALL INFORMATION

A lot of essential information is molded into the sidewall of the average passenger car and light truck tire, including the tire rating. There are many dimensions related to tires and rims; not all this information is listed on the sidewall of the tire but is no less important

(**Figure 3-17**). The tire rating is a group of letters and numbers that identify the tire type, section width, aspect ratio, construction type, rim diameter, load capacity, and speed symbol. The tire in **Figure 3-18** has a P215/55R17 95H rating.

Figure 3-17 Various dimensions associated with tire and rim.

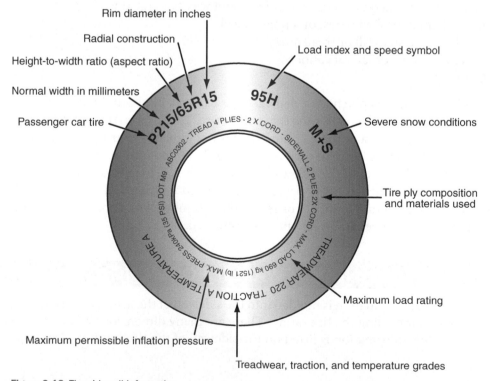

Figure 3-18 Tire sidewall information.

P indicates a passenger car tire

LT indicates a light truck tire

T indicates a temporary tire most commonly a compact spare tire

The complete DOT code must be stamped on one side of the tire sidewall. This code is a combination of letters and numbers and indicates the manufacturer plant code where the tire was manufactured, the batch code, and the week and year the tire was manufactured (**Figure 3-19**). It is not a unique number to a specific tire but rather an identification for a batch of tires produced. Beginning in 2000, the week and year the tire was manufactured are indicated by the last four digits of the DOT number (**Figure 3-20**). The first two digits indicate the week while the last two digits indicate the year of production. As an example, if the last four digits are 2514, then the tire was produced during the 25th week of 2014. On tires produced before 2000, the week and year the tire was manufactured are indicated by the last three digits of the DOT number (**Figure 3-21**). The first two digits indicate the week, while the last digit indicates the year of production. As an example, if the last three digits are 254, then the tire was produced during the 25th week of 1994.

Section Width, Aspect Ratio, and Ply Design

In the tire size P215/55R17, the number 215 is the section width of the tire in millimeters measured from sidewall to sidewall with the tire mounted on the recommended rim width. The section width, also referred to as the cross-section width, is the width of the tire from the widest point measured from one sidewall to the opposite sidewall.

The aspect ratio is a number that indicates the section height of the tire sidewall. In the tire size P215/55R17, the number 55 indicates the aspect ratio, which is the ratio of the height to the width. With a 55 aspect ratio, the tire's height is 55 percent of its width

Figure 3-19 The tire DOT code contains batch information about where and when a tire was manufactured.

In the example above:
DOT U2LL LMLR 5107

51   Manufactured during the 51st week of the year
07   Manufactured during 2007

Figure 3-20 On tires produced beginning in 2000 the last four digits of the DOT code indicates among other things the week and year the tire was produced.

In the example above:
DOT EJ8J DFM 408

40	Manufactured during the 40th week of the year
8	Manufactured during the 8th year of the decade

Figure 3-21 On tires produced before 2000, the last three digits of the DOT code indicates, among other things, the week and year the tire was produced.

P225/60R16 97T

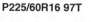

$$\text{Aspect ratio} = \frac{\text{Section height}}{\text{Section width}} = 60\%$$

Figure 3-22 The aspect ratio is determined by dividing section height by section width and is represented as a percentage.

(**Figure 3-22**). The aspect ratio is also referred to as the profile of the tire. Low-profile tires are those with an aspect ratio of 50 and below. Currently, more than 50 percent of the vehicles manufactured are fitted with low-profile tires. The letter R indicates radial-ply tire construction. Tires used on passenger cars and light trucks have been of radial-ply construction since the late 1970s. If the tire construction was indicated by the letter B, the tire has a belted bias-ply construction. The letter D represents diagonal bias-ply tire design. Belted bias-ply tires can still be found on some utility and recreational trailers, but are not suitable for modern passenger vehicles.

Rim Diameter and Load Rating

In the tire size P215/55R17 95H, the number 17 is the rim diameter in inches. The load index is represented by the number 95. The tire in Figure 3-9 has load rating of 1521 pounds. The load index number correlates to the maximum rated load the tire can carry (**Figure 3-23**). A higher number represents the tire has a higher maximum load capacity. The maximum load is shown on the tire in pounds (lb) and kilograms (kg) together with the maximum inflation pressure in pounds per square inch (psi) and kilograms (kg). As an example, MAX LOAD 730 kg (1609 lb) and 240 kPa (35 psi) MAX PRESSURE COLD may be listed on the sidewall of the tire. This indicates the maximum load at a specified cold inflation pressure. Never exceed the maximum inflation pressure listed on the sidewall, or tire failure could result. Some tire manufacturers have used the letters B, C, D, or E to indicate the **load rating** most commonly on LT tires. The letter B indicates the lowest

The **load rating** indicates the tire's load-carrying capability.

Load Index	Maximum Load (lbs.)	Load Index	Maximum Load (lbs.)
74	827	88	1235
75	853	89	1279
76	882	90	1323
77	908	91	1356
78	937	92	1389
79	963	93	1433
80	992	94	1477
81	1019	95	1521
82	1047	96	1565
83	1074	97	1609
84	1102	98	1653
85	1135	99	1709
86	1168	100	1764
87	1201		

Figure 3-23 The load index number correlates to a number assigned to the maximum load the tire can carry.

load rating and the letter C represents a higher load rating. A tire with a D load rating is designed for light-duty trucks, and this tire will safely carry a load of 2623 pounds when inflated to 65 psi. The ply rating is not the actual number of sidewall plies used but rather an equivalent strength compared to earlier bias-ply tires. An E load range light truck tire may have the ply rating of a 10-ply tire but, in actuality, only have a 4-ply sidewall.

> Aspect ratio is the percentage of a tire's height in relation to its width.

Speed Rating

In the tire size P215/55R17 95H, the letter H indicates the tire has a 130 miles per hour (mph) **speed rating**. Other speed ratings are as follows:

> The **speed rating** indicates the tire's capability to withstand high speed.

Q – 99 mph
S – 112 mph
T – 118 mph
U – 124 mph
V – above 130 mph without service description
V – 149 mph with service description
Z – above 149 mph
W – 168 mph
Y – 186 mph

> The aspect ratio may be referred to as the tire profile or series number.

Tire speed ratings do not suggest that vehicles can always be driven safely up to the maximum designated speed rating, because many different road and weather conditions may be encountered. Tire manufacturers do not endorse the operation of a vehicle in an unlawful or unsafe manner. Speed ratings are based on laboratory tests, and these ratings are not valid if tires are worn out, damaged, altered, underinflated, or overloaded. The condition of the vehicle may also affect high-speed operation.

Department of Transportation (DOT) Tire Grading

The United States Department of Transportation (DOT) and the National Highway Traffic Safety Administration (NHTSA) developed the **Uniform Tire Quality Grading (UTQG)** system to provide customers with standardized information related to a tire's tread wear, traction, and temperature capabilities to aid them with purchasing decisions. It is required that tire manufacturers grade most of their passenger car tires, except deep-treaded light truck tires, winter tires, temporary spare tires, and trailer tires. It should be noted that the DOT does not test the tires. It is left to the manufacturer to assign a grade based on their test results and those of independent testing companies they have hired.

Tread Wear Rating

The UTQG for **tread wear ratings** allows customers to compare tire tread life expectancies. The tread wear rating is based on the tire tread wear when tested under specific conditions on a regulated test track. Tire tread wear is monitored for a total of 7200 miles and then the data are extrapolated out for total tread life. A baseline tire has a tread wear rating of 100. A tire with a 150 tread wear rating would last 1.5 times as many miles on the test track as the baseline tire with a 100 tread wear rating. Tread wear ratings are valid only for comparison within a manufacturer's product line, and these ratings are not as valid for comparisons between tire manufacturers. This is one of the major flaws of the UTQG as it relates to tread wear. But you can expect a tire with a tread wear number of 600 to last longer than one rated at 500.

Traction Rating

The UTQG for **traction ratings** indicates the tire's straight line wet coefficient of traction and braking on both wet asphalt and wet concrete surfaces. The test does not evaluate the dry braking performance, wet cornering, dry cornering, or high-speed hydroplaning resistance. So while the information it provides is limited it does offer some ability to cross compare tires as long as the limits are understood.

To determine the traction rating, a properly inflated tire is placed on the axle of a skid trailer that is electronically monitored. Test conditions are carefully controlled to maintain test uniformity. A trailer speed of 40 mph is maintained over both wet concrete and wet asphalt. The brakes are momentarily applied on both surfaces while sensors gather information related to the tire's coefficient of friction, or braking g-force, as the tire slides, maintaining a constant speed of 40 mph. By design this test places less emphasis on tread design and more emphasis on the tread compound.

The results of the skid tests are averaged, and the traction rating is designated as AA, A, B, or C (**Figure 3-24**). An AA traction rating indicates the best traction, while a C rating indicates acceptable traction.

Temperature Rating

The UTQG for **temperature rating** indicates the tire's ability to generate or dissipate heat during tire operation. A tire that is unable to dissipate heat efficiently, or is unable to resist the destructive effects of internal heat buildup, will have a reduced ability to run at higher speeds.

Traction Grades	Asphalt g-Force	Concrete g-Force
AA	Above 0.54	0.41
A	Above 0.47	0.35
B	Above 0.38	0.26
C	Less than 0.38	0.26

Figure 3-24 The UTQG Traction Grade is determined by the tire's average coefficient of brake traction on both wet asphalt and wet concrete at a constant 40 mph in a straight line.

Temperature Grades	Speeds in mph
A	Over 115
B	Between 100 and 115
C	Between 85 and 100

Figure 3-25 The UTQG Temperature Grade indicates the tire's ability under a load to withstand heat buildup at various speeds.

To obtain the temperature rating, tires are tested on a laboratory test wheel under a load. The temperature rating indicates the tires' ability to operate at high speeds without failure. Extremely high temperatures may cause tire materials to degenerate and thus reduce tire life or cause tire failure. The temperature ratings indicate the tires' ability to withstand heat at various speeds and are A for best, B for intermediate, and C for acceptable (**Figure 3-25**). In laboratory testing a B-rated tire can withstand speeds between 100 and 115 mph without failure. The test is similar to those used to confirm a tire's speed rating, though it is not an exact test.

The UTQG is a good indicator of what can be expected from a particular tire. But it falls short of reflecting what actual real-world performance will be. It should be viewed as a useful tool for comparison only.

DOT regulations also require tire manufacturers to place the following information on tire sidewalls:

1. Size
2. Load range
3. Maximum load
4. Maximum pressure
5. Number of plies under the tread and in the sidewalls
6. Manufacturer's name
7. Tubeless or tube-type construction
8. Type of carcass construction
9. DOT approval number, including the manufacturer's code number and date of manufacture

Some tires have a **tire performance criteria (TPC)** number molded on the sidewall (**Figure 3-26**). This number represents that the tire meets the car manufacturer's performance standards for traction, endurance, dimensions, noise, handling, and rolling

TIRE IDENTIFICATION

Figure 3-26 Tire performance criteria (TPC) number.

resistance. Most car manufacturers assign a different TPC number to each tire size. Replacement tires should have the same ratings and TPC number as the original tires.

TIRE CONTACT AREA

The **tire contact area**, or contact patch, refers to the area of the tire that is in contact with the road surface when the tire is supporting the vehicle weight. The **tire free diameter** is the distance of a horizontal line through the center of the spindle and wheel to the outer edges of the tread. The **tire rolling diameter** is the distance of a vertical straight line through the center of the spindle to the outer edges of the tread when the tire is supporting the vehicle weight. The rolling diameter is always less than the free diameter. The difference between the free diameter and the rolling diameter is referred to as deflection. Tire tread grooves take up excess rubber and prevent scrubbing as the tire deflects in the contact area. The rolling diameter, free diameter, and contact area are shown in **Figure 3-27**. The contact patch is a measurement of the tire area that is in contact with the road (**Figure 3-28**).

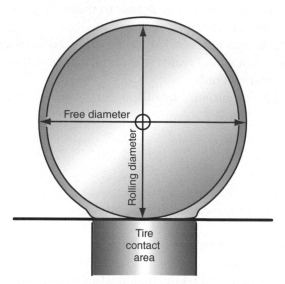

Figure 3-27 Tire rolling diameter, free diameter, and contact area.

Figure 3-28 The contact patch is a measurement of the tire area that is in contact with the road.

| 40PSI | 35PSI | 30PSI | 25PSI | 20PSI | 15PSI |

Figure 3-29 As air pressure is increased or decreased in a tire, the contact patch area will inversely change, decreasing as pressure is increased and increasing as pressure is decreased.

As air pressure is increased or decreased in a tire, the contact patch area will inversely change, decreasing as pressure is increased and increasing as pressure is decreased (**Figure 3-29**).

TIRE PLACARD AND INFLATION PRESSURE

The vehicle weight is supported by the correct air pressure exerted evenly against the entire interior tire surface, which produces tension in the tire carcass. Therefore, **tire pressure** is extremely important. Underinflation decreases the rolling diameter and increases the contact area, which results in excessive sidewall flexing and tread wear. Overinflation decreases the contact area, increases the rolling diameter, and stiffens the tire. This action results in excessive wear on the center of the tread. Tire pressure should be checked when the tires are cool. Since tire pressure normally increases at high tire temperatures, air pressure should not be released from hot tires. Excessive heat buildup in a tire may be caused by underinflation. This condition may lead to severe tire damage.

Tire pressure is the amount of air pressure in the tire.

> **AUTHOR'S NOTE** There are many causes of excessive tire tread wear, such as improper wheel balance or alignment. However, it has been my experience that one of the most common causes of excessive tire tread wear is improper inflation pressure. Many vehicle owners or drivers do not check tire pressures regularly.

On many vehicles, the **tire placard** is permanently attached to the rear face of the driver's door. This placard provides tire information such as maximum vehicle load; tire size, including spare; and cold inflation pressure, including spare (**Figure 3-30**).

Tire pressure is carefully calculated by the vehicle manufacturer to provide satisfactory tread life, handling, ride, and load-carrying capacity. Most vehicle manufacturers recommend that tire pressures be checked cold once a month or prior to any extended trip. The manufacturer considers the tires to be cold after the vehicle has sat for three hours, or when the vehicle has been driven less than one mile. The tires should be inflated to the pressure indicated on the tire placard. Tire pressures may be listed in metric or English system values. Conversion charts provide pressures in either of these systems (**Table 3-2**). The tire pressure that is listed on the sidewall is the maximum pressure and weight load that the tire can safely handle and is not the recommended tire pressure (**Figure 3-31**). Vehicle manufacturer's recommended tire pressure is always less than the pressure stated on the tire sidewall.

Shop Manual
Chapter 3, page 114

TABLE 3-2 Tire Inflation Pressure Conversion Chart

Inflation Pressure Conversion Chart (Kilopascals to psi)			
kPa	psi	kPa	psi
140	20	215	31
145	21	220	32
155	22	230	33
160	23	235	34
165	24	240	35
170	25	250	36
180	26	275	40
185	27	310	45
190	28	345	50
200	29	380	55
205	30	415	60

Conversion: 6.9 kPa = 1 psi

Front, rear, and
spare tire pressures

Figure 3-30 Tire placard.

Figure 3-31 The tire pressure that is listed on the sidewall is the maximum pressure and weight load that the tire can safely handle and is not the recommended tire pressure.

Nitrogen Tire Inflation

Some shops are presently equipped with **nitrogen tire inflation** equipment. When tires are inflated with compressed air, the air slowly passes through the tire walls and tread. Tires inflated with compressed air can lose as much as 12 psi in a 6-month period. When tire inflation pressure is lower than specified, tire tread wear is increased, while fuel economy and vehicle stability are decreased.

Nitrogen molecules are considerably larger than air molecules (**Figure 3-32**). Therefore, when a tire is inflated with nitrogen, the nitrogen passes through the tire walls and tread more slowly when compared with air. When tires are inflated with nitrogen, the tire inflation pressure remains more stable over a longer time period, which provides reduced tire tread wear, improved fuel economy, and increased vehicle stability. Another benefit from using nitrogen to inflate tires is reduced aluminum rim corrosion because the nitrogen does not contain any oxygen or moisture (water molecules). When tires are filled with air, the oxygen and water vapor in the air tends to react with the aluminum in the rims to cause corrosion.

SPECIALTY TIRES

Sport Utility Vehicle (SUV) and 4 × 4 Tires

SUV tires may be classified for use on pavement or off-road. SUV tires have greater load-carrying capacity compared to passenger car tires, because of the extra loads that may be carried in an SUV. A typical pavement SUV tire has these features:

1. A silica tread compound that provides low noise levels and exceptional wet braking capability.
2. An enhanced casing system and a stable tire contact area that supplies even tread wear and responsive handling quality.
3. Larger cables in the steel belts to provide increased strength and durability.
4. Full-depth, interlocking sections in the tread that supply excellent wet and snow traction.

Figure 3-32 Oxygen and nitrogen leakage through tire walls and tread.

Compared to SUV tires for use on pavement, off-road SUV tires have stronger tread rubber to prevent cutting and chipping. Off-road SUV tires may also have thicker belt wire and more belt strands.

Tires designed for 4 × 4 vehicles may have some of the same features as SUV tires. A typical 4 × 4 tire has these additional features:

1. Two wide circumference grooves with a stepped profile to reduce hydroplaning.
2. Staggered shoulder blocks in the tread to improve lateral grip on slopes.
3. Gradual profile changes in the shoulder area of the tread to provide progressive breakaway during hard cornering.
4. Additional rubber at the base of the tread to reduce the possibility of tire damage.

Puncture sealing tires are available as an option on certain car lines, and some rubber companies sell these tires in the replacement tire market. These tires contain a special rubber sealing compound applied under the tread area during the manufacturing process. When a nail or other object up to 3/16 in. (4.76 mm) in diameter punctures the tread area, it picks up a coating of sealant. If the object is removed, the sealant sticks to the object and is pulled into the puncture. This sealant completely fills the puncture and forms a permanent seal to maintain tire inflation pressure (**Figure 3-33**). Puncture sealing tires usually have a special warranty. These tires can be serviced with conventional tire changing and balancing equipment. When repairing tires, the maximum repairable puncture size is ¼ in.

When snow tires are installed on a vehicle, these tires should be the same size and type as the other tires on the vehicle.

REPLACEMENT TIRES

Shop Manual
Chapter 3, page 110

Most tires have tread wear indicators built into the tread. When the tread wears down to 2/32 in. (1.5 mm) the minimum allowable tread depth, the wear indicators appear as bands across the tread (**Figure 3-34**). Car manufacturers typically recommend tire replacement when the wear indicators appear in one or more tread grooves at three locations around the tire. It is generally advisable that tires be replaced before they reach the minimum of 2/32 in. (1.5 mm) of tread depth. A tread depth gauge (**Figure 3-35**) should be used to determine the actual amount of serviceable tread life left on a tire. Technicians typically recommend replacing tires when there is only 4/32 in. (3 mm) of tread depth available. This is especially true in areas where wet road traction is a consideration. In areas where

Figure 3-33 Puncture sealing tire.

Figure 3-34 Tread wear indicators.

Figure 3-35 A tread depth gauge should be used for an accurate measurement of the serviceable tread available.

snow is a concern technicians often recommended all-season tire replacement when there is only 5/32 in. (4 mm) of tread depth available.

If **replacement tires** have a different size or construction type than the original tires, vehicle handling, ride quality, speedometer/odometer calibration, and antilock brake system (ABS) operation may be seriously affected. When replacement tires are a different size than the original tires, the vehicle ground clearance and tire-to-body clearance may be altered. Steering and braking quality may be seriously affected if different sizes or types of tires are installed on a vehicle. This does not include the compact spare tire, which is intended for temporary use. All vehicles manufactured after 2012 are equipped with ABS due to the requirement of stability control, and most vehicles manufactured since the early 1990s are equipped with ABS. When different-sized tires are installed on these vehicles, the ABS operation is abnormal, which may result in serious braking defects. When selecting replacement tires, the following precautions must be observed to maintain vehicle safety:

1. Replacement tires must be installed in pairs on the same axle. Never mix tire sizes or designs on the same axle. If it is necessary to replace only one tire, it should be paired with the tire having tread depth within 4/32 in. (3 mm) of the replacement tire.
2. The tire load rating must be adequate for the vehicle on which the tire is installed. Light-duty trucks, station wagons, and trailer-towing vehicles are examples of vehicles that require tires with higher load ratings than passenger car tires.
3. Snow tires should be the same type and size as the other tires on the vehicle.
4. A four-wheel-drive and all-wheel drive vehicle should have the same type and size of tires on all four wheels, and tread depth between tires should be within manufacturer specifications.
5. Do not install tires with a load rating less than the car manufacturer's recommended rating.
6. Replacement tire ratings should be equivalent to the original tire ratings in all rating designations.
7. When combining different tires front to rear on a vehicle, consult the car manufacturer's or tire manufacturer's recommendations.

Vehicles that are four-wheel-drive or all-wheel-drive, whether they are light trucks, SUVs, or passenger cars should have four matched tires. This includes tread depth. All tires should be within 2/32 in.–4/32 in. (1.5 mm–3 mm) of tread depth with one another depending on manufacturer's recommendations. The reason for this is that four-wheel-drive and all-wheel-drive vehicles are equipped with viscous couplings or differentials that are designed for momentary differences in wheel speeds. If all drive tires are not exactly the same size, the differentials and viscous couplings will be always forced to compensate for the different rolling diameters, resulting in unwanted heat buildup and potential wear to expensive drivetrain components. If a tire requires replacement because it was damaged, it may be necessary to replace all four tires due to too great a variation in tread depth created by installing one new tire. Refer to vehicle manufacturer's service information for specific details and allowable tread depth differences.

Tires should be rotated every 6000 miles as a general rule following manufacturer's recommendations for both frequency and rotation pattern (**Figure 3-36**). It is also important to torque the wheel lug nuts to specification following the recommended torque sequence or star pattern (**Figure 3-37**). Beginning in the 2008 model year, tire pressure monitoring systems (TPMS) became mandatory. Many vehicles with TPMS require that

Shop Manual
Chapter 3, page 102

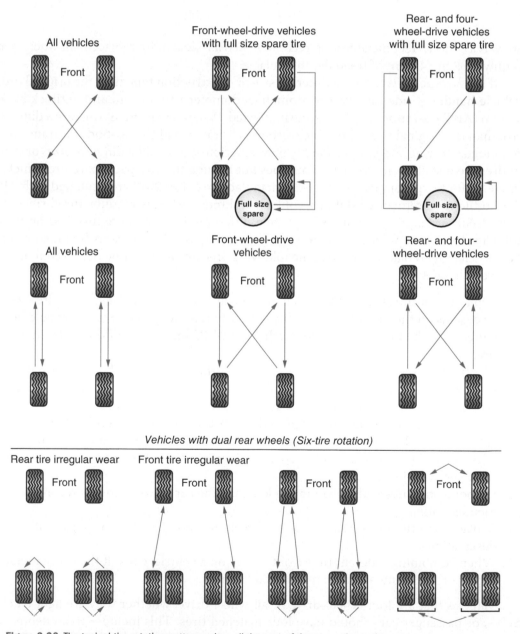

Figure 3-36 The typical tire rotation patterns when all tires are of the same size and type.

Figure 3-37 The typical rim torquing sequence.

the sensor position be relearned if any wheel and tire positions are changed (such as a tire rotation) since the last relearn procedure. Indirect TPMS systems that do not have a pressure sensor mount internally in the tire-and-wheel assembly do not require a relearn procedure. Consult the vehicle manufacturer's service information for exact procedures required following a tire rotation or replacement.

PLUS SIZING

Sometimes a customer does not like the appearance of their vehicle with the stock tire-and-wheel assembly and they would like to install a larger rim diameter to improve the looks and performance of the vehicle. This is referred to as plus sizing. It should be noted that a stock vehicle is designed to operate with a specific tire-and-wheel combination and deviating from allowed OEM combinations may adversely affect vehicle operation. Increasing a tire's contact patch may have a negative impact on wheel bearing loads and suspension component wear. It may also cause harsher ride quality due to less road isolation, which is caused by a narrower and stiffer tire side wall.

When the decision to plus size is made, you must remember that the overall height of the tire-and-wheel assembly must be maintained (**Figure 3-38**). By using a larger diameter wheel with a lower profile and a lower aspect number, it is possible to properly maintain the overall diameter of the tire and height of the assembly, keeping odometer and speedometer changes negligible. In addition, it is important to maintain the overall diameter of the tire to maintain the effectiveness of the antilock braking system, traction control, and vehicle stability system. It should be noted that when the decision is made to increase the rim diameter, it may also be necessary to change rim width and wheel offset. Refer to the section on rims for more detail regarding rim dimensions.

Now that we know what the tire size numbers mean, we can calculate the overall height of a tire-and-wheel assembly. We multiply the tire width by the aspect ratio to get the height of the tire:

$$\text{Tire sidewall height} = 235 \text{ mm (width)} \times 0.75 \text{ (aspect ratio percentage)}$$
$$= 176.25 \text{ mm (64.94 in.)}$$

Figure 3-38 Plus sizing must maintain the overall height of the tire-and-wheel assembly.

It will be necessary to convert rim diameter from inches to millimeters (15 in. × 25.4 = 381mm). Then we add twice the tire sidewall height to the rim diameter.

Overall tire and wheel height = 2 × 176.25 mm + 381 = 733.5 mm (28.9 in.)

To convert from inches to mm multiply by 25.4 and to convert from mm to inches divide by 25.4.

This is the unloaded height of the assembly; as soon as any weight is put on the tire, the height will decrease.

Then use this information when comparing optional tire sizes. It should be within 0.25 of original assembly height.

TIRE VALVES

The **tire valve** allows air to flow into the tire, and it is also used to release air from the tire. The core in the center of the valve is spring loaded and allows air to flow inward while the tire is inflated (**Figure 3-39**). Once the tire is inflated, the valve core seals and prevents air flow out of the tire. The small pin on the outer end of the valve core may be pushed to unseat the valve core and release air from the tire. An airtight cap on the outer end of the valve keeps dirt out of the valve, and provides an extra seal against air leakage. A deep groove is cut around the inner end of the tire valve. When the rubber valve assembly is pulled into the wheel opening, this groove seals the valve in the opening. This type of valve stem is used on vehicles with indirect TPMS or on vehicles and trailers without TPMS. Steel valve stems are installed in some wheels. The lower end of the valve stem is threaded, and a nut retains the valve stem in the wheel (refer to Figure 3-39). Steel washers and sealing washers are located on the valve stem on the inside and outside of the wheel. The sealing washers are positioned next to the wheel.

On many vehicles equipped with TPMS, the rubber valve stem is mounted directly to the tire pressure sensor with a screw (**Figure 3-40**). The rubber valve stem and core can be served separately from the sensor, but care must be taken during both tire dismounting

Groove seals valve to rim

Wheel rim

One-way valve core

Steel valve stem

Figure 3-39 Valve stems and cores.

Figure 3-40 On some TPMSs, the rubber valve stem and core can be served separately from the tire pressure sensor.

1. Metal washer
2. Sensor-to-wheel seal
3. Valve stem nut (with pressed-in washer)
4. Valve stem cap
5. Wheel
6. Sensor

Figure 3-41 On some TPMSs, the metal valve stem is an integral part of the tire pressure sensor, and only the valve core and sensor-to-wheel seal can be serviced separately.

and mounting, as well as valve stem replacement, to avoid sensor damage. On some tire pressure monitoring systems, the metal valve stem is an integral part of the tire pressure sensor, and only the valve core and sensor to wheel seal can be serviced separately (**Figure 3-41**).

TIRE PRESSURE MONITORING SYSTEMS

The Transportation Recall Enhancement Accountability and Documentation Act passed by the U.S. government in October 2000 requires all new vehicles to have tire pressure monitoring systems no later than the 2008 model year. A TPMS illuminates a warning light in the instrument panel to inform the driver if one tire has low pressure (**Figure 3-42**).

Most direct TPMSs have a pressure sensor mounted to a rubber valve stem, or the sensor and valve stem are integrated into one assembly as was described earlier. Regardless of the sensor's mounting location, it transmits radio frequency (RF) signals. A vehicle with

Shop Manual
Chapter 3, page 114

Figure 3-42 When the TPMS determines that a tire is low on air pressure, a warning image is illuminated on the instrument panel cluster to warn the driver.

direct pressure monitoring uses one of two frequencies, either 315 MHz and 433 MHz sensors. These RF signals will change if any tire is underinflated a specific amount. The RF signals are sent to a receiver that is usually mounted under the dash (**Figure 3-43**). On some systems, the receiver illuminates the low tire pressure warning light in the instrument panel if the pressure in any tire drops below 25 psi (172 kPa). On other systems, the tire pressure warning light is illuminated with a smaller reduction in tire pressure. Some TPMSs have a sensor strapped in the drop center on each rim (**Figure 3-44**). These sensors must be mounted on the rim directly opposite to the valve stem.

Some TPMSs provide continuous monitoring with the vehicle stopped and the ignition switch on. Other systems will not monitor tire pressure until the vehicle is moving above a specific speed, such as 25 mph (40 km/h). Some TPMSs have two warning lights: the low/flat tire warning light informs the driver regarding low tire pressure, and the service low tire pressure warning system (LTPWS) warning light informs the driver if there is a defect in the system (**Figure 3-45**). On some vehicles the low tire pressure receiver is connected to the data link connector (DLC) under the dash. A scan tool may be connected to the DLC to diagnose the TPMS. If a defect occurs in the system, a diagnostic trouble code (DTC) is stored in the receiver memory. A scan tool may be connected to the DLC to obtain the DTCs from the receiver.

RECEIVER

SENSOR

Figure 3-43 Receiver in a TPMS.

Pressure sensor

Valve stem

Figure 3-44 Tire pressure sensor.

Location	Color code
Right front	Blue
Left front	Green
Right rear	Orange
Left rear	Yellow

WARNING: Pressure sensor inside tire. Avoid contacting sensor with tire changing equipment tools or tire bead.

Service note: Pressure sensor must be mounted directly across from valve stem.

Figure 3-45 Warning lights in a TPMS.

Figure 3-46 Antenna in a tire sidewall.

Other TPMSs use the wheel speed sensor signals on four-channel ABS to detect low tire pressure. A tire with low air pressure has a smaller diameter, and the wheel speed sensor generates a higher-frequency signal.

Some TPMSs have a miniature pressure sensor and computer chip about the size of a watch battery. This sensor and chip assembly is imbedded in the tire. The pressure sensor senses the actual tire pressure and sends voltage signals in relation to the tire pressure to a transceiver via a 360-degree circumferential antenna mounted in the tire sidewall close to the rim (**Figure 3-46**). A transceiver is mounted in each wheel well. The pressure sensor and computer chip do not require a battery, because they receive energy from the transceiver-generated field. Each transceiver is connected by data links to a central electronic control unit (ECU) (**Figure 3-47**). The transceivers relay voltage signals from the pressure sensors to the ECU. The computer chip calculates the recommended tire pressure based on data such as air temperature, tire pressure, and vehicle speed. Vehicle speed information is transmitted via data links from the powertrain control module (PCM) to

Figure 3-47 Transceivers and ECU in a TPMS.

the ECU. If a low tire pressure signal is received by the ECU, the instrument panel displays a warning message together with a calculated number of miles of safe travel.

Tire Pressure Monitoring Systems (TPMS) with Valve Stem Sensors

Most current TPMSs have pressure sensors that are attached to each valve stem inside the tire, and these sensors require a battery (**Figure 3-48**). Average battery life is approximately 10 years. A retention nut is threaded onto the valve stem on the outside of the wheel rim, and a sealing grommet is positioned between this nut and the rim (**Figure 3-49**). The internal threads on the retention nut are positioned near the bottom of the nut, and matching threads are located near the bottom end of the valve stem. A nickel-plated valve core is threaded into the internal threads in the valve stem, and a special valve cap is threaded onto the external threads at the top of the valve stem.

Many vehicles have a TPMS sensor in the spare tire. The valve stem acts as a radio antenna, and the TPMS module is the receiver. Each sensor transmits radio signals to the TPMS module every 60 seconds. On some SUVs, the TPMS module is located behind the right-hand C pillar. The TPMS module compares the data from each tire pressure sensor to the low tire pressure limits. When the TPMS module determines that any of the five tires has a pressure below the specified pressure, this module sends voltage signals through the data links to the vehicle message center. When the message center receives voltage signals indicating low pressure in one or more tires, a warning message is displayed

Figure 3-48 Tire pressure sensor mounted on the valve stem.

Figure 3-49 Tire pressure sensor retention nut, valve cap, valve core, and grommet.

in the message center, and the TPMS warning light is illuminated. The possible messages related to the TPMS are as follows:

1. WARNING TIRE VERY LOW—On some systems, this message is displayed if any tire has pressure less than 25 psi. An audible warning chime is heard when this message is displayed, and the TPMS warning light is illuminated.
2. CHECK TIRE PRESSURE—On some systems this message is displayed if any tire has pressure less than 30 psi. An audible warning chime is heard when this message is displayed, and the TPMS warning light is illuminated.
3. TIRE PRESSURE SENSOR FAULT—This message is displayed if one or more sensors are malfunctioning. If this message is displayed, the TPMS warning light flashes for 20 seconds.
4. TIRE PRESSURE MONITOR FAULT—This message is displayed if the TPMS module is defective or if all four sensors have failed. The TPMS warning light flashes for 20 seconds when this message is displayed.

Some TPMSs have the capability to sense a loss or an increase in tire pressure, and these systems also allow the driver to display the individual tire pressures and their locations on the driver information center (DIC) while the vehicle is being driven. This type of TPMS has conventional radio frequency–transmitting pressure sensors in each of the four valve stems. These sensors transmit signals to the antenna module, dash integration module (DIM), instrument panel cluster (IPC), and the DIC via the serial data circuit. The sensor's pressure accuracy from 14°F to 158°F (10°C to 70°C) is plus or minus 1 psi (7 kPa).

When the vehicle speed is less than 20 mph (32 km/h), the system remains in the stationary mode. In this mode, the sensors transmit data every 60 minutes to provide longer sensor battery life. When the vehicle speed is 20 mph (32 km/h) or greater, centrifugal force closes a roll switch in each sensor, and this action causes the sensors to enter the drive mode in which the sensors transmit signals every 60 seconds. The antenna module receives and translates each sensor signal into sensor presence, sensor mode, and tire pressure. The temperature and speed ratings of the TPMS system vary depending on the vehicle manufacturer. Always consult the vehicle manufacturer's specifications. The antenna module then transmits this information from each sensor to the DIC via the serial data circuit. The DIC displays an overhead view of the vehicle in which tire locations and pressures are displayed. If the TPMS senses a specific pressure gain or loss in any tire, a CHECK TIRE PRESSURE warning message is displayed on the DIC, and the low tire pressure warning indicator is illuminated in the IPC. Any defect in the TPMS system causes a SERVICE TIRE MONITOR warning message to be displayed on the DIC.

TPMS with Electronic Vehicle Information Center (EVIC) Display

Some TPMSs provide graphic tire pressure displays and warning messages on an EVIC display in the instrument panel. These systems may have four-tire or five-tire monitoring capabilities. A TPMS with five-tire monitoring capabilities also monitors the spare tire. These systems have valve stem sensors that broadcast tire pressure once per minute when the vehicle is moving at 25 mph (40 km/h) or faster. If the vehicle has a five-tire TPMS, the sensor in the spare tire transmits a signal every hour. Each valve stem sensor transmits a unique code so the EVIC module can determine the location of the sensor. If the wheels are rotated on the vehicle, or the spare tire and wheel are installed on the vehicle, the EVIC must be reprogrammed to recognize the new wheel and tire locations.

This type of TPMS provides a warning display in the EVIC if the tire pressure drops below a specific value or increases above a certain threshold. Typically, if the tire pressure in any tire decreases below 25 psi (172 kPa), the EVIC requests a chime warning, displays a LOW PRESSURE message, and indicates the location of the tire with low pressure. If the pressure in any tire exceeds 45 psi (310 kPa), the EVIC requests a chime warning and displays HIGH PRESSURE while indicating the location of the tire with high pressure. After a few seconds the EVIC reverts to a blinking display of the tire pressure. This blinking display continues for the entire ignition cycle or until an EVIC button is pressed. When an EVIC button is pressed, the blinking display returns after 60 seconds without the chime warning. If high or low pressure is detected in the spare tire on a five-tire system, SPARE HIGH PRESSURE or SPARE LOW PRESSURE appears on the EVIC display for 60 seconds during each ignition cycle.

WHEEL RIMS

Wheel rims are circular devices on which the tires are mounted. Wheel rims are bolted to the wheel hubs or axles.

Wheel rims may be manufactured from stamped or pressed steel discs that are riveted or welded together to form the circular rim, or may be cast or forged aluminum (**Figure 3-50**). Currently, less than 50 percent of the wheel rims installed by the original equipment manufacturers (OEMs) are stamped steel type.

A rim is equipped with a drop center, which allows tires to be more easily removed and installed (**Figure 3-51**). The closer the drop center is to the outside mounting flange the easier it is to remove and install the tire (**Figure 3-52**). Some performance rims and run-flat tire rims are equipped with a bead lock ridge, which aids in keeping the tire bead locked or secured to the rim (**Figure 3-53**). This is important especially during cornering maneuvers or in the case of a run-flat tire when it has lost all air pressure. These bead lock rims also commonly have a drop center that is located in the center of the rim farther from the outer mounting flange. Rims of this design make tire service much more challenging (**Figure 3-54**).

Shop Manual
Chapter 3, page 120

Figure 3-50 Wheel rim design.

Specialized wheel designs

Outboard

Must be inverted
drop center up for
mounting/demounting

Inboard

Reverse drop center

Conical

Outboard

Inboard

Long drop center

Racing application
not recommended
for side shovel tire
changers

Cylindrical

Outboard

Wheel disassembly
required.
Do not use on
tire changer

Inboard

Cylindrical, no drop center

Figure 3-51 Various rim drop center designs.

Easy
mountability

Drop center

Outer flange

Easy mountable wheel

Figure 3-52 The closer the drop center is to the outer mounting flange, the
easier it will be to dismount and mount tires on the rim assembly.

BMW Z3 "AH2" bead locking wheel.

Figure 3-53 The bead lock on a rim assembly aids in keeping the tire bead secure to the rim.

Figure 3-54 A rim with a drop center located in or close to the center of the assembly can make it very challenging to dismount and mount tires on the rim assembly.

There are many dimensions that are associated with the rim assembly (**Figure 3-55**), and all must be considered if a customer wants a replacement rim of a different design. The rim offset locates the tire-and-wheel assembly in relationship to the suspension and the lines of force on the wheel bearing-and-spindle assembly (**Figure 3-56**). When the mounting face of the rim aligns directly with the rim's centerline, then the rim has zero offset. If a rim is designed with positive offset, the rim centerline is inboard of the mounting face. Stated another way the mounting face is closer to the street side of the wheel. The rim with negative offset has the centerline outboard of the mounting face. Stated another way the mounting face is closer to the brake side of the wheel. Front-wheel-drive vehicles tend to have positive offset rims to allow for proper clearance of the hub assembly within the vehicle wheel well. The rim offset affects front suspension loading and operation. As a general rule, if a replacement set of rims is being considered, they should meet all the dimensional characteristics of the original equipment rim.

A large hole in the center of the rim fits over a flange on the mounting surface. The rim has a small hole for the valve stem. The wheel stud mounting holes in the rim are tapered to match the taper on the wheel nuts. Some late-model Ford trucks have non-self-centering wheel nuts. On these nuts, a swivel washer is attached to the inner side of each nut (**Figure 3-57**). When the wheel and nuts are installed, the flat side of the swivel nut fits against the wheel rim. The wheels are hub-centered, and an O-ring mounted on the

Figure 3-55 Various dimensions associated with the wheel rim assembly.

hub provides improved wheel centering (**Figure 3-58**). If this O-ring is missing or damaged, the wheel may not be properly centered, and this condition results in wheel vibrations. The O-ring also provides a seal to prevent corrosion.

Cars and light-duty trucks have four, five, six, or eight mounting stud openings and an equal number of matching studs in the hub or axle (**Figure 3-59**). The number of wheel studs depends on the vehicle weight and load. Heavier vehicles and/or increased vehicle load usually require more wheel mounting studs. These nuts usually have a taper on the inner side of the nuts, and this taper fits against a matching taper on the wheel stud openings (**Figure 3-60**). These tapers on the nuts and wheel openings center the wheel on the hub or axle; this is referred to as a stud piloted rim. Wheel stud openings are listed by the number of studs and the circle through the stud centers. For example, if wheel openings are listed as 5–4.5, there are five wheel studs and stud openings, and the centers of the wheel studs and wheel openings are on a 4.5-in. (11.4-cm) circle. If a wheel has an even number of studs, measure the distance between opposite stud centers to check the stud position and wheel circle. On a five-bolt wheel, measure the distance between two adjacent stud centers. The specified distances between adjacent wheel stud centers and the corresponding wheel circles are the following:

4.5-in. wheel circle	2.64 in. between adjacent stud centers
4.75-in. wheel circle	2.79 in. between adjacent stud centers
5.0-in. wheel circle	2.93 in. between adjacent stud centers
5.5-in. wheel circle	3.23 in. between adjacent stud centers

Figure 3-56 If a rim is designed with positive offset, the rim centerline is inboard of the mounting face. The rim with negative offset has the centerline outboard of the mounting face.

Figure 3-57 Non-self-centering wheel nut design.

Figure 3-58 Hub O-ring that provides wheel centering.

Figure 3-59 Wheel stud configurations.

Templates are available to measure the stud position on wheels that do not have an even number of studs.

The width of the wheel is measured between the vertical bead seats on the rim flanges. Rim diameter is the distance between the horizontal part of the bead seats measured through the rim center bottom. A drop center in the rim makes tire changing easier

Figure 3-60 Wheel lug nuts generally have a tapered or ball seat to locate and secure the rim to the hub assembly.

Figure 3-61 Wheel rim with drop center and safety ridges.

(**Figure 3-61**). Rims have a safety ridge or bead lock ridge behind the tire bead locations, which help prevent the beads from moving into the drop center area if the tire blows out. If the tire blows out and a bead enters the drop center area, the tire may come off the wheel.

Cast aluminum wheels are commonly used on many vehicles by OEMs. These wheels may be polished, chromed, or painted. A cast aluminum wheel costs about twice as much as a stamped steel wheel.

Forged aluminum wheels are installed on some vehicles. They are more durable than cast aluminum wheels and have a bright finish. Forged aluminum wheels are used on heavy-load applications such as trucks. Some expensive sports cars are also equipped with forged aluminum wheels. A forged aluminum wheel is lighter than a steel wheel, but the cost of a forged aluminum wheel is considerably higher than a stamped steel or cast aluminum wheel.

Some vehicles are equipped with pressure-cast aluminum wheels that are designed to be chrome plated. Pressure-cast wheels may also be called squeeze-cast wheels. During the manufacture of a pressure-cast wheel, the molten metal is squeezed under pressure into the mold. This action squeezes all the air out of the metal and reduces the metal porosity, which provides an improved substrate for chrome plating. The chrome plating increases the cost of the pressure-cast aluminum wheel so it is considerably more expensive than other types of wheels.

Race cars may be equipped with **magnesium alloy wheels**, which are lighter than either steel or aluminum wheels.

Replacement wheel rims must be the same as the original equipment wheels in load capacity, offset, width, diameter, and mounting configuration. An incorrect wheel can affect tire life, steering quality, wheel bearing life, vehicle ground clearance, tire clearance, and speedometer/odometer calibrations.

Magnesium alloy wheels may be called "mag" wheels.

TIRE CHAINS

Tire chains may be used in emergency situations, such as driving on snow-covered or ice-covered mountain roads. Most tire manufacturers do not recommend the use of tire chains. Use only SAE class "S" tire chains unless specified by the vehicle and tire manufacturer. Chain types include class S, type P, and type RP (**Figure 3-62**). These chains must be the proper size and must be installed tightly with the ends secured. Always follow the chain manufacturer's recommended installation procedure. While using chains, driving speed should be reduced. If the chains are heard striking the vehicle body, stop immediately and tighten the chains to prevent body damage.

COMPACT SPARE TIRES

Because cars have been downsized in recent years, space and weight have become major concerns for vehicle manufacturers. For this reason, many car manufacturers have marketed cars with **compact spare tires** to save weight and space. The high-pressure mini-spare tire is the most common type of compact spare (**Figure 3-63**), and the first letter in the tire size will be a T for temporary. This compact spare rim is usually four inches wide, but is one inch larger in diameter than the other rims on the vehicle. The compact spare rim should not be used with standard tires, snow tires, wheel covers, or trim rings. Any of these uses may result in damage to these items or other parts of the vehicle. The compact spare should be used only on vehicles that offer it as original equipment. Inflation pressure in the compact spare should be maintained at 60 psi (415 kPa). The compact spare tire is designed for very temporary use until the conventional tire can be repaired or replaced. Limit driving speed to 50 mph (80 km/h) when the high-pressure mini-spare tire is installed on a vehicle.

The space-saver spare tire must be inflated with a special compressor. Battery voltage is supplied to the compressor from the cigarette lighter. This type of compact spare should be inflated to 35 psi (240 kPa). After the tire is inflated, be sure there are no folds in the sidewalls.

The lightweight-skin spare tire is a bias-ply tire with a reduced tread depth to provide an estimated 2000 mi. (3200 km) of tread life. Always inflate the lightweight-skin spare tire to the pressure specified on the tire placard.

TYPE "PL"
1100 series, SAE class "S"

TYPE "P"
1200 series, SAE class "U"

TYPE "RP"
1800 series, lug-reinforced

Figure 3-62 Types of tire chains.

Figure 3-63 Compact, high-pressure, mini-spare tire.

RUN-FLAT TIRES

Some tire manufacturers utilize **run-flat tires** on some of their vehicle platforms. Run-flat tires may be called extended mobility tires (EMT). These tires are standard equipment on the 1995 and late-model Corvette. Run-flat tires eliminate the need for a spare tire and a jack, saving both weight and space. Run-flat tires are designed with stiffer sidewalls that will partially support the vehicle weight without air pressure in the tire.

Run-flat tires must provide acceptable levels of inflated performance in the areas of comfort, ride, handling, and adequate deflated mobility. Run-flat tires share these same basic design objectives:

1. Minimize the difference between run-flat tires and conventional tires when the tires are inflated.
2. Enhance the handling and riding capabilities of run-flat tires when inflated.
3. Provide acceptable handling when run-flat tires have zero pressure on various vehicles.
4. Enhance low-pressure and zero-pressure bead retention on run-flat tires.
5. Provide sufficient zero-pressure durability so the vehicle can be driven a reasonable distance to a repair facility.

> **Run-flat tires** are designed to operate at zero air pressure for limited mileage. Run-flat tires may be called self-supporting, extended mobility, or zero-pressure tires.

Run-Flat Tires with Sidewall Reinforcements

One of the most important requirements for run-flat tires is bead retention when running with low or zero pressure. Run-flat tires have improved beads to meet this requirement (**Figure 3-64**). Some run-flat tires have sidewall reinforcements that may be manufactured from flexible, low-hysteresis rubber, thermal-resistive materials, or metallic and/or textile tissues. Run-flat tires with a high aspect ratio require increased sidewall stiffness compared to run-flat tires with a low aspect ratio. Because of these sidewall reinforcements, run-flat tires are 20 to 40 percent stiffer compared to conventional tires. Therefore, run-flat tires may increase ride harshness and vertical firmness especially when the tire strikes a large road irregularity. This increase in ride harshness is not as noticeable when the tire contacts a smaller road irregularity. The increased stiffness in run-flat tires usually provides a small improvement in vehicle handling.

Run-flat tires with a low aspect ratio usually have the same or less rolling resistance compared to equivalent-size conventional tires. Because run-flat tires with a high aspect ratio have increased sidewall stiffness, these tires tend to have increased rolling resistance compared to conventional tires.

> **Shop Manual**
> Chapter 3, page 105

Special bead design:
* Enhanced retention after pressure loss
* Acceptable seating pressure

Sidewall reinforcement:
* Flexible low-hysteresis rubber
* Thermal-resistive material
* Metallic and/or textile tissues

Appropriate summit adjustments:
*Maintain inflated performance (comfort & handling like std. tires)

Figure 3-64 Run-flat tire with sidewall reinforcement.

⚠ **Caution**

On a vehicle with PAX tires as original equipment, conventional tires and rims should not be substituted for the PAX tires and rims. This action will decrease ride quality and vehicle steering characteristics.

In moderate cornering and lane-change maneuvers, a run-flat tire with zero pressure provides a slight decrease in vehicle handling. The key word in the previous sentence is moderate. Run-flat tires are used with a tire pressure monitoring system, which has a warning light in the instrument panel to inform the driver when one tire has low pressure. Once the low tire pressure warning light is illuminated with the engine running, the driver should avoid high-speed driving and high-speed cornering, because run-flat tires provide reduced handling capabilities during high-speed cornering.

The zero-pressure durability of a run-flat tire varies depending on the vehicle weight, atmospheric temperature, and the tire aspect ratio. For example, a run-flat tire with a 40 aspect ratio supporting 880 lb (400 kg) provides 186 mi. (300 km) of driving at 55 mph (89 km/h) (**Figure 3-65**). A run-flat tire with a 60 aspect ratio supporting 1322 lb (600 kg) provides 50 mi. (80 km) of driving at 55 mph (89 km/h).

Run-Flat Tires with Support Ring

Michelin PAX run-flat tires were first introduced in 1998, but later discontinued in 2009. PAX run-flat tires were installed as original equipment on some Honda Odyssey, Toyota Sienna, Nissan Quest, and Acura RL models, as well as Audi A8 and A4 models in Europe. The letters "PAX" translate to the values of peace of mind, safety, and the future. The technology never gained traction on the world market, and has now become part of automotive history.

⚡ **Caution**

Using conventional tire changers that do not have PAX capabilities on PAX run-flat tires and rims may damage the tire and rim. Tire deflection is the difference between the free diameter and the rolling diameter of the tire.

The run-flat tires in the PAX system have a flexible support ring mounted on a special rim to support the tire if deflation occurs. PAX tires do not have a stiffer sidewall, and so they provide excellent ride quality. When PAX tires are installed on a vehicle as original equipment, all the suspension system components such as springs and shock absorbers are engineered to go with the PAX tires.

PAX run-flat tires have a vertical, mechanical locking bead system that "latches" the bead to the rim. The outer tire bead has a smaller diameter compared with the inside tire bead. Therefore, the outer edge of the rim has a smaller diameter compared with the inner edge (**Figure 3-66**). When mounted on the rim, the tire beads lock into vertical grooves in the outer edges of the rim. The outer edges of the inner and outer tire beads extend slightly outside the wheel rim edges. This design helps protect the rim. PAX tires may be driven for 125 mi. (201 km) at 55 mph (88 km/h) with zero air pressure. When mounting the tire on the rim, the tire is first positioned such that the tire beads are outside the rim. The outer (smaller-diameter) bead is then mounted onto the rim, followed by the inner (larger-diameter) bead. The disadvantages of this type of run-flat tire include the special rim that is required and the extra weight of the support ring. A TPMS is used with this type of run-flat tire.

Shop Manual

Chapter 3, page 110

P225/60R16
(135-mm sidewall)

Representative situation:

* Zero pressure 600 kg
* Front-wheel-drive luxury vehicle
* 89 km of zero-pressure durability at 80 km/h

P275/40ZR17
(110-mm sidewall)

Representative situation:

* Zero pressure 400 kg
* Rear-wheel-drive high-performance vehicle
* 300 km of zero-pressure durability at 89 km/h

Figure 3-65 Run-flat tire zero-pressure durability.

Figure 3-66 Run-flat tire with support ring.

Labels: Support ring; Minimum intrusion well; Support ring locked by the external bead of tire; Protection of the external rim edge; Single piece wheel

STATIC WHEEL BALANCE THEORY

When a tire-and-wheel assembly has proper **static balance**, it has the weight equally distributed around its axis of rotation, and gravity will not force it to rotate from its rest position. If a vehicle is raised off the floor and a wheel is rotated in 120-degree intervals, a statically balanced wheel will remain stationary at each interval. When a wheel and tire are statically unbalanced, the tire has a heavy portion at one location. The force of gravity acting on this heavy portion will cause the wheel to rotate until the heavy portion is located near the bottom of the tire (**Figure 3-67**).

Static balance refers to the balance of a wheel in the stationary position.

Results of Static Unbalance

Centrifugal force may be defined as the force that tends to move a rotating mass away from its axis of rotation. As we have explained previously, a tire and wheel are subjected to very strong acceleration and deceleration forces when a vehicle is in motion. The heavy portion of a statically unbalanced wheel is influenced by centrifugal force. This influence attempts to move the heavy spot on a tangent line away from the wheel axis. This action tends to lift the wheel assembly off the road surface (**Figure 3-68**).

The wheel-lifting action caused by static unbalance may be referred to as **wheel tramp** (**Figure 3-69**). Wheel tramp action allows the tire to slip momentarily when it is lifted vertically. When the wheel and tire move downward as the heavy spot decelerates, the tire strikes the road surface with a pounding action. This repeated slipping and pounding action causes severe tire scuffing and cupping (**Figure 3-70**).

The vertical wheel motion from static unbalance is transferred to the suspension system and then absorbed by the chassis and body. This action causes rapid wear on

Wheel tramp may be defined as rapid upward and downward wheel and tire oscillations.

Figure 3-67 Static wheel imbalance.

Labels: Gravity; Heavy spot

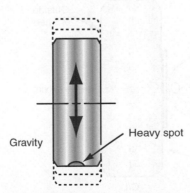

Figure 3-68 Effects of static imbalance.

Labels: Gravity; Heavy spot

Figure 3-69 Wheel tramp.

Scuffing or cupping

Figure 3-70 Cupping tire wear caused by static imbalance.

suspension and steering components. The wheel tramp action resulting from static unbalance is also transmitted to the passenger compartment, which causes passenger discomfort and driver fatigue.

When a vehicle is traveling at normal highway cruising speed, the average wheel speed is 850 revolutions per minute (rpm). A statically unbalanced tire-and-wheel assembly is an uncontrolled mass of weight in motion. When a vehicle is traveling at 60 mph (97 km/h) and a tire has 2 ounces (oz), or 57 grams (g), of static unbalance, the resultant pounding force is approximately 15 lb, or 6.8 kg, against the road surface.

DYNAMIC WHEEL BALANCE THEORY

Dynamic balance refers to the balance of a wheel in motion.

When a tire-and-wheel assembly has correct **dynamic balance**, the weight of the assembly is distributed equally on both sides of the wheel center viewed from the front. Dynamic wheel balance may be explained by dividing the tire into four sections (**Figure 3-71**). In Figure 3-71, if sections A and C have the same weight and sections B and D also have the same weight, the tire has proper dynamic balance. If a tire has dynamic unbalance, section D may have a heavy spot; thus, sections B and D have different weights (**Figure 3-72**).

From our discussion of dynamic balance, we can understand that a tire and wheel assembly may be in static balance, but have dynamic unbalance. Therefore, wheels must be in balance statically and dynamically. When a tire-and-wheel assembly is placed on a

Shop Manual
Chapter 3, page 125

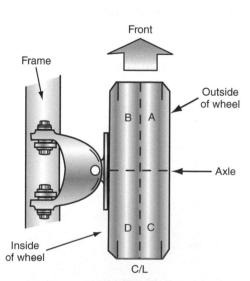

Figure 3-71 Dynamic wheel balance.

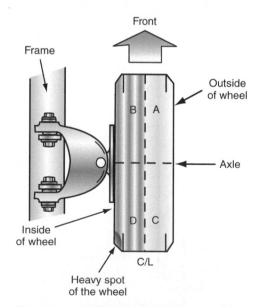

Figure 3-72 Dynamic wheel unbalance.

computer wheel balancer to check and correct for an imbalance condition, it is generally necessary to attach a wheel weight to the rim assembly to offset this condition. Wheel weights are generally either clip-on or stick-on weights. Clip-on wheel weights are attached to the outboard flange of the rim and come in an assortment of sizes and shapes depending on the shape of the rim flange (**Figure 3-73**).

Alloy Wheels: Passenger Cars & Light Trucks

ALC® Aluma-Guard™ Coated — 10 AL 26
Domestic (GM, Ford, & Chrysler) 12 sizes: 0.25 to 3 oz.

ALCMC Aluma-Guard™ Coated — 10 AL-MC 20
Domestic (GM, Ford, & Chrysler) 12 sizes: 0.25 to 3 oz.

ALCLH Aluma-Guard™ Coated — 10 LH 28
Newer Chrysler 12 sizes: 0.25 to 3 oz.

TTC Aluma-Guard™ Coated
New Domestic Lt. Truck (GM, Ford, and Chrysler) 12 sizes: 0.25 to 3 oz.

ALCIW Aluma-Guard™ Coated — 10G AL-IW 30
European Subaru 11 sizes: 10 g to 60 g

ALCTW Aluma-Guard™ Coated — 1.G TW 2G
European Subaru & Mitsubishi 12 sizes: 0.25 to 3 oz.

ALCFN Aluma-Guard™ Coated — 1.06 ALFN 30
Newer Japanese 12 sizes: 5 g to 60 g

ALCEN Aluma-Guard™ Coated — 1.06 AL-EN 30
Mercedes & older Japanese 11 sizes: (in 5 g incr.) 10 g to 60 g

MBC Aluma-Guard™ Coated — 20
Mercedes, 2 pc. 12 sizes: 5 g to 60 g

BMC Aluma-Guard™ Coated — F / BMW
BMW & newer Jaguar, 2 pc. 11 sizes: 5 g to 55 g

AVW Aluma-Guard™ Coated — 2B
Audi & VW 2 pc. 12 sizes: 5 g to 60 g

Steel Wheels: Passenger Cars & Light Trucks

TC-THINLINE® Coated — 1.0 MICRO 28
All import and domestic passenger cars, 1/2 ton trucks and all mini pickups & vans with standard 0.125" flange. 20 sizes: 0.25 to 6 oz.

T-THINLINE® Uncoated — 1.0 MICRO 28
All import and domestic passenger cars, 1/2 ton trucks, mini pickups and vans with standard 0.125" flange. NASCAR sanctioned. 20 sizes: 0.25 to 6 oz.

GRAMLINE® Uncoated
All import and domestic passenger cars, 1/2 ton trucks, mini pickups and vans with standard 0.125" flange. 12 sizes: 5 g to 60 g

Note: See application chart and use rim gauge to make actual weight selections for a particular wheel.

Figure 3-73 Wheels weights come in various dimensions depending on wheel rim assembly design.

Results of Dynamic Wheel Unbalance

When a dynamically unbalanced wheel is rotating, centrifugal force moves the heavy spot toward the tire centerline. The centerline of the heavy spot arc is at a 90° angle to the spindle. This action turns the true centerline of the left front wheel inward when the heavy spot is at the rear of the wheel (**Figure 3-74**). When the wheel rotates until the heavy spot is at the front of the wheel, the heavy spot movement turns the left front wheel outward (**Figure 3-75**).

From these explanations, we can understand that dynamic wheel unbalance causes lateral wheel shake, or shimmy (**Figure 3-76**). This action causes steering wheel oscillations at medium and high speeds with resultant driver fatigue and passenger discomfort. Wheel shimmy and steering wheel oscillations also cause unstable directional control of the vehicle.

Earlier in this chapter, when discussing wheel rotation, we mentioned that a tire stops momentarily where it contacts the road surface. A wheel with dynamic unbalance pivots on the contact area, which results in excessive tire scuffing and wear. Dynamic wheel unbalance causes premature wear on steering linkage and suspension components.

Figure 3-74 Dynamic wheel unbalance with heavy spot at the rear of the left front tire.

Figure 3-75 Dynamic wheel unbalance with heavy spot at the front of the left front tire.

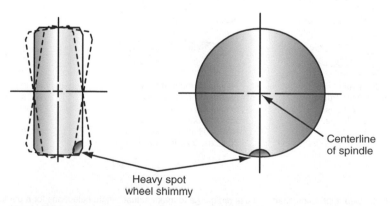

Figure 3-76 Dynamic wheel unbalance causes wheel shimmy.

Therefore, dynamic wheel balance is extremely important to provide normal tire life, reduce steering and suspension component wear, increase directional control, and decrease driver fatigue. The main purposes of proper wheel balance are as follows:

1. Maintains normal tire tread life.
2. Provides extended life of suspension and steering components.
3. Helps provide directional control of the vehicle.
4. Reduces driver fatigue.
5. Increases passenger comfort.
6. Helps maintain the life of body and chassis components.

TIRE MOTION FORCES

When a vehicle is in motion, wheel rotation subjects the tires to centrifugal force. The tires are also subjected to accelerating and decelerating forces because of their path of travel. If a vehicle is traveling at 55 mph, or 88.5 km/h, the part of the tire exactly fore and aft of the spindle is also traveling at 55 mph (88.5 km/h). At the exact top of the tire, the tire speed is 110 mph (177 km/h). The tire speed actually drops to zero at the exact bottom of the tire where the arc of deceleration ends and the arc of acceleration begins. Since the tires are subjected to strong acceleration and deceleration forces, the tire construction must be uniform. A soft spot in a tire will deflect farther than the surrounding area, and this area will be subjected to rapid wear as it strikes the road surface (**Figure 3-77**).

Tires must have equal stiffness in the sidewalls. If a tire does not have equal stiffness in all areas around the tire sidewalls, it may cause a vibration when driving. This vibration is called **road force variation** and produces a vibration similar to the vibration caused by improper wheel and tire balance. Imagine the tire as a collection of springs around the circumference of the tire. If the sidewall stiffness is not uniform a varied force is exerted, which creates a vibration. Hunter road force wheel balancers have a roller that is forced against the tire tread during the balance procedure, and this type of balancer detects road force variation (**Figure 3-78**).

The radial force variation can be determined by examining the radial run out. Loaded radial runout of one thousands of an inch is equivalent to approximately one pound of

Road force variation is a condition that occurs when tire sidewalls do not have equal stiffness around their complete area.

Shop Manual
Chapter 3, page 128

Figure 3-77 Think of a tire as sets of springs around its diameter. The variation in sidewall stiffness is referred to as road force variation and if excessive can create harmonic vibration and tire wear.

Figure 3-78 The roller on some wheel balancers senses road force variation.

Figure 3-79 The roller on some wheel balancers senses concentricity.

radial force variation. As an example, at 50 mph a wheel imbalance of 1.5 oz (42 g) will result in the same amount of vibration as 0.30 in. (0.76 mm) of loaded radial runout or 30 lb of road force variation.

If excessive road force variation is detected, match mounting the rim and tire should be considered. In match mounting, the high point on the tire is matched with the low point on the rim runout or vice versa to decrease the rolling vibration in the wheel assembly.

Tires must be manufactured so the cords are concentric with the center of the tire. Tire conicity is a term used to describe a tire in which the cords are off-center in relation to the tire center. If the tire cords are not concentric with the tire center, the tire may cause a steering pull condition when driving. Modern wheel balancers use a roller pressed against the tire tread to detect conicity during the balance procedure (**Figure 3-79**).

NOISE, VIBRATION, AND HARSHNESS ANALYSIS

OEMs have made significant improvements in reducing vehicle noise, vibration, and harshness (NVH), because many customers prefer a quieter-running vehicle with reduced vibration and ride harshness. The suspension system plays a significant role in reducing NVH. Many of the suspension features described in this chapter are designed to reduce NVH. These features include track bars and braces, proper car riding height, hydraulic suspension system mounts, independent rear suspension systems, and large rubber insulating bushings on suspension system mountings.

However, the technician must understand that reducing NVH involves many mechanical and body or chassis components on the vehicle. For example, some vehicles have an optimized body structure, which is the major reason for reduced NVH. In this body design, lateral tie bars that connect the front longitudinal rails provide a stiffer front end. At the rear of the car, one-piece side rings with integral quarters eliminate rear pillar seams and provide a more precise door fit. The instrument panel and steering column are integrated solidly into the body structure by a cast magnesium beam. The door hinges are through bolted and thick spacer blocks are installed on these bolts to provide a very solid door attachment. Because the entire body is stronger and more rigid, the suspension can provide excellent ride quality and steering control without having to compensate for unwanted body flexing. In addition, appropriate body cavities like the dash panel are filled with expandable baffles to eliminate noise. The five-layer noise buffer in the dash panel contains these materials:

1. Fiberglass insulation mat
2. Viscoelastic energy-absorbing layer
3. Double steel panel
4. Single one-piece dash mat

A cast foam floor carpet system reduces noise transmitted through the floor pan and wheel wells. The door pillar and rocker panel cavities contain over 20 noise blockers.

Many engine refinements reduce vibration. For example, many engines now have a deep skirt block with the main bearings bolted through the sides of the block and vertically. Many engines now have cast oil pans and rocker covers rather than stamped steel components. The main bearing caps are contained in a one-piece casting on many engines to increase bottom end strength and reduce engine vibration. Some V-8 engines have a rubber intake manifold valley stuffer attached to the underside of the intake manifold to reduce vibration. Other vehicles have a slip yoke vibration damper mounted on the front of the drive shaft to reduce driveline vibration. Therefore, reducing NVH is a total vehicle concept.

Vibration Theory

Vibrations have these three elements:

1. Source—the cause of the vibration
2. Path—where the vibration travels through the vehicle
3. Responder—the component where the vibration is felt

For example, if the vehicle has an unbalanced tire, this is the vibration source. The vibration path is the steering and suspension system through which the vibration travels. The responder is the steering wheel, because this component is where the customer feels the vibration (**Figure 3-80**). When diagnosing vibration problems, locate and correct the source of the vibration. In the previous example of the unbalanced tire, installing a rigid brace from the steering column to the instrument panel and chassis may reduce the vibration experienced by the customer, but this does not solve the problem. To eliminate the problem, also diagnose the unbalanced tire condition and then balance the tire-and-wheel assembly.

Vibration may also produce noise. If a vehicle has a broken or improperly positioned tailpipe hanger that allows the tailpipe to contact the chassis, the customer may complain about a vibrating noise. The floor panel acts as a large speaker and amplifies the vibrating noise. The vibration path is through the exhaust system and chassis to the floor panel. The responder is the chassis and floor panel. In this case, after diagnosing the broken or improperly positioned tailpipe hanger, replace or reposition it to eliminate the vibration transfer path.

Clamp a yardstick to a table top with 18 in. (45 cm) hanging over the edge of the table (**Figure 3-81**). If the outer end of the yardstick is pulled upward or downward and then released, the end of the yardstick repeatedly vibrates up and down. Each vibration **cycle** begins

A **cycle** begins and ends at the same point.

Figure 3-80 Vibration source, path, and responder.

Midpoint

Figure 3-81 Theory of vibration cycles.

at the midpoint with the yardstick straight. From this point, the vibration cycle continues to the lowest point of travel and then moves up through the midpoint to the highest point of travel. The vibration cycle then returns to the midpoint where it begins over again.

If the end of the vibrating yardstick completed 10 cycles per second, this is called the **frequency** of the vibration. To calculate the cycles per minute multiply the cycles per second by 60. In this example it is $10 \times 60 = 600$ cycles per minute.

If the yardstick is clamped to the table top with an 8-in. (20-cm) overhang, the end of the yardstick will vibrate much faster when the end of the yardstick is pulled upward or downward. Under this condition, the end of the yardstick may vibrate at a frequency of 30 cycles per second or 1800 cycles per minute.

Many vehicle vibrations are caused by an out-of-balance rotating component or engine firing pulses improperly isolated from the passenger compartment. Customers usually complain about vibrations that are felt in the steering wheel, instrument panel, frame or chassis, or the front or rear seat.

Vehicle vibrations may be tested with an **electronic vibration analyzer (EVA)** (**Figure 3-82**). The EVA has a vibration sensor mounted on the suspected vibration source. The EVA senses and records vibration cycles. A vibration cycle begins and ends at the same point and is continually repeated (**Figure 3-83**). Cycles per second are measured in **Hertz (Hz)**, and the Hz may be multiplied by 60 to obtain the cycles per minute. The **amplitude** of a vibration is the maximum value of the varying vibration. In **Figure 3-84**, 1 represents the maximum amplitude, 2 is the minimum amplitude, 3 is the zero-to-peak amplitude, and 4 indicates the peak-to-peak amplitude. The vibration amplitude may vary with the rotating speed of a component. For example, if a tire-and-wheel assembly is unbalanced, the amplitude of the resulting vibrations increases with wheel speed.

All objects have **natural vibration frequencies**. The natural frequency of a front suspension system is 10 to 15 Hz. The suspension system design determines the natural frequency. The natural frequency of the suspension system is the same at all vehicle speeds.

> The **frequency** of a vibration is the number of cycles per second or cycles per minute.

> **Natural vibration frequency** is the frequency at which the object tends to vibrate.

Figure 3-82 Electronic vibration analyzer (EVA).

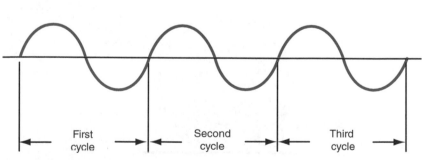

| First cycle | Second cycle | Third cycle |

Figure 3-83 Vibration cycles.

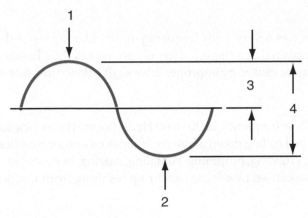

Figure 3-84 Amplitude of vibration cycles.

The vibration frequency caused by an unbalanced tire-and-wheel assembly increases with wheel speed. When the vibration frequency caused by the wheel balance problem intersects with the natural frequency of the suspension system, the suspension system begins to vibrate. This intersection point is called the **resonance**.

Vibration Classifications and Terminology

Vibrations that can be felt are:

1. Shake
2. Roughness
3. Buzz
4. Tingling

Vibrations that result in noise may be classified as:

1. Boom
2. Moan or drone
3. Howl
4. Whine

Shake

A shake is a vibration with a low frequency of 5 to 20 Hz, and is sometimes experienced in the steering wheel, seat, or console. Customers may refer to shake as shimmy, wobble, shudder, waddle, or hop. Most cases of a shake vibration are caused by brake rotors and drums or unbalanced tire-and-wheel assemblies. These defects cause a shake-type vibration that is vehicle-speed sensitive. The engine, clutch, or transmission may cause a shake-type vibration that is engine-speed sensitive.

Roughness

Roughness is a vibration with a higher frequency of 20 to 50 Hz. Holding a jigsaw to cut a piece of wood produces a roughness-type vibration. A roughness vibration may be caused by a defective wheel bearing. Prior to causing a roughness vibration, the defective wheel bearing would cause a howl.

Buzz

A buzz has a faster frequency of 50 to 100 Hz. Holding a vibrator-type electric razor produces a feeling similar to a buzz vibration. This type of vibration is usually felt in the vehicle floor or the seat. A buzz-type vibration is usually caused by defects in the exhaust system hangers, A/C compressor, or the engine.

Resonance is the point where the frequency of a vibration from a defective component intersects with the natural frequency of the component or system where the component is located.

Shop Manual
Chapter 3, page 133

Tingling

A tingling vibration has a very high frequency much like a pins-and-needles sensation. Customers may complain that a tingling-type vibration puts their hands and feet to sleep. A tingling vibration may be caused by improper drive shaft balance in a rear-wheel-drive vehicle.

Boom

A boom noise has a low frequency of 20 to 60 Hz. A boom-type vibration produces a noise similar to a bowling ball rolling down a bowling alley. A customer may describe a boom-type vibration as droning, growling, humming, rumbling, roaring, or moaning. A boom-type noise and vibration may be caused by engine backfiring resulting from ignition defects.

Moan or Drone

A moan or drone is a tone with a higher frequency of 60 to 120 Hz. A moan or drone produces a noise similar to a bumblebee in flight. Moan or drone may be caused by defects in the exhaust system or defective engine or transmission mounts.

Howl

A howl is a noise with frequency of 120 to 300 Hz, and this sound is much like the wind howling. A howling noise may be caused by worn differential bearings.

Whine

A whine is a high-pitched sound with a frequency of 300 to 500 Hz, and this sound is similar to a vacuum cleaner. A whine problem may be caused by meshing gears in the transmission of differential. When diagnosing vibration problems, the technician should match the vibration frequency to the rotating speed of a component, to help locate the component responsible for the vibration.

SUMMARY

- Tires are extremely important because they provide ride quality, support the vehicle's weight, provide traction for the drive wheels, and contribute to steering quality and directional stability.
- Tires have many design features and are carefully engineered to meet specific driving requirements.
- Tires may be bias-ply, belted bias-ply, or radial construction design.
- Tire ratings provide information regarding tire type, section width, aspect ratio, construction type, rim diameter, load capacity, and speed rating.
- Numeric tire ratings provide information about tire type, aspect ratio, construction type, and rim diameter.
- The tire performance criteria (TPC) number represents that the tire meets the car manufacturer's performance standards for traction, endurance, dimensions, noise, handling, and rolling resistance.
- The Uniform Tire Quality Grading (UTQG) designation includes tread wear, traction, and temperature ratings.
- Replacement tires must be the same type and size as the original tires to maintain vehicle safety.
- Replacement tires must have the same ratings as the original tires to maintain vehicle safety.
- The tire placard provides valuable information regarding the tires on the vehicle.
- Tires are subjected to severe acceleration and deceleration forces during normal operation.
- Replacement wheel rims must have the same width, diameter, offset, load capacity, and mounting configuration as the original rims to maintain vehicle safety.
- Static wheel unbalance causes wheel tramp and severe tire cupping.
- Dynamic wheel unbalance causes wheel shimmy, increased tire wear, unstable directional control, driver fatigue, and increased wear on suspension and steering components.

REVIEW QUESTIONS

Short Answer Essays

1. State general tire functions.

2. Interpret what each item of a tire size means, using P225/45R16 as your example.

3. What is the meaning of the service description 88H on the tire sidewall following the tire size?

4. Describe the structural difference between a bias-ply and radial-ply tire.

5. List and define the three major uniform tire quality grading system ratings.

6. What is the tire DOT code and how do you know the date the tire was manufactured?

7. Explain the difference between an A and C tire temperature rating.

8. Define tire contact area and tire deflection.

9. Explain the purpose of the wheel rim drop center and safety ridges.

10. Describe the results of static imbalance.

Fill-in-the-Blanks

1. To calculate the tire aspect ratio, the tire section height is divided by the _____.

2. Tires equipped on today's passenger cars and light trucks are of the _____ ply design.

3. In _____ _____, the ply cords are arranged radially at right angle to the tire centerline.

4. Replacement tires should be installed in pairs on the same _____.

5. When a high-pressure mini-spare tire is installed on a vehicle, driving speed should not exceed _____ mph.

6. Run-flat tires eliminate the need for a _____ _____ and a _____.

7. Radial runout refers to variations in tire _____.

8. Car manufacturers recommend that tire inflation pressures should be checked when the tires are _____.

9. _____ occurs when water on the pavement is allowed to remain between the pavement and the tire tread contact area.

10. _____ _____ vehicles tend to have positive offset rims to allow for proper clearance of the hub assembly within the vehicle wheel well.

Multiple Choice

1. All of the following statements about rim offset are correct EXCEPT:
   A. Zero offset, the rim centerline aligns with the mounting face of the rim.
   B. Positive offset, the rim centerline is inboard of the mounting face of the rim.
   C. Positive offset, the centerline is outboard of the mounting face of the rim.
   D. Negative offset, the centerline is outboard of the mounting face of the rim.

2. Tire pressure monitoring systems (TPMS) became required equipment on all vehicles sold in the United States beginning with what model year?
   A. 1996   C. 2006
   B. 2000   D. 2008

3. What is the maximum speed rating for a T rated tire?
   A. 112 mph   C. 124 mph
   B. 118 mph   D. 130 mph

4. Dynamic wheel unbalance can result in:
   A. Lateral wheel shimmy.
   B. Increased steering effort.
   C. Tire and wheel tramp.
   D. Normal tire tread life.

5. When a tire has a conicity problem:
   A. The tire has excessive lateral runout.
   B. The tire has too much radial runout.
   C. The tire is cone shaped and not level across the tread area.
   D. The tire has separation between the cord plies in the sidewall.

6. The Uniform Tire Quality Grading (UTQG) system provides consumers with useful information to help them purchase tires based on all of the following EXCEPT:

   A. Temperature rating.

   B. Traction rating.

   C. Speed rating.

   D. Tread wear rating.

7. Nitrogen tire inflation provides:

   A. Increased tire rolling resistance.

   B. Improved puncture resistance.

   C. More stable tire pressure.

   D. Improved ride quality.

8. The tread wear indicator bar becomes visible when tread depth is less than:

   A. 4/32 in.   C. 2/32 in.

   B. 2/16 in.   D. 3/16 in.

9. Static wheel unbalance causes:

   A. Cupped tire tread wear.

   B. Even wear on one edge of the tire tread.

   C. Even wear on the center of the tire tread.

   D. Feathered tire tread wear.

10. When installing replacement tires:

    A. Different tire sizes can be installed on the same axle.

    B. Snow tires can be a different type or size than the original tires on the vehicle.

    C. If the tire size is different than the original tire size, antilock brake system operation can be adversely affected.

    D. A four-wheel-drive vehicle can have different size tires on the front and rear wheels.

CHAPTER 4
WHEEL BEARINGS

Upon completion and review of this chapter, you should be able to understand and describe:

- The purposes of a bearing.
- Three different types of bearing loads.
- The basic parts in ball bearings or roller bearings, and describe the purpose of each part.
- The action between the balls and the race when a ball bearing is rotating.
- The purposes of bearing snap rings, shields, and seals.
- The load-carrying capabilities of ball bearings, roller bearings, tapered roller bearings, and needle roller bearings.

- The advantage of tapered roller bearings compared to other types of bearings.
- Seal design and purpose.
- The purpose of the garter spring behind a lip seal.
- The purpose of flutes on seal lips.
- Two different types of rear axle bearings in rear-wheel-drive cars, and give the seal location for each bearing type.
- Grease classifications.

Terms To Know

American Petroleum Institute (API)	Inner race	Single-row ball bearing
Angular bearing load	Lithium-based grease	Snap ring
Ball bearings	Needle roller bearing	Society of Automotive Engineers (SAE)
Bearing hub unit	Outer race	Sodium-based grease
Bearing seals	Radial bearing load	Springless seals
Bearing shields	Rear-axle bearings	Spring-loaded seal
Double-row ball bearings	Roller bearing	Tapered roller bearing
Fluted lip seal	Rolling elements	Thrust bearing load
	Separator	

INTRODUCTION

Many different types of bearings are used in the automobile. A bearing may be defined as a component that supports and guides one of these parts:

1. Pivot
2. Wheel
3. Rotating shaft
4. Oscillating shaft
5. Sliding shaft

While a bearing is supporting and guiding one of these components, the bearing is designed to reduce friction and support the load applied by the component and related assemblies. Since the bearing reduces friction, it also decreases the power required to rotate or move the component. Bearings are precision-machined assemblies, which provide smooth operation and long life. When bearings are properly installed and maintained, bearing failure is rare.

> **AUTHOR'S NOTE** Only 21 percent of the power developed by a vehicle engine actually gets to the drive wheels. Much of the engine energy is lost in overcoming friction and wind resistance. The U.S. Department of Energy (DOE) is working with a major bearing manufacturer to explore the use and advantage of roller bearings to reduce friction in vehicle engines. The DOE estimates that if all vehicle engines used roller-bearing, low-friction technology, 100 million barrels of oil could be saved in a year.

Shop Manual
Chapter 4, page 166

A **thrust bearing load** may be referred to as an axial load.

BEARING LOADS

When a bearing load is applied in a vertical direction on a horizontal shaft, it is called a **radial bearing load**. If the vehicle weight is applied straight downward on a bearing, this weight is a radial load on the bearing. A **thrust bearing load** is applied in a horizontal direction (**Figure 4-1**). For example, while a vehicle is turning a corner, horizontal force is applied to the front wheel bearings. When an **angular bearing load** is applied, the angle of the applied load is somewhere between the horizontal and vertical positions. This can also be referred to as a combination load.

BALL BEARINGS

Ball bearings have round steel balls between the inner and outer races.

Front and rear wheel bearings may be **ball bearings** or roller bearings. Either type of bearing contains these basic parts:

1. Inner race, or cone
2. Rolling elements, balls, or rollers
3. Separator, also called a cage or retainer
4. Outer race, or cup

The **inner race** supports the inner side of the rolling elements in a bearing.

The **rolling elements** are the precision-machined balls or rollers between the inner and outer bearing races.

The **inner race** is an accurately machined component. The inner surface of the race is mounted on the shaft with a precision fit. The **rolling elements** are mounted on a very

Radial load

Angular contact load

Thrust load

Figure 4-1 Types of bearing loads.

smoothly machined surface on the inner race. The surfaces of the rolling elements and the inner and outer races are case hardened to provide long bearing life. Positioned between the inner and outer races, the **separator** retains the rolling elements and keeps them evenly spaced. The rolling elements have precision-machined surfaces, and these elements are mounted between the inner and outer races. The **outer race** is the bearing's exterior ring. Both sides of this component have precision-machined surfaces. The outer surface of this race supports the bearing in the housing, and the inner surface is in contact with the rolling elements.

A **single-row ball bearing** has a crescent-shaped machined surface in the inner and outer races in which the balls are mounted (**Figure 4-2**). When a ball bearing is at rest, the load is distributed equally through the balls and races in the contact area. When one of the races and the balls begin to rotate, the bearing load causes the metal in the race to bulge out in front of the ball and flatten out behind the ball (**Figure 4-3**). This action creates a certain amount of friction within the bearing, and the same action is repeated for each ball, while the bearing is rotating. If metal-to-metal contact is allowed between the balls and the races, these components will experience very fast wear. Therefore, bearing lubrication is extremely important to eliminate metal-to-metal contact in the bearing and reduce wear.

A ball bearing is designed primarily to handle radial loads. However, this type of bearing can also withstand a considerable amount of thrust load in either direction, even at high speeds. A maximum capacity ball bearing has extra balls for greater radial load-carrying capacity. Ball bearings are available in many different sizes for various applications (**Figure 4-4**).

Double-row ball bearings contain two rows of balls side by side. As in the single-row ball bearing, the balls in the double-row bearing are mounted in crescent-shaped grooves in the inner and outer races. The double-row ball bearing can support heavy radial loads and withstand thrust loads in either direction.

Ball Bearing Seals, Shields, and Snap Rings

For some applications, a ball bearing is held in place with a **snap ring**. A groove is cut around the outside surface of the outer race, and the snap ring is mounted in this groove. The snap ring may fit against a machined housing surface, or the outer circumference of the snap ring may be mounted in a groove in the housing. Ball bearings retained with a snap ring are not used on wheels because they are not designed to withstand high thrust loads encountered by wheel bearings.

> The **outer race** supports the outer side of the rolling elements and positions the bearing properly in the housing.

> A **single-row ball bearing** has one row of balls between the inner and outer races.

> A **snap ring** is made from spring steel and fits into a groove to retain a bearing on a shaft.

Figure 4-3 When a load is applied to a ball bearing, the metal in the race bulges out in front of the ball and flattens out behind the ball.

Figure 4-2 Parts of a ball bearing.

Figure 4-4 A ball bearing.

Bearing shields cover the space between the two bearing races on one, or both, sides of the bearing. These shields are usually attached to the outer race, but space is left between the shield and the inner race. Bearing shields prevent dirt from entering the bearing, but excess lubrication can still flow through the bearing.

A **bearing seal** is a circular metal ring with a sealing lip on the inner edge. The seals are usually attached to the outer bearing race on each side of the bearing, and the lip surface contacts the inner race. The seal lip may have single, double, or triple lips made of synthetic or nonsynthetic rubber or elastomers. Lubricant is retained in the bearing by the seal, and the seal also keeps moisture, dirt, and contaminants out of the bearing. Some rear axle bearings on rear-wheel-drive cars are sealed on both sides and retained on the axle with a retainer ring (**Figure 4-5**).

Shop Manual
Chapter 4, page 172

Figure 4-5 Some rear axle bearings are sealed on both sides and retained on the axle with a retainer ring.

ROLLER BEARINGS

Roller Bearing

A **roller bearing** contains precision-machined rollers that have the same diameter at both ends. These rollers are mounted in square-cut grooves in the outer and inner races (**Figure 4-6**). In the roller bearing, the races and rollers run parallel to one another. Roller bearings are designed primarily to carry radial loads, but they can withstand some thrust load. Since ball bearings do not withstand high thrust loads, they are usually not used in wheel bearings.

A **roller bearing** is sometimes called a nontapered roller bearing.

Tapered Roller Bearing

In a **tapered roller bearing**, the inner and outer races are cone-shaped. If imaginary lines extend through the inner and outer races, these lines taper and eventually meet at a point extended through the center of the bearing (**Figure 4-7**). The most important advantage of the tapered roller bearing compared to other bearings is an excellent capability to carry radial, thrust, and angular loads. In the tapered roller bearing, the rollers are mounted on cone-shaped precision surfaces in the outer and inner races. The bearing separator has an open space over each roller (**Figure 4-8**). Grooves cut in the side of the separator roller openings match the curvature of the roller. This design allows the rollers to rotate evenly

A **tapered roller bearing** has tapered rollers between the inner and outer races. Tapered surfaces on the races match the tapered roller surfaces.

Shop Manual
Chapter 4, page 167

Figure 4-6 Parts of a roller bearing.

Figure 4-7 Imaginary lines extending from tapered roller bearing races eventually meet at a point extending from the bearing center.

Figure 4-8 Tapered roller bearings.

Figure 4-9 Needle roller bearing.

without interference between the rollers and the separator. Lubrication and proper end play adjustment are critical on tapered roller bearings. A tapered roller bearing outer and inner race may be called a cup and a cone, respectively.

Needle Roller Bearings

A **needle roller bearing** contains many small-diameter steel rollers in a thin outer race. This type of bearing is very compact, and it is used in steering gears where mounting space is limited. Most needle roller bearings do not have a separator, but the steel rollers push against each other and maintain the roller position. Rather than having an inner race, a machined surface on the mounting shaft contacts the inner surface of the rollers (**Figure 4-9**). The needle roller bearing is designed to carry radial loads; it does not withstand thrust loads.

⚠ **WARNING** **Never strike a roller bearing with a steel hammer. This action may cause the bearing to shatter, resulting in severe personal injury.**

⚠ **WARNING** **Spinning a roller bearing with compressed air may rotate the bearing at extremely high speed and cause the bearing to disintegrate, resulting in serious personal injury.**

SEALS

Seals are designed to keep lubricant in the bearing and prevent dirt particles and contaminants from entering the bearing. Static seals are used between two surfaces that do not move and dynamic seals are used between two surfaces where at least one surface moves. Wheel bearing seals are mounted in front and rear wheel hubs, and in rear axle housings on rear-wheel-drive cars. The metal seal case has a surface coating that resists corrosion and rust and acts as a bonding agent for the seal material. Seals have many different designs, including single lip, double lip, and fluted. A lip seal is usually encased in metal. The seal material is usually made of a synthetic rubber compound such as nitrile, silicon, polyacrylate, or a fluoroelastomer such as Viton. The actual seal material depends on the lubricant and contaminants that the seal encounters. All seals may be divided into two groups, springless and spring loaded. **Springless seals** are used in some front or rear wheel hubs, where they seal a heavy lubricant into the hub (**Figure 4-10**).

Springless seals do not have a garter spring behind the seal lip.

Figure 4-10 Springless seal.

Figure 4-11 Spring-loaded seal.

Figure 4-12 Fluted lip seal redirects oil back into the housing.

Figure 4-13 Sealer painted on the seal case prevents leaks between the case and the housing.

In a **spring-loaded seal**, the **garter spring** behind the seal provides additional force on the seal lip to compensate for lip wear, shaft movement, and bore eccentricity (**Figure 4-11**). A **fluted lip seal** may be used to direct oil back into a housing. This seal design provides a pumping action to redirect the oil back into the housing (**Figure 4-12**).

Some seals have a sealer painted on the outside surface of the metal seal housing. When the seal is installed, this sealer prevents leaks between the seal case and the housing (**Figure 4-13**). Seals should be replaced whenever it is necessary to remove one during a service procedure.

A **spring-loaded seal** has a garter spring behind the seal lip.

A **garter spring** is a circular coil spring mounted behind a seal lip.

A **fluted lip seal** has small ridges on the outer lip surface.

WHEEL BEARINGS

Some front-wheel-drive vehicles have front wheel bearing and hub assemblies that are bolted to the steering knuckles (**Figure 4-14**). The bearings are lubricated and sealed, and the complete bearing and hub assembly is replaced as a unit. The **bearing hub unit** is more compact than other types of wheel bearings mounted in the wheel hub. This type of bearing contains two rows of ball bearings with an angular contact angle of 32° (**Figure 4-15**). The inner bearing assembly bore is splined, and the inner ring extends to the outside to form a flange and spigot. The flange attached to the outer ring contains bolt holes, and bolts extend through these holes into the steering knuckle. This type of bearing attachment allows the bearing to become a structural member of the front suspension.

A **bearing hub unit** is a one-piece bearing and hub assembly that supports a front or rear wheel.

Figure 4-14 Wheel bearing and hub assembly.

Figure 4-15 Double-row, sealed wheel bearing hub unit.

Since the bearing outer ring is self-supporting, the main concern in knuckle design is fatigue strength rather than stiffness. The drive axle shaft transmits torque to the inner bearing race. This shaft is not designed to hold the bearing together. This type of wheel bearing is designed for mid-sized front-wheel-drive cars.

Shop Manual
Chapter 4, page 174

Each front drive axle has splines that fit into matching splines inside the bearing hubs (**Figure 4-16**). A hub nut secures the drive axle into the inner bearing race. Some wheel bearing hubs contain a wheel speed sensor for the antilock brake system (ABS).

Some front-wheel-drive vehicles have a sealed bearing unit that is pressed into the steering knuckle (**Figure 4-17**). The wheel hub is pressed into the inner bearing race, and the drive axle is splined into the hub. This type of bearing is designed for smaller front-wheel-drive cars. These bearings may contain two rows of ball bearings, or two tapered

Figure 4-16 Front drive shaft installed in wheel bearing hub.

Figure 4-17 Steering knuckle with pressed-in bearing.

Figure 4-18 Cutaway view of front wheel bearing that is pressed into the steering knuckle.

roller bearings and a split inner race (**Figure 4-18**). The bearing containing two tapered roller bearings has more radial load capacity than the double-row ball bearing. However, the tapered roller bearing is more sensitive to misalignment. Both sides of the bearing are sealed, and a seal is positioned behind the bearing in the steering knuckle to keep contaminants out of the bearing area.

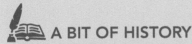

A BIT OF HISTORY

Nearly all rear-wheel-drive cars have two serviceable tapered roller bearings or ball bearings in the front wheel hubs. Most front-wheel-drive cars have sealed bearing hubs on the front wheels, and some have these units on the rear wheels. These sealed bearing hub units have largely replaced the tapered roller bearings or ball bearings in the front wheel hubs.

Front Steering Knuckles with Two Separate Tapered Roller Bearings

Other front-wheel-drive vehicles have two separate tapered roller bearings mounted in the steering knuckles. The bearing races are pressed into the steering knuckle, and seals are located in the knuckle on the outboard side of each bearing (**Figure 4-19**). Correct bearing endplay adjustment is supplied by the hub nut torque. The wheel hub is pressed into the inner bearing races, and the drive axle splines are meshed with matching splines in the wheel hub.

Wheel Hubs with Two Separate Tapered Roller Bearings

This type of bearing system was very popular on rear-wheel-drive cars and light trucks during the 1980s and earlier. A serviceable hub assembly uses two tapered roller bearings that support the hubs and wheels on the spindles. This is still a popular bearing configuration on towable utility, camper, and boat trailers. Serviceable bearings require routine maintenance and inspection, which is covered in the Shop Manual. This type of wheel bearing has the bearing races pressed into the hub. A grease seal is pressed into the inner end of the hub to prevent grease leaks and keep contaminants out of the bearings. The hub and bearing assemblies are retained on the spindle with a washer, adjusting nut, nut lock, and cotter pin. The adjusting nut must be adjusted properly to provide the correct bearing endplay. A grease cap is pressed into the outer end of the hub to prevent bearing contamination (**Figure 4-20**).

Some older front-wheel-drive cars from the 1980s and early 1990s have two tapered roller bearings in the rear wheel hubs that are very similar to the wheel bearings in **Figure 4-20**. The two tapered roller bearings in the rear wheel of a front-wheel-drive car are shown in **Figure 4-21**.

Figure 4-19 Steering knuckle with two separate tapered roller bearings.

Cotter pin

Washer

Spindle

Grease seal

Nut and retainer

Tapered roller bearings

Hub

Figure 4-20 Front wheel bearing assembly, rear-wheel-drive car.

Brake support

DRUM BRAKES

Brake drum

Spindle

Caliper support

Seal

Bearing cone (inner)

Bearing cone (outer)

Washer

Nut

Lock

Cotter pin

Grease cap

Hub

DISC BRAKES

Braking disc

Figure 4-21 Rear wheel hub with tapered roller bearings for drum or disc brakes, front-wheel-drive car.

AUTHOR'S NOTE An understanding of wheel bearings and related service procedures is critical to maintain vehicle safety! If the technician's knowledge of wheel bearings and appropriate service procedures is inadequate, bearing failure may occur. Wheel bearing failure may cause a wheel to fly off a vehicle with disastrous results!

Shop Manual
Chapter 4, page 182

REAR-AXLE BEARINGS

On many rear-wheel-drive cars, the rear axles are supported by roller bearings mounted near the outer ends of the axle housing. The outer bearing race is pressed into the housing, and a machined surface on the axle contacts the inner roller surface. A seal is mounted in the axle housing on the outboard side of each bearing (**Figure 4-22**). This type of axle bearing is usually not sealed, and lubricant in the differential and rear axle housing provides axle bearing lubrication. The seals prevent lubricant leaks from the outer ends of the axle housing and keep dirt out of the bearings.

Other **rear-axle bearings** on rear-wheel-drive vehicles have sealed roller bearings pressed onto the rear axles. These axle bearings are sealed on both sides, and an adapter ring is pressed onto the axle on the inboard side of the bearing (**Figure 4-23**). The outer bearing race is mounted in the rear axle housing with a light press fit, and a seal is positioned in the housing on the inboard side of the bearing and adapter ring. A retainer plate is mounted between the bearing and the outer end of the axle. This retainer plate is bolted to the axle housing to retain the axle in the housing.

Rear-axle bearings support the rear axles in the rear axle housing on rear-wheel-drive vehicles.

Many wheel bearings require lithium-based or sodium-based grease.

A **lithium-based grease** has a specific amount of lithium mixed with the lubricant.

A **sodium-based grease** has a specific amount of sodium mixed with the lubricant.

BEARING LUBRICATION

Proper bearing lubrication is extremely important to maintain bearing life. Bearing lubricant reduces friction and wear, dissipates heat, and protects surfaces from dirt and corrosion. Sealed or shielded bearings are lubricated during the manufacturing process, *and no attempt should be made to wash these bearings or pack them with grease.*

Bearings that are not sealed or shielded require cleaning and repacking at intervals specified by the vehicle manufacturer. *Always use the bearing grease specified by the*

Figure 4-22 Rear-axle bearing, rear-wheel-drive car.

Figure 4-23 Rear-axle bearing and retainer.

vehicle manufacturer. Bearing lubricants may be classified as greases or oils. Grease is oil with a thickening agent added. Greases are named for the thickening agent:

- Aluminum
- Barium
- Calcium
- Lithium
- Sodium

ASTM International sets the standards for testing for grease. Classification ASTM D 4950 covers greases designed for the lubrication of wheel bearings and chassis components for passenger cars, trucks, and other vehicles. The National Lubricating Grease Institute (NLGI) classifies automotive greases into two main groups. The prefix designation for wheel bearing grease is G, and the prefix designation for chassis lubrication grease is L (**Figure 4-24**).

Service category L chassis grease:

- LA-Mild duty, non-critical applications with frequent relubrication intervals of less than 2000 mi. (3200 km). Performance offers oxidation resistance, shear stability, and corrosion and wear protection.
- LB-Mild to severe duty, high loads, and vibration with a usable temperature range of $-40 - 248°\pmb{F}$. Relubrication intervals of greater than 2000 mi. (3200 km). Performance offers oxidation resistance, shear stability, and corrosion and wear protection even under heavy loads and in the presence of water contamination.

Service category G wheel bearing grease:

- GA-Mild duty
- GB-Mild to moderate duty, with a usable temperature range of $-40 - 248°\pmb{F}$ and occasional spikes to 320°F. Performance offers oxidation and evaporation resistance, shear stability, and corrosion and wear protection.
- GC-Mild to severe duty, with a usable temperature range of $-40 - 320°\pmb{F}$ and occasional spikes to 392°F. Performance offers oxidation and evaporation resistance, shear stability, and corrosion and wear protection.

The most common grease designation used in automotive shops is GC and LB. A designation label GC/LB indicates that the grease is acceptable for bearings and chassis parts. Number 2 grease is the most common rating with the higher the number indicating the stiffer the consistency of the grease.

The Society of Automotive Engineers (SAE) is responsible for the establishment of many automotive standards.

New bearings usually have a protective coating to prevent rust and corrosion. This coating should not be washed from the bearing. On rear-wheel-drive vehicles, where the rear-axle bearings are lubricated from oil in the differential housing, the type and level of oil in the housing are important.

Vehicle manufacturers usually recommend a **Society of Automotive Engineers (SAE)** No. 90 or SAE No. 140 hypoid gear oil in the differential. In very cold climates, the manufacturer may recommend an SAE No. 80 differential gear oil. The **American Petroleum Institute (API)** classifies gear lubricants as GL-1, GL-2, GL-3, GL-4, and GL-5. The GL-4

The American Petroleum Institute (API) is responsible for establishing standards related to oils and lubricants.

NLGI SYMBOLS

Figure 4-24 The National Lubricating Grease Institute (NLGI) classifies automotive greases into two main groups, L and G.

lubricant is used for hypoid gears under normal conditions. The GL-5 lubricant is used in heavy-duty hypoid gears. Always use the vehicle manufacturer's specified differential gear oil.

The differential should be filled until the lubricant is level with the bottom of the filler plug opening in the differential housing. If the differential is overfilled, the bearings and seals may have excessive lubricant. Under this condition, the lubricant may leak past the seal. When the lubricant level is low in the differential, the lubricant may not be available in the axle housings. When this condition exists, the bearings do not receive enough lubrication and bearing life is shortened.

> **⚠ Caution**
>
> If a bearing is operated without proper lubrication, bearing life will be very short.

SUMMARY

- A bearing reduces friction, carries a load, and guides certain components such as pivots, shafts, and wheels.
- Radial bearing loads are applied in a vertical direction.
- Thrust bearing loads are applied in a horizontal direction.
- Angular bearing loads are applied at an angle between the vertical and the horizontal.
- The inner bearing race is positioned at the center of the bearing and supports the rolling elements.
- The rolling elements in a bearing are positioned between the inner and outer races.
- The bearing separator keeps the rolling elements evenly spaced.
- The outer bearing race forms the outer ring on a bearing.
- A cylindrical ball bearing is designed primarily to withstand radial loads, but these bearings can handle a considerable thrust load.
- A snap ring can be mounted in a groove in the outer bearing race, and the snap ring retains the bearing in the housing.
- A bearing shield prevents dirt from entering the bearing, but it is not designed to keep lubricant in the bearing.

- Bearing seals keep lubricant in the bearing and prevent dirt from entering the bearing.
- Roller bearings are designed primarily to carry radial loads, but they can handle some thrust loads.
- Tapered roller bearings have excellent radial, thrust, and angular load-carrying capabilities.
- Needle roller bearings are very compact and are designed to carry radial loads. They will not carry thrust loads.
- Springless seals are used for wheel bearing seals in some wheel hubs.
- The garter spring provides additional force on the seal lip to compensate for lip wear, shaft movement, and bore eccentricity.
- Flutes on seal lips provide a pumping action to direct oil back into a housing.
- Bearing hub units are compact compared to bearings that are mounted in the wheel hub. This compactness makes bearing hub units suitable for front-wheel-drive cars.
- Some bearing hub units are bolted to the steering knuckle; other bearing hub units are pressed into the steering knuckle.
- Some steering knuckles contain two separate tapered roller bearings
- Rear-axle bearings are mounted between the drive axles and the housing on rear-wheel-drive cars.

REVIEW QUESTIONS

Short Answer Essays

1. Define a radial bearing load.

2. Define a thrust bearing load, and give another term for this type of load.

3. Explain an angular bearing load.

4. Describe the main parts of a bearing, including the location and purpose of each part.

5. Describe the difference between a maximum-capacity ball bearing and an ordinary ball bearing.

6. Explain the design and purpose of bearing seals.

7. Explain the purpose of the sealer on the outside surface of a seal housing.

8. What does ASTM and NLGI stand for, and what is their function?

9. Explain how the proper bearing endplay adjustment is obtained when two tapered roller bearings are mounted in the steering knuckle.

10. Describe two different types of rear-axle bearings on rear-wheel-drive cars.

Fill-in-the-Blanks

1. A bearing can be described as a component that supports a _____, _____, or _____.

2. A bearing is designed to support a load and _____.

3. An angular bearing load is applied at an angle between the _____ and the _____.

4. A ball bearing is designed primarily to withstand _____ loads.

5. Greases are named for the _____ agent used.

6. Lubrication and proper_____ adjustment are important on tapered roller bearings.

7. A needle roller bearing is not designed to carry _____ loads.

8. A springless seal may be used to seal a _____ lubricant into a hub.

9. When a vehicle is turning a corner, the front wheel bearings must carry a _____ load.

10. When a rear-wheel-drive vehicle has cylindrical roller bearings mounted in the rear axle housing, the inner surface on the rollers contacts a machined surface on the _____.

Multiple Choice

1. A tapered roller bearing has:
   A. Rollers that have the same diameter at both ends.
   B. Excellent ability to carry radial, thrust, and angular loads.
   C. Horizontal inner surfaces of the inner and outer races.
   D. A separator that allows the rollers to lightly contact each other.

2. A spring-loaded seal compensates for all of these conditions EXCEPT:
   A. Seal lip wear.
   B. Shaft movement.
   C. Bore eccentricity.
   D. Wear on the seal bore.

3. When two separate tapered roller bearings are located in a front steering knuckle on a front-wheel-drive car, the wheel bearing endplay adjustment is provided by:
   A. The bearing race position.
   B. Wheel nut torque.
   C. Hub nut torque.
   D. Wheel hub position.

4. Greases are named for the thickening agent. All of the following are common greases EXCEPT:
   A. Lithium   B. Sodium
   C. Teflon   D. Aluminum

5. All of the following statements are correct EXCEPT:
   A. Greases are named for their thickening agent.
   B. NLGI classifies gear lubricants.
   C. The prefix GC is the highest designation for automotive wheel bearing greases.
   D. The prefix LB is the highest designation for automotive wheel chassis greases.

6. When inspecting and servicing bearings:
   A. Some rear-wheel-drive vehicles have rear-axle bearings that are sealed on both sides.
   B. In a sealed bearing, the seal is usually attached to the inner race.
   C. Endplay adjustment is not critical on tapered roller bearings.
   D. A roller bearing can be removed by striking it with a steel hammer.

7. All of these statements about bearings are true EXCEPT:
   A. Tapered roller bearings have excellent capability to carry radial, thrust, and angular loads.
   B. Ball bearings are used in most wheel bearings.
   C. A roller bearing has rollers with the same diameter at each end.
   D. In a tapered roller bearing, the openings in the separator grooves match the curvature of the rollers.

8. While inspecting and servicing wheel hub seals:

   A. Many wheel hub seal lips are made from a plastic compound.

   B. Springless seals may be used in applications where they must seal a light fluid.

   C. A spring-loaded seal helps compensate for wheel hub bore eccentricity.

   D. The sealer on the outer surface of a seal housing improves seal and hub bore alignment.

9. When inspecting and servicing front wheel bearings on front-wheel-drive cars:

   A. Some one-piece front wheel bearing hubs are bolted to the front strut.

   B. In a one-piece front wheel bearing hub, the outer end of the drive axle is splined to the outer bearing race.

   C. When the steering knuckle contains two tapered roller bearings, the torque on the drive axle nut does not affect bearing endplay.

   D. Some front wheel bearings mounted in the steering knuckle contain two tapered roller bearings with a split inner race.

10. When inspecting and servicing rear wheel bearings on rear-wheel-drive cars:

   A. On some rear wheel bearings, a threaded adapter ring retains the bearing on the axle shaft.

   B. Roller-type rear-axle bearings may be lubricated from the oil supply in the differential.

   C. A GL-2 lubricant is recommended in hypoid-type differentials.

   D. An SAE 50 gear oil is recommended in many differentials.

CHAPTER 5
SHOCK ABSORBERS AND STRUTS

Upon completion and review of this chapter, you should be able to understand and describe:

- The three purposes of shock absorbers.
- How shock absorbers contribute to vehicle safety.
- Wheel jounce and rebound.
- Spring operation during wheel jounce and rebound.
- Shock absorber operation during wheel jounce.
- Shock absorber operation during wheel rebound.
- The advantages of nitrogen-gas-filled shock absorbers and struts.
- Shock absorber ratios.
- Travel-sensitive shock absorber operation.
- The operation of an adjustable shock absorber.
- The operation of load-leveling struts and shock absorbers.

Terms To Know

Active suspension systems
Adjustable struts
Gas-filled shock absorber
Heavy-duty shock absorbers
Jounce travel

Load-leveling shock absorbers
Magneto-rheological fluid
Rebound travel
Shock absorbers
Shock absorber ratios

Spring insulator
Struts
Travel-sensitive struts
Upper strut mount

INTRODUCTION

Two front **shock absorbers** or **struts** are connected from the front suspension to the chassis, and two rear shock absorbers or struts are attached between the rear suspension and the chassis. The internal strut design is very similar to shock absorber design, and struts perform the same control functions as shock absorbers. Shock absorbers have three main purposes:

Shock absorbers are devices used to control, or dampen, spring oscillations.

1. They control spring action and oscillations (bounce) to provide the desired ride quality and noise reduction.
2. They help prevent body sway (roll) and lean while cornering.
3. They reduce brake dive and acceleration squat, thereby reducing the tendency of a tire tread to lift off the road, which improves tire life, traction, and directional stability.

The kinetic energy of the suspension system spring action is converted to thermal energy as the shock absorber piston travels though the hydraulic fluid. This heat is then dissipated into the air flowing around the outer steel tube of the shock absorber. Because shock absorbers control spring action, spring oscillations, and chassis oscillations, they contribute to vehicle safety and passenger comfort. If the shock absorbers are worn out, excessive chassis oscillations may occur, particularly on rough road surfaces. These

Worn-out shocks increase stopping distance

Worn shocks — 112 ft.

New shocks — 91 ft.

0 20 40 60 80 100 120
Feet

Figure 5-1 Worn-out shock absorbers can increase stopping distance.

excessive chassis oscillations may result in loss of steering control. Worn-out shock absorbers also cause excessive body lean and sway while cornering, which may cause the driver to lose control of the vehicle and can contribute to increased stopping distance (**Figure 5-1**). Shock absorbers are extremely important to providing longer tire life and improving vehicle handling, steering quality, and ride quality. A faulty shock absorber can cause tire cupping to occur. The two most common causes of tire cupping are a static imbalance condition of the wheel assembly and a weak hydraulic shock absorber.

Shock absorber design is matched to the deflection rate of the suspension spring to control the spring action. Several different types of shock absorbers are available for special service requirements.

SHOCK ABSORBER AND STRUT RATIOS

Most automotive shock absorbers and struts are a double-acting-type that controls spring action during jounce and rebound wheel movements. The piston and valves in many shock absorbers are designed to provide more extension control than compression control. An average shock absorber may have 70 percent of the total control on the extension cycle, and thus 30 percent of the total control is on the compression cycle. Shock absorbers usually have this type of design because they must control the heavier sprung body weight on the extension cycle. The lighter unsprung axle, wheel, and tire weight are controlled by the shock absorber on the compression cycle. A shock absorber with this type of design is referred to as a 70/30 type. **Shock absorber ratios** vary from 50/50 to 80/20.

A shock absorber is mounted between the rear axle and the chassis on a front-wheel-drive car (**Figure 5-2**). Mounting bolts extend through hangers on the rear axle and

Shock absorber ratios indicate the amount of control on the extension and compression cycles.

Chassis

Track bar

Shock absorber

Axle assembly

Figure 5-2 Rear shock absorber mounting on a front-wheel-drive car.

chassis. These bolts also pass through the isolating bushings on each of the shock absorbers. The isolating bushings are very important for preventing vibration and noise. Front shock absorbers may be mounted in a similar way between the lower control arms and the chassis.

TWIN-TUBE SHOCK ABSORBER DESIGN

The lower half of a shock absorber is a twin-tube steel unit filled with hydraulic oil and nitrogen gas (**Figure 5-3**). In some shock absorbers, the nitrogen gas is omitted. The outer tube is the reservoir tube, which stores excess hydraulic fluid and the inner tube is the pressure or working tube. A relief valve is located in the bottom of the unit to control fluid movement during compression, and a circular lower mounting is attached to the lower

Figure 5-3 Shock absorber filled with hydraulic oil and nitrogen gas.

tube. This mounting contains a rubber isolating bushing, or grommet, to reduce the transmitted road noise and suspension vibration. The rubber bushings are flexible to allow for minimal movement as road forces are transmitted through the assembly. The upper mount is connected to the vehicle frame and the lower mount is typically connected to the lower control arm or axle tube of the suspension system. A piston and rod assembly is connected to the upper half of the shock absorber. This upper portion of the shock absorber has a dust shield that surrounds the lower twin-tube unit. The piston is precision fit in the inner cylinder of the lower unit. A piston rod guide and seal are located in the top of the lower unit. A circular upper mounting with a rubber bushing is attached to the top of the shock absorber. Shock absorber bore size refers to the diameter of the working piston. Generally, the larger the piston diameter, the higher the potential energy and damping control level of the shock absorber because of larger piston displacement and pressure area. As a general rule, the larger the piston area, the lower the internal operating temperature and pressure. To achieve optimal ride characteristics of stability and balance under a wide variety of driving conditions, ride engineers select piston size and valve rates for specific vehicle applications.

Shop Manual
Chapter 5, page 196

The advantages of the twin-tube shock absorber are as follows:

- Combines both comfort and control.
- Adjusts more rapidly than a single tube design shock absorber to changing weight and road conditions.

A disadvantage of the twin-tube design is that it can only be mounted in one direction. The twin-tube shock absorber is the most common original equipment manufacturer (OEM) design used on cars, SUVs, and light trucks.

SHOCK ABSORBER OPERATION

Shock absorbers are usually mounted between the lower control arms and the chassis. When a vehicle wheel strikes a bump, the wheel and suspension move upward in relation to the chassis. Upward wheel movement is referred to as **jounce travel**. This jounce action causes the spring to deflect or compress. Under this condition, the spring stores energy and springs back downward with all the energy absorbed when it deflected upward. This downward spring and wheel action is called **rebound travel**. If this spring action were not controlled, the wheel would strike the road with a strong downward force, and the wheel jounce would occur again. Therefore, some device must be installed to control the spring action, or the wheel would bounce up and down many times after it hit a bump, causing passenger discomfort, directional instability, and suspension component wear along with severe tire cupping.

Shock absorbers are installed on suspension systems to control spring action. When a wheel strikes a bump and jounce travel occurs, the shock absorber lower tube unit is forced upward. This action forces the piston downward in the lower tube unit. Because oil cannot leak past the piston, the amount of resistance a shock absorber develops depends on the piston size and the number of holes passing through it. The oil in the lower unit is forced through the piston orifices or valves to the upper oil chamber (**Figure 5-4**). These valves provide precise oil flow control and control the upward action of the wheel and suspension, which is referred to as a shock absorber compression stroke (**Figure 5-5**). Modern shock absorbers are velocity-sensitive damping devices that increase resistance the faster the suspension moves in effect adjusting themselves to road conditions.

When the spring expands downward in rebound travel, the lower shock absorber unit is also forced downward. When this occurs, the piston moves upward in the lower tube unit, and hydraulic oil is forced through the piston orifices or valves from the upper oil

Figure 5-4 As the shock absorber is compressed (jounce) oil flows from the lower chamber to the upper chamber through piston passages and a one-way valve. As the shock absorber expands (rebound), oil flows from the upper chamber to the lower chamber through piston passages and the one-way valve. The ratio of fluid flow through the valved passages determines the shock absorbers rate or ratio.

Twin-tube shock absorbers

Cylinder tube

Work piston

Base valve

Compression phase   Rebound phase

Figure 5-6 Fluid flow during the jounce and rebound travel is restricted through valve passages in the working piston and base valve to achieve proper shock rate.

Figure 5-5 Shock absorber action.

chamber to the lower oil chamber. Because the valves restrict oil flow with precise control, the downward suspension and wheel movement is controlled. Fluid flow during the jounce and rebound travel is restricted through valve passages in the working piston and base valve on a twin-tube design to achieve proper shock rate or ratio (**Figure 5-6**).

When the shock absorber piston moves, oil is forced through the piston. Because the piston valves and orifices resist the flow of oil, friction and heat are created. The resistance of the oil moving through the piston must be calibrated as closely as possible to the spring's deflection rate or strength. Wheels and suspension systems deflect at many different speeds, depending on the type and size of bump and the vehicle speed. The resistance of a shock absorber piston increases with the square of its speed. For example, if the wheel deflection speed increases 4 times, the piston resistance is 16 times as great. Therefore, if a wheel strikes a large bump at high speed, the wheel deflection and rebound can be effectively locked by the shock absorber. Shock absorber engineers prevent this action by precisely designing shock absorber valves and orifices to provide enough friction to prevent the spring from overextending on the rebound stroke. These piston valves and orifices must not create excessive friction, which slows the wheel from returning to its original position.

Shock absorber pistons have many different types of valves and orifices. Valving is selected to achieve optimal ride characteristics of a specific vehicle platform. In some pistons, small orifices control the oil flow during slow wheel and suspension movements. Stacked steel valves control the oil flow during medium speed wheel and suspension movements. During maximum wheel and suspension movements, larger orifices between the piston valves provide oil flow control (**Figure 5-7**). On other shock absorber pistons, the stacked steel valves alone provide oil flow control (**Figure 5-8**). As a simple example, think of a shock absorber with a multistage damping stack of three steel valve plates or discs (**Figure 5-9**). The first stage valve is for light bumps experienced during normal flat road travel and is constantly opening and closing as the vehicle is driven. The second stage valve is for moderate suspension movement that may occur during corning maneuvers. The third stage valve is for extreme suspension movement such as large bumps, pot holes, speed bumps, and it does not open as frequently for each mile driven as the first and second stage valve. When enough force is built up inside the shock absorber each valve progressively opens to allow fluid to bypass the piston seal offering variable damping. A shock absorber piston assembly may make as many as 1500 to 1900 strokes per mile or 75 million strokes in 50,000 mi. (80,467 km), which means that a first stage valve may open and close 37.5 million times in that same distance driven. In this design the constant flexing of the valve plate eventually causes metal fatigue of one or more of the valve plates

Piston support washer

Orifice

Piston

Notched bleed

Deflective valve disc

Valve spring

Piston rod nut

Figure 5-7 A combination of valve discs and orifice passages are used on some shocks to achieve proper shock rate.

Figure 5-8 Stacked shims on the piston can be used to control fluid flow rates on both the rebound and compression cycles.

Figure 5-9 A simple stacked steel valve disc used to achieve proper shock damping rate.

effecting damping performance related to the pressure valve that has failed. Regardless of the piston orifice and valve design, the shock absorber must be precisely matched to absorb the spring's energy.

During fast upward wheel movement on the compression stroke, excessive pressure in the lower oil chamber forces the base valve open and thus allows oil to flow through this valve to the reservoir. A rebound rubber is located on top of the piston. If the wheel drops downward into a hole, the shock absorber may become fully extended. Under these conditions, the rebound rubber provides a cushioning action.

AUTHOR'S NOTE Of all the suspension components, shock absorbers and struts contribute the most to ride quality. As the vehicle is driven over road irregularities, the shock absorbers or struts are continually operating to control the spring action and provide acceptable ride quality. It has been my experience that shock absorbers and struts usually wear out first in suspension systems, because they are working every time a wheel strikes a road irregularity. Therefore, you must understand not only shock absorber and strut operation but also how ride quality is affected if these components are not functioning properly. Thus, you must know shock absorber and strut diagnosis and service procedures.

GAS-FILLED SHOCK ABSORBERS AND STRUTS

Gas-filled units are identified with a warning label. Many shock absorbers and struts today contain a nitrogen gas charge (**Figure 5-10**). A gas charge shock absorber may contain either a low-pressure in twin-tube designs or a high-pressure charge in a mono-tube design. The nitrogen gas provides a compensating space for the oil that is displaced into the reservoir on the compression stroke and when the oil is heated. The primary function of the gas charge is to minimize aeration of the hydraulic fluid (**Figure 5-11**). Because the gas exerts pressure on the oil, cavitation, or foaming of the oil, is eliminated, which in turn reduces fade or loss of damping capability. If foaming occurs, shock absorber performance would be reduced because of the compressibility of air bubbles suspended in the hydraulic fluid, reducing efficiency and transfer of energy. As you recall from Chapter 2, fluids are noncompressible. When oil bubbles are eliminated by the pressurized gas charge, the shock absorber provides continuous damping for wheel deflections as small as 0.078 in. (2.0 mm). By eliminating aeration the shock absorber is more responsive and predictable under a broader performance range. The gas charge also creates a mild boost in spring rate assisting in the reduction of body roll, acceleration squat, and brake dive. This mild boost in spring rate is achieved because of the increased surface area below the working piston than above (**Figure 5-12**). With an increased surface area below the piston as a

Faulty shock absorber   Intact shock absorber

Additional oil flow in worn part of cylinder

Controlled oil flow through the piston

Oil   Oil

Gas   Gas

A   B

Figure 5-10 The nitrogen gas charge maintains pressure on the oil charge to eliminate aeration. As the cylinder bore and piston seal wear, oil will bypass the piston, reducing shock absorber effectiveness.

Aeration–hydraulic fluid mixes with air Shock operation suffers when fluid becomes aerated

Pressureless twin-tube shock absorber

Piston rod

Outer tube

Inner tube

Working piston

Compensating chamber

Base valve

Figure 5-11 Shock absorbers that lose their gas charge or non-gas charged shocks are subject to aeration which decreases the absorbers effectiveness.

Above the piston

Below the piston

Figure 5-12 The surface area of the shock absorber working piston is greater at the bottom than it is at the top because of the loss of area due to the attachment area of the piston rod. This causes applied force to be greater below the piston than above the piston. Though, this does not take valving and orifices into account.

result of the piston rod attachment to the upper piston surface more pressurized fluid contacts the bottom of the piston causing more force to be transferred. The force applied to the piston is equal to the pressure applied to the piston times the surface area of the piston ($F = P \times A$). This is the reason that gas charged shocks will extend on their own. If a low-pressure (100–150 psi) **gas-filled shock absorber** is removed and compressed to its shortest length, it should reextend when it is released. Failure to reextend indicates that shock absorber or strut replacement is necessary. A high-pressure shock absorber should not be compressible by hand because it contains a 360 psi gas charge.

A **gas-filled shock absorber** contains a nitrogen gas charge to maintain pressure on the oil in the shock.

⚡ **WARNING** New gas-filled shock absorbers are wired in the compressed position for shipping purposes. Exercise caution when cutting this wire strap because shock absorber extension may cause personal injury. After the upper shock absorber attaching bolt is installed, the wire strap can be cut to allow the unit to extend. Front gas-filled struts have an internal catch that holds them in the compressed position. This catch is released when the strut rod is held and the strut rotated 45 degrees counterclockwise.

⚡ **WARNING** Do not throw gas-filled shock absorbers or struts in the fire or apply excessive heat or flame to these units. These procedures may cause the unit to explode, resulting in personal injury.

⚡ **WARNING** Never apply heat to a shock absorber or strut chamber with an acetylene torch. This action may cause a shock absorber or strut explosion resulting in personal injury.

⚡ **Caution**

When drilling worn-out shock absorbers or struts to relieve the gas pressure prior to disposal, drill the shock absorber only at the vehicle manufacturer's specified location.

HEAVY-DUTY MONO-TUBE SHOCK ABSORBER DESIGN

Some **heavy-duty mono-tube shock absorbers** have a dividing piston in the lower oil chamber. The area below this piston is pressurized with nitrogen gas to 360 pounds per square inch (psi), or 2.482 kilopascals (kPa). Hydraulic oil is contained in the oil chamber above the dividing piston. The other main features of the heavy-duty shock absorber are:

Heavy-duty mono-tube shock absorbers have several design features that provide improved durability compared with conventional shock absorbers.

1. High-quality seal for longer life.
2. Single tube design to prevent excessive heat buildup. This design may be called a mono-tube shock absorber.
3. Rising rate valve to provide precise spring control under all conditions.

The operation of the heavy-duty shock absorber is similar to that of the conventional type (**Figure 5-13**). A dent in the oil chamber will affect shock absorber operation on this type of design.

The advantages of the mono-tube shock absorber are as follows:

- Heat may be dissipated more rapidly.
- The shock assembly may be mounted in any position including upside down, reducing unsprung weight.
- Larger piston diameter allows for lower working pressures.

The disadvantages of the mono-tube shock absorber are as follows:

- Piston rod seal is subjected to internal damping pressure.
- Sufficient room must be provided by the suspension system layout to provide for its larger diameter.
- The single wall working chamber is susceptible to damage from rocks and other road debris. A dent will destroy the unit.
- Larger than a twin-tube design.

Upper mounting

Rod guide

Dust shield

Rising rate valve system dampens all bumps

Working piston with valves and rebound rubber

Single tube design prevents excessive heat buildup

Dividing piston with O-ring separates oil and nitrogen gas to eliminate foaming

Nitrogen gas pressure principle gives better control

Lower mounting

Figure 5-13 Heavy-duty shock absorber.

STRUT DESIGN, FRONT SUSPENSION

A strut-type front suspension is used on most front-wheel-drive cars and some rear-wheel-drive cars. Internal strut hydraulic design is essentially the same as a shock absorber design, and the **struts** perform the same functions as shock absorbers damping spring oscillations. Some struts have a replaceable cartridge. In many strut-type suspension systems, the coil spring is mounted on the strut. The coil spring is largely responsible for proper curb riding height. A weak or broken coil spring reduces curb riding height and provides harsh riding. The lower end of the front suspension strut is bolted to the steering knuckle (**Figure 5-14**). An **upper strut mount** is attached to the strut, and this mount is bolted into the chassis

Struts are similar to shock absorbers, but struts are usually positioned between the knuckle and the chassis to provide knuckle support.

Figure 5-14 Front strut assembly.

strut tower. A lower spring seat is part of the strut assembly, and a lower insulator is positioned between the coil spring and the spring seat on the strut. Another **spring insulator** is located between the coil spring and the upper strut mount. The two insulators prevent metal-to-metal contact between the spring and the strut, or mount. These insulators reduce the transmission of noise and harshness from the suspension to the chassis. A rubber spring bumper is positioned around the strut piston rod. When a front wheel strikes a large road irregularity and the strut is fully compressed, the jounce bumper provides a cushioning action between the top of the strut and the upper support. The jounce bumper stops the upward wheel and suspension movement before the spring is completely compressed. If the spring becomes completely compressed and the coils strike each other, ride quality is very harsh. Therefore, the jounce bumper in the strut improves ride quality. Most jounce bumpers are made from butyl rubber. Some late-model vehicles have microcellular urethane (MCU) jounce bumpers, which are lighter than rubber and provide more progressive cushioning to improve ride quality. MCU jounce bumpers are also 20–40 percent lighter than rubber jounce bumpers, which reduces road noise transmission to the passenger compartment. In relation to temperature changes, MCU jounce bumpers remain more stable and provide improved ride quality regardless of the temperature. The upper strut mount contains a bearing, upper spring seat, and jounce bumper (**Figure 5-15**).

 A BIT OF HISTORY

For many years, rear-wheel-drive cars were equipped with front and rear shock absorbers. Most front-wheel-drive cars are equipped with front struts, and some of these cars also have rear struts. Because massive numbers of front-wheel-drive cars have been introduced in the 1980s and 1990s, front and rear struts are now very common.

Figure 5-15 Upper strut mount.

When the front wheels are turned, the front strut and coil spring rotate with the steering knuckle. The strut-and-spring assembly rotates on the upper strut mount bearing.

Some cars have a multilink front suspension with an upper link connected from the chassis to the steering knuckle. The strut is connected from the upper link to the strut tower (**Figure 5-16**). A bearing is mounted between the upper link and the steering knuckle, and the wheel and knuckle turn on this bearing and the lower ball joint. Therefore, the coil spring and strut do not turn when the front wheels are turned, and a bearing in the upper strut mount is not required (**Figure 5-17**).

When strut replacement is necessary on many vehicles, complete strut assemblies are available that include a premium strut, the coil spring, upper strut mount and bearing plate (if equipped), upper spring seat, protective boot, and spring insulators in a fully-assembled unit ready for installation in the vehicle (**Figure 5-18**). These assemblies offer several benefits over strut servicing, which include greater safety (since there is no need to compress the coil spring), and ease of installation (since now special spring compressing tools are required). Monroe calls their assembly a Quick-Strut and Gabriel refers to theirs as ReadyMount. In addition to the previously mentioned benefits, these assemblies greatly

Figure 5-16 Multilink front suspension with strut connected between the upper link and the strut tower.

Figure 5-17 Multilink front suspension with knuckle and wheel pivots at the upper link bearing and lower ball joint, and a nonrotating upper strut mount.

Figure 5-18 Several aftermarket strut manufactures offer complete replacement strut service assemblies that are ready to be installed without the need for specialty tools.

reduce replacement time and are a cost-effective choice for consumers when mounts and springs show signs of wear. While the complete assembly is more expensive than a strut or cartridge alone, the labor savings achieved by eliminating the labor charge for disassembly and reassembly brings the cost to a comparable level with the added benefits of all-new complete assembly components.

Shop Manual
Chapter 5, page 200

SHOCK ABSORBER AND STRUT DESIGN, REAR SUSPENSION

In some rear suspension systems, the lower end of the strut is bolted to the spindle, and the top of the strut is connected through a strut mount to the chassis. The rear coil springs are mounted separately from the struts. These springs are mounted between the lower control arms and the chassis (**Figure 5-19**).

Figure 5-19 Rear suspension system with coil springs mounted separately from the struts.

In other rear suspension systems, the coil springs are mounted on the rear struts (**Figure 5-20**). An upper insulator is positioned between the top of the spring and the upper spring support, and a lower insulator is located between the bottom of the spring and the spring mount on the strut. A rubber spring bumper is positioned on the strut piston rod. If a rear wheel strikes a severe road irregularity and the strut is fully compressed, the spring bumper provides a cushioning action between the top of the strut and the upper support.

On some rear-wheel-drive light-duty trucks and sport utility vehicles (SUVs), the rear shock absorbers are slanted rearward and inward (**Figure 5-21**). On other light-duty trucks and SUVs, the rear shock absorbers are staggered on each side of the rear axle (**Figure 5-22**). Either of these shock absorber mountings improves ride quality and reduces noise, vibration, and harshness (NVH).

Shop Manual
Chapter 5, page 208

Figure 5-20 Rear suspension system with coil springs mounted on the struts.

Figure 5-21 Rear shock absorber mounting, sport utility vehicle (SUV).

Figure 5-22 Shock absorbers staggered on each side of the rear axle.

TRAVEL-SENSITIVE STRUT

Some **travel-sensitive struts** contain narrow longitudinal grooves in the lower oil chamber (**Figure 5-23**). These grooves are parallel to the piston orifices, and some oil flows through the grooves as well as the orifices. Under normal driving and road conditions, the orifices and grooves are calibrated to provide normal spring damping and control. If the front wheel drops suddenly, such as when it strikes a large hole, the piston moves into the narrow portion of the oil chamber. Under this condition, all the oil must flow through the piston orifices, which greatly increases the strut's resistance to movement and the suspension damping action. This strut action prevents harsh impacts against the internal strut rebound rubber.

ADJUSTABLE STRUTS

Some adjustable struts have a manual adjustment that allows the vehicle owner or technician to adjust the struts to suit driving conditions (**Figure 5-24**). The strut adjusting knob varies the strut orifice opening. This knob has eight possible settings. The factory setting

Figure 5-23 Travel-sensitive strut.

Figure 5-24 Manual adjustable strut.

is No. 3, which provides average suspension control. The No. 1 setting provides reduced spring control and the softest ride, whereas a No. 8 adjustment gives increased spring control and the hardest ride. The adjustment knob is usually accessible without raising the vehicle.

LOAD-LEVELING SHOCK ABSORBERS

Load-leveling rear shock absorbers or struts are used with an electronic height control system. An onboard air compressor pumps air into the rear shocks to raise the rear of the vehicle, and an electric solenoid releases air from the shocks to lower the rear chassis. An electromagnetic height sensor may be contained in the shock absorber, or an external sensor may be used (**Figure 5-25**). This sensor sends a signal to an electronic control module in relation to the rear suspension height. The module controls the air compressor and the exhaust solenoid to control air pressure in the shock absorbers. This action maintains a specific rear suspension trim height regardless of the load on the rear suspension. If a heavy package is placed in the trunk, the vehicle chassis is forced downward. However, the **load-leveling shock absorbers** extend to restore the original rear suspension height.

Aftermarket air shock absorbers are available. These shock absorbers contain an air valve connection. A shop air hose may be used to supply the desired pressure in these air shock absorbers.

Aftermarket spring-assisted shock absorbers are also available. These are conventional shock absorbers with a small coil spring mounted over them. Upper and lower spring seats are attached near the top and bottom of these shock absorbers to support and retain the spring. The coil springs on the shock absorbers help the springs in the suspension system support the vehicle weight.

Load-leveling shock absorbers use air pressure supplied to the shock absorbers to maintain rear suspension height. Load-leveling shock absorbers may be called air shocks.

Traveling
magnet

Electromagnetic
sensor

Inflatable
shock absorber

Sealed wiring
connector

Figure 5-25 Load-leveling strut.

ELECTRONICALLY CONTROLLED SHOCK ABSORBERS AND STRUTS

Computer controlled
suspension systems
may be referred to as
**active suspension
systems**.

Many cars are equipped with computer-controlled **active suspension systems**. In these systems, a computer-controlled actuator is positioned in the top of each shock absorber or strut (**Figure 5-26**). The shock absorber or strut actuators rotate a shaft inside the piston rod, and this shaft is connected to the shock valve. Many of these systems have two modes: soft and firm. In the soft mode, the actuators position the shock absorber valves so there is less restriction to the movement of oil. When the computer changes the

Computer-controlled
actuator

Hex drive
shaft

Strut
shaft

Figure 5-26 Computer-controlled strut actuator.

actuators to the firm mode, the actuators position the shock valves so they provide more restriction to oil movement, which provides a firmer ride.

Some electronically controlled shock absorbers and struts contain a synthetic **magneto-rheological fluid** that contains numerous small, suspended metal particles.

Delphi's trademarked name for the system is called MagneRide. There are no small moving parts or electromechanical valves to fail. The electro-rheological fluid viscosity changes when electrical current passes through the fluid itself, which contains magnetically soft particles of iron microspheres in a synthetic hydrocarbon-based fluid. Each shock absorber contains an electromagnetic coil winding inside the piston itself that is energized by the suspension computer (**Figure 5-27**). Only the fluid in the oil passage of the piston is involved in the viscosity change. The wires to the coil are routed through the piston rod to a connector at the top of the shock housing. If the shock absorber winding is not energized, the metal particles in the fluid align randomly in the fluid. Under this condition, the fluid has a mineral-oil-like consistency and the fluid moves easily through the shock absorber orifices. If the suspension computer energizes a shock absorber coil winding, the metal particles in the fluid are aligned into fibrous structures. When this occurs, the fluid has a jelly-like consistency for a firm ride (**Figure 5-28**). The computer can change the shock absorber damping characteristics almost instantaneously, in 1 millisecond, and can also supply a varying amount of current through the shock absorber windings to provide a wide variety of shock absorber damping characteristics. Depending on the amount of current supplied, the fluid's viscosity or resistance to flow can be varied from thinner than water to a near plastic or solid state or any consistency in-between. This makes the fluid infinitely adjustable in viscosity and achieves continuously variable real-time damping to suit almost every driving condition instantly.

Magneto-rheological fluid A fluid that when subjected to a magnetic field will increase viscosity as the magnetic field is increased, to the point of becoming an elastic solid.

Figure 5-27 Some of the major electrical components that are integrated into a magneto-rheological damping system.

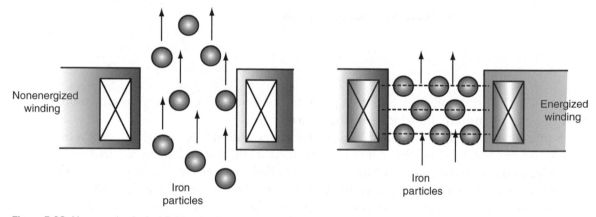

Figure 5-28 Magneto-rheological fluid action in a strut or a shock absorber.

The advantages of magneto-rheological-fluid-controlled shock absorber are as follows:

- Ability to smooth out the action of each tire.
- Reduced noise, bounce, and vibration giving a flatter ride by controlling body motions.
- Exceptional roll control during evasive steering maneuvers.
- Excellent handling by controlling weight transfer during lateral and longitudinal maneuvers. Enhanced road isolation by reducing high-frequency road noise through the dampers.
- When integrated with ABS and traction control, ensures maximum stability and balance on slippery surfaces and gravel.

SUMMARY

- Shock absorbers or struts play a very important role in ride quality, steering control, and tire life.
- Tire and wheel jounce travel occurs when a tire strikes a bump in the road surface and the tire and wheel move upward.
- Rebound tire and wheel travel occurs when the tire and wheel move downward after jounce travel.
- When a spring is deflected upward during jounce travel, it stores energy. The spring then expands downward in rebound travel with all the energy it stored during the jounce travel. If the spring action is not controlled, the energy in the spring during rebound travel drives the tire against the road surface with excessive force. This action drives the tire and wheel back upward in jounce travel and the wheel continues oscillating up and down.
- The shock absorbers control spring action and prevent excessive tire and wheel oscillations.
- During jounce travel, the piston moves downward in the lower shock absorber chamber; during rebound travel, this piston moves upward. Because the lower shock absorber chamber is sealed and filled with oil, this oil must flow past the piston during any piston movement.
- Valves and openings in the shock absorber piston provide precision control of the oil flow past the piston to control spring action. Shock absorber

- valves are matched to the amount of energy that may be stored in the spring.
- A nitrogen gas charge is located in the oil reservoir of many shock absorbers and struts to prevent oil cavitation or foaming, which provides more positive shock absorber action.
- Shock absorber ratio refers to the difference between the shock absorber control on the compression and extension cycle. Many shock absorbers provide more control on the extension cycle.
- Internal design is similar in shock absorbers and struts, but struts also support the coil spring.
- Most front struts are connected between the steering knuckle and the upper strut mount.
- Many rear struts are connected between the spindle and the upper support. A travel-sensitive shock absorber provides increased resistance to piston movement as the shock absorber is extended.
- Adjustable shock absorbers and struts have a manual adjustment that allows the technician or owner to adjust the strut orifice opening.
- Load-leveling shock absorbers have air pumped into the shock absorbers from an onboard compressor to maintain a specific rear suspension height regardless of the rear suspension load.
- Most front struts rotate on the upper strut mount bearing as the front wheels are turned.

REVIEW QUESTIONS

Short Answer Essays

1. Describe spring action during jounce and rebound travel.

2. Describe uncontrolled spring action without a shock absorber.

3. Explain shock absorber operation.

4. Describe the vehicle safety hazards created by worn-out shock absorbers.

5. Explain shock absorber ratio.

6. Explain the purpose of the nitrogen gas charge in shock absorbers and struts.

7. Explain the differences between heavy-duty mono-tube and conventional twin-tube shock absorbers.

8. Identify the purposes of an upper strut mount.

9. Explain the purpose of strut-and-spring insulators.

10. Describe the purpose of the rubber spring bumper on a strut piston rod.

Fill-in-the-Blanks

1. Shock absorbers control spring action and prevent excessive spring _____.

2. Shock absorber design is matched to the _____ rate of the spring.

3. Modern shock absorbers are _____ _____ damping devices that increase resistance the faster the suspension moves.

4. The nitrogen gas charge in a shock absorber prevents oil _____ and _____.

5. The single tube design in a heavy-duty shock absorber prevents excessive_____
_____.

6. A shock absorber with a 70/30 ratio provides more control on the _____ cycle.

7. The lower end of many front struts is bolted to the steering _____.

8. When a higher number is selected on the adjustment knob of an adjustable strut, the strut provides a _____ ride.

9. Load-leveling shock absorbers may have an internal _____ _____.

10. Travel-sensitive shock absorbers provide increased spring control when the shock absorber is _____.

Multiple Choice

1. The main purpose of shock absorbers and struts is to:
   A. Control spring action.
   B. Prevent fore-and-aft wheel movement.
   C. Reduce lateral wheel movement.
   D. Prevent wheel shimmy.

2. All of these statements about shock absorber design and operation are true EXCEPT:
   A. The oil flow through the orifices and valves is matched to the spring's strength and deflection rate.
   B. A typical shock absorber may have 70 percent of the total control on the extension cycle.
   C. During jounce wheel travel the piston moves upward in the shock absorber lower tube unit.
   D. A nitrogen gas charge in a shock absorber prevents oil foaming.

3. Compared to a conventional shock absorber, a heavy-duty shock absorber has:
   A. Triple layers of steel in the lower tube unit.
   B. Higher viscosity oil.
   C. A larger diameter piston rod.
   D. A high-quality seal for longer life.

4. All of these statements about gas-filled shock absorbers are true EXCEPT:
   A. A hydrogen gas charge is used in most shock absorbers and struts.
   B. The gas charge creates a mild boost in spring rate.
   C. The primary function of the gas charge is to minimize aeration of the hydraulic fluid.
   D. A gas-filled shock absorber provides continuous damping action with very little wheel movement.

5. The magneto-rheological fluid used in some shock absorbers and struts contains:
   A. Transmission fluid.
   B. Small, suspended metal particles.
   C. Antifoaming agents.
   D. A dye for lead detection purposes.

6. During normal strut operation:
   A. Downward wheel movement is called wheel jounce.
   B. The strut prevents excessive wheel oscillations.
   C. When the wheel moves upward the strut piston also moves upward.
   D. The strut prevents wheel shimmy.

7. Travel-sensitive struts:
   A. Have an adjustment on the outside of the strut.
   B. Provide more resistance to oil movement when compressed.
   C. Provide more resistance to oil movement when driving on a smooth road surface.
   D. Bypass some of the oil past the piston when compressed.

8. Adjustable struts:
   A. May be adjusted with a control knob in the instrument panel.
   B. Provide five different settings.
   C. Provide softest ride with a No. 1 setting.
   D. Rotate the upper part of the strut to provide adjustable ride quality.

9. All of the following are benefits of replace faulty struts with a complete preassembled replacement strut assembly EXCEPT:

    A. Quicker, no need to disassemble existing strut.

    B. Safer, no exposure to dangers of compressing coil spring.

    C. Easier, reduce replacement time.

    D. Lower cost than strut assembly alone.

10. Electronically controlled shock absorbers and struts:

    A. Contain an electronically controlled actuator in the top of each strut.

    B. Usually have 12 different modes of strut operation from soft to very firm.

    C. Have an electronically controlled actuator that rotates the strut piston rod.

    D. Have an actuator that positions the strut valves to provide more oil restriction in the soft mode.

CHAPTER 6
FRONT SUSPENSION SYSTEMS

Upon completion and review of this chapter, you should be able to understand and describe:

- The causes of coil spring failure.
- The design and spring rate of a linear-rate coil spring.
- The difference between a regular-duty coil spring and a heavy-duty coil spring.
- The functions of three different types of coils in a variable-rate coil spring.
- Basic torsion bar action as a front wheel strikes a road irregularity.
- How friction and noise problems are reduced in a multiple-leaf spring.
- The advantages of a front suspension system with ball joints compared to earlier I-beam front suspension systems.
- Two types of load-carrying ball joints, and explain the location of the control arm in each type.
- How to recognize a worn ball joint from a visual inspection of the ball joint wear indicator.

- The mounting location and purpose of a stabilizer bar.
- The purpose of a strut rod.
- The two steering knuckle pivot points in a MacPherson strut front suspension system.
- The advantage of a short-and-long arm front suspension system compared with a suspension system with equal-length upper and lower control arms.
- Design of the twin I-beam front suspension.
- Vehicle ride height.
- The effect of sagged front springs on caster angle and directional stability.
- The effect of sagged front springs on camber angle.
- The advantages of hydraulic control arm bushings.
- The advantages of aluminum control arms.

Terms To Know

Compression loaded
Full-wire open-end springs
Gussets
Heavy-duty coil springs
Inactive coils
Linear-rate coil springs
Load-carrying ball joint
MacPherson strut front suspension systems

Mono-leaf springs
Multilink front suspension
Multiple-leaf springs
Non-load-carrying ball joint
Pigtail spring ends
Regular-duty coil springs
Short-and-long arm front suspension systems
Stabilizer bar

Taper-wire closed-end
Tension loaded
Transitional coils
Unsprung weight
Variable-rate coil springs
Wear indicators

INTRODUCTION

The front and rear suspension systems must perform several extremely important functions to maintain vehicle safety and owner satisfaction. The suspension system must provide steering control for the driver under all road conditions. Vehicle owners expect the suspension system to provide a comfortable ride. The suspension, together with the frame, must maintain proper vehicle tracking and directional stability. Another important purpose of the suspension system is to provide proper wheel alignment and minimize tire wear.

The impact of the front tires striking road irregularities must be absorbed and dissipated by the front suspension system. These impacts are distributed throughout the suspension, and this action isolates the vehicle passengers from road shock. The vehicle's ride characteristics are determined by the amount of impact energy that the suspension can absorb and by the rate at which the suspension dissipates these tire impacts. Ride characteristics are designed into the suspension system, and these characteristics are not adjustable. For example, the front suspension may be designed to provide a very soft ride in which all the tire impacts are absorbed and dissipated quickly by the suspension system. This type of suspension system provides a very comfortable ride, but it also allows excessive body lean during cornering, which reduces high-speed cornering and handling capabilities. Vehicles such as sports cars and sport utility vehicles (SUVs) are usually designed with a suspension system that provides a firm ride and absorbs and dissipates tire impacts more slowly. Although high-speed cornering and handling capabilities are improved, this type of suspension may transfer some road shock to the passenger compartment.

When the vehicle is driven over road irregularities, the front suspension system must allow the front wheels to move vertically while maintaining the tire's proper horizontal position in relation to the road surface. To provide this wheel action, the steering knuckle must be mounted between the upper and lower control arms on short-and-long arm (SLA) suspension systems or between the lower control arm and the lower end of the strut on MacPherson strut suspension systems. The upper and lower control arms must be pivoted on the inner ends; the steering knuckle pivots on the ball joints in SLA suspensions or on the lower ball joint and upper strut mount on MacPherson strut suspensions.

Springs absorb much of the shock from tire impacts with the road surface. When a front wheel strikes a road irregularity and moves upward, the coil spring compresses and absorbs energy during this movement. The coil spring immediately dissipates this energy as the spring moves back to its original state. Shock absorbers are installed in the front suspension system to dampen the oscillations of the coil springs.

A stabilizer bar is mounted on rubber insulating bushings and bolted to the chassis. The outer ends of the stabilizer bar are attached through links to the lower control arms. The stabilizer bar controls the amount of independent lower control arm movement. When one front wheel strikes a road irregularity and moves upward, the stabilizer bar transfers part of the lower control arm movement to the opposite lower control arm, which reduces and stabilizes body roll. Therefore, the stabilizer bar helps to define the suspension characteristics related to body roll.

All suspension components, including the frame and chassis, are designed or "tuned" to provide the ride and handling qualities that the manufacturer believes the average driver will desire. For example, the late-model Explorer Sport Trac frame has been stiffened 40 percent more than the Explorer frame. To achieve this increased stiffness, the frame side rails are thicker than the ones on the Explorer, and a new tubular crossmember has been added to the frame. **Gussets** have also been welded into the corners where the crossmembers meet the side rails. This frame design is matched to the torsion bar suspension to provide improved vehicle agility on and off the road.

Gussets are pieces of metal welded into the corner where two pieces of metal are attached together, which provide increased strength and stiffness.

SUSPENSION SYSTEM COMPONENTS

Coil Springs

Shop Manual
Chapter 6, page 230

The coil spring is the most commonly used spring for front and rear suspension systems. Coil springs are actually a coiled-spring steel bar. When a vehicle wheel strikes a road irregularity, the coil spring compresses to absorb shock, and then recoils back to its original installed height. Many coil springs contain a steel alloy that contains different types of steel mixed with other elements such as silicon or chromium. Coil springs may be manufactured by a cold or hot coiling process. The hot coiling process includes procedures for tempering and hardening the steel alloy. Coil springs are designed to carry heavy loads, but they must be light in weight. Many coil springs have a vinyl coating, which increases corrosion resistance and reduces noise.

Coil spring failures may be caused by these conditions:

1. Constant overloading
2. Continual jounce and rebound action
3. Metal fatigue
4. A crack or nick on the surface layer or coating

Coil springs do not have much ability to resist lateral movement. However, when coil springs are used on the drive wheels, the suspension usually has special bars to prevent lateral movement.

Linear-Rate Coil Springs. Coil springs are classified into two general categories: linear rate and variable rate. **Linear-rate coil springs** have equal spacing between the coils and one basic shape with a consistent wire diameter. When the load is increased on a linear-rate spring, the spring compresses and the coils twist or deflect. As the load is removed from the spring, the coils unwind, or flex, back to their original position. The spring rate is the load required to deflect the spring 1 in. Linear-rate coil springs have a constant spring rate, regardless of the load. For example, if 200 lb deflect the spring 1 in., 400 lb deflect the spring 2 in. The spring rate on linear springs is usually calculated between 20 and 60 percent of the total spring deflection.

Variable-Rate Coil Springs. **Variable-rate coil springs** have a variety of wire sizes and shapes. The most common variable-rate coil springs have a consistent wire diameter with a cylindrical shape and unequally spaced coils (**Figure 6-1**).

The **inactive coils** at the end of the spring introduce force into the spring when the wheel strikes a road irregularity. When the **transitional coils** are compressed to their point of maximum load-carrying capacity, these coils become inactive. The **active coils** operate during the complete range of spring loading. When a stationary load is applied to

Inactive coils are positioned at each end of a coil spring.

Transitional coils are positioned between the inactive coils and the active coils in the center of the spring.

The **active coils** are located in the center of a coil spring.

Variable-rate spring   Conventional spring

Figure 6-1 Variable-rate coil springs have consistent wire diameter and unequally spaced coils.

a variable-rate coil spring, the inactive coils theoretically support the load. If the load is increased, the transitional coils support the load until they reach maximum load-carrying capacity, and the active coils carry the remaining overload. This spring action provides automatic load adjustment while maintaining vehicle height.

Some variable-rate coil springs have a tapered wire, in which the active coils have the larger diameter and the inactive coils have the smaller diameter. Other variable-rate spring designs include truncated cone, double cone, and barrel shape. A variable-rate spring does not have a standard spring rate. This type of spring has an average spring rate based on the load at a predetermined spring deflection. *It is impossible to compare variable spring rates and linear spring rates because of this difference in spring rates.* Variable-rate coil springs usually have more load-carrying capacity than linear-rate coil springs in the same application.

Lightweight Coil Springs

A few sports cars are presently equipped with titanium coil springs. This type of coil spring reduces the weight on the front springs by 39 percent and 28 percent on the rear springs compared to steel coil springs. Decreasing the weight of the coil springs reduces the **unsprung weight** and improves ride control. The unsprung vehicle weight tends to force the wheel downward during rebound wheel travel. Higher unsprung weight drives the wheel downward with greater force and increases the impact force between the tire and the road surface, resulting in a harsh ride. To compensate for the high unsprung weight, the shock absorber damping rate must be increased. This suspension design reduces ride quality. When the unsprung weight is reduced, the wheel is forced downward with less force, reducing the shock absorber damping rate and improving ride quality.

> **Unsprung weight** is the vehicle weight that is not supported by the coil springs, and sprung weight is the vehicle weight that is supported by the coil springs.

Heavy-Duty Coil Springs   **Heavy-duty coil springs** are designed to carry 3 to 5 percent greater loads than regular-duty coil springs. The wire diameter may be up to 0.10 in. greater in a heavy-duty spring than in a regular-duty coil spring. This larger-diameter wire increases the load-carrying capacity of the spring. The free height of a heavy-duty coil spring is shorter than a regular-duty coil spring for the same application (**Figure 6-2**).

Selecting Replacement Springs   When replacement coil springs are required, the technician must select the correct spring. The original part number is usually on a tag wrapped around one of the coils. However, this tag may have fallen off if the spring has been in service for very long. Some aftermarket suppliers stamp the part number on the end of the coil spring. If the original part number is available, the replacement springs may be ordered with the same part number. Most vehicle manufacturers recommend that both front or rear springs be replaced at the same time. The replacement springs must have the same type of ends as the springs in the vehicle. Coil spring ends may be square tapered, square nontapered, or tangential (**Figure 6-3**).

Figure 6-2 Comparison of heavy-duty and regular-duty coil springs.

SQUARE END (tapered)   SQUARE END (untapered)   TANGENTIAL

Figure 6-3 Types of coil spring ends.

If a wheel moves upward in relation to the chassis, this action is referred to as jounce travel. Downward wheel movement is called rebound travel.

Full-wire open-end springs have the ends cut straight off, and sometimes these ends are flattened, squared, or ground to a D-shape. **Taper-wire closed-end springs** are ground to a taper and wound to ensure squareness. **Pigtail spring ends** are wound to a smaller diameter. Springs are generally listed for front or rear suspensions.

Regular-duty coil springs are a close replacement for the original spring in the vehicle, and these springs may replace several different original equipment (OE) springs in the same vehicle. Linear-rate coil springs are usually found in regular-duty, heavy-duty, and sport suspension packages. Heavy-duty coil springs are required when the vehicle is carrying a continuous heavy load, such as trailer towing.

Variable-rate coil springs are generally used when automatic load leveling is required under increased loads. Variable-rate coil springs maintain the correct vehicle height under various loads and provide increased load-carrying capacity compared to heavy-duty coil springs. The technician must select the correct spring to meet the requirements of the vehicle and load conditions.

⚠ WARNING **A very large amount of energy is stored in a compressed coil spring. Always follow the spring service procedures in the vehicle manufacturer's service manual to avoid personal injury.**

⚠ WARNING **Never disconnect a suspension component that quickly releases the tension on a compressed coil spring. This action may cause personal injury.**

Torsion Bars

In some front suspension systems, torsion bars replace the coil springs. During wheel jounce, the torsion bar twists. During wheel rebound, the torsion bar unwinds back to its original position. A torsion bar may be thought of as a straight, flattened coil spring. One end of the heat-treated alloy steel torsion bar is attached to the vehicle frame, and the opposite end is connected to the lower control arm. A few older vintage vehicles had the end of the torsion bars connected to the upper control arm. Transversely mounted torsion bars were used in some front suspensions on older cars. Some light-duty trucks and SUVs are presently equipped with longitudinal torsion bars in the front suspension (**Figure 6-4**). On four-wheel-drive trucks, the use of torsion bars replaces coil springs, which allows more space for the front drive axles. The front end of the torsion bar is anchored in the lower control arm and the rear end of the torsion bars are anchored to a chassis

Figure 6-4 Longitudinal torsion bar.

Figure 6-5 Torsion bar mounting.

crossmember (**Figure 6-5**). Since the twisting of the torsion bars supports the vehicle weight, the torsion bars replace the coil springs.

Because the lower control arm moves up and down as the wheel strikes road irregularities, this control arm action twists the torsion bar. The bar's natural resistance to twisting causes it to return to its original position. During the manufacturing process, torsion bars are prestressed to provide fatigue strength. These bars are directional. Torsion bars are marked right or left, and they must be installed on the appropriate side of the vehicle. Left and right on a vehicle are always viewed from the driver's seat.

A torsion bar is capable of storing a higher maximum energy compared with a loaded coil or a leaf spring. Shorter, thicker torsion bars have increased load-carrying capacity as compared to longer, thinner bars. Since torsion bars require less space than coil or leaf springs, they are usually found on front suspensions.

On a torsion bar front suspension system with upper and lower control arms (**Figure 6-6** and **Figure 6-7**) the front of the torsion bars has hex-shaped ends that are

Figure 6-6 Four-wheel-drive light-duty truck torsion bar front suspension.

ITEM	DESCRIPTION
1	Final drive unit
2	Drive axle
3	Upper control arm
4	Upper ball joint
5	Steering knuckle
6	Lower ball joint
7	Lower control arm
8	Torsion bar adjuster
9	Torsion bar
10	Crossmember

Figure 6-7 Component identification light-duty four-wheel-drive truck torsion bar front suspension.

Figure 6-8 Torsion bar adjusting arm and adjusting bolt.

mounted in a matching hex in the lower control arm. The hex-shaped rear ends of the torsion bars are mounted in adjusting arms that are retained in the torsion bar crossmember. Torsion bars have a riding height adjustment screw at the end where they are attached to the frame. An adjusting bolt in this crossmember contacts the outer end of each adjusting arm (**Figure 6-8**). Rotation of this adjusting bolt changes the tension on the torsion bar to adjust the riding height. On this type of torsion bar front suspension system, the front wheel bearing hubs are a sealed, non-serviceable assembly (**Figure 6-9**).

Vehicle ride height is controlled by the torsion bar anchor adjusting bolts in the crossmember. Front suspension heights must be within specifications for correct wheel alignment, tire wear, satisfactory ride, and accurate bumper heights. A conventional stabilizer bar is connected between the lower control arms and the crossmember. Ball joints are located in the upper and lower control arms, and both ball joints are bolted into the steering knuckle. The shock absorbers are connected between the lower control arms and the crossmember support, and the inner ends of the lower control arms are bolted to the crossmember through an insulating bushing (**Figure 6-10**).

Figure 6-9 Four-wheel-drive torsion bar front suspension with sealed non-serviceable front wheel bearing hubs.

Figure 6-10 Front suspension system with longitudinal torsion bars.

Multiple-Leaf Springs

Leaf springs may be multiple-leaf or mono-leaf. **Multiple-leaf springs** have a series of flat steel leaves of varying lengths that are clamped together. A center bolt extends through all the leaves to maintain the leaf position in the spring. The upper leaf is called the main leaf, and this leaf has an eye on each end. An insulating bushing is pressed into each main leaf eye. The front bushing is attached to the frame, and the rear bushing is connected through a shackle to the frame. The shackle provides fore-and-aft movement as the spring compresses (**Figure 6-11**).

The main leaf is the longest leaf in the spring, and the other leaves get progressively shorter. Each spring leaf is curved in the manufacturing process. If this curve were doubled, it would form an ellipse. Therefore, leaf springs are referred to as semielliptical or quarter elliptical. Most leaf springs are semielliptical. The ellipse designation refers to how much of the ellipse the spring actually describes.

As a leaf spring compresses, it becomes progressively stiffer. When a leaf spring is compressed, the length of the leaves changes, and the leaves slide on each other. This sliding action could be a source of noise and friction. These noise and friction problems are reduced by interleaves, or spacers, made from zinc and plastic placed between the steel leaves. The

Figure 6-11 Leaf spring design.

head on the spring center bolt fits into an opening in the axle to position the axle properly and provide proper vehicle tracking. If the center bolt is broken, the axle position may shift and alter vehicle tracking and alignment. Leaf springs are usually mounted at right angles to the axle. They provide excellent resistance to lateral movement.

Mono-leaf Springs

Some leaf springs contain a single steel leaf, and these springs may be referred to as **mono-leaf springs**. The single leaf is thicker in the center and becomes gradually thinner toward the outer ends. This design provides a variable spring rate for a smooth ride and adequate load-carrying capacity. Mono-leaf springs do not have a friction and noise problem as the spring compresses. Some cars, such as the Corvette and front-wheel-drive Oldsmobile Cutlass Supreme, use a fiberglass-reinforced plastic mono-leaf spring in place of a steel spring to reduce weight (**Figure 6-12**). The mono-leaf spring may be mounted longitudinally or transversely, and this type of spring may be used in front or rear suspensions. Older model Chevrolet Astro vans have longitudinally mounted fiberglass mono-leaf rear springs.

Shop Manual
Chapter 6, page 213

A BIT OF HISTORY

The transverse leaf-spring front and rear suspension is one of the oldest suspension systems in the automotive industry. It was used on Ford products from the early 1900s until 1948.

Figure 6-12 Transversely mounted fiberglass mono-leaf spring.

Ball Joints

The ball joints act as pivot points that allow the front wheels and spindles, or knuckles, to turn between the upper and lower control arms. Compared with the earlier I-beam and kingpin-type front suspension systems, ball-joint suspension systems are much simpler. A front suspension with ball joints reduces the number of load-carrying bearing surfaces. Compared with an I-beam front suspension with kingpins, a front suspension with ball joints has these advantages:

1. Reduced space requirements
2. Reduced unsprung weight
3. Easier alignment
4. More dependable steering control
5. Improved safety
6. Improved tire life
7. Reduced steering effort
8. Improved ride quality
9. Simplified service—no kingpin reaming or honing
10. Simplified lubrication

⚡ **WARNING Worn ball joints may suddenly pull apart, resulting in loss of steering control and possibly a collision! Refer to the procedures in the Shop Manual Chapter 6 for measuring ball joint wear.**

Load-Carrying Ball Joint

Ball joints may be grouped into two classifications, load carrying and non-load carrying. Ball joints may be manufactured with forged, stamped, cold-formed, or screw-machined housings. The coil spring is seated on the control arm to which the load-carrying ball joint is attached. For example, when the coil spring is mounted between the lower control arm and the chassis, the lower ball joint is a load-carrying joint (**Figure 6-13**). In a torsion bar suspension, the **load-carrying ball joint** is mounted on the control arm to which the torsion bar is attached. A load-carrying ball joint supports the vehicle weight.

In a load-carrying ball joint, the vehicle weight forces the ball stud into contact with the bearing surface in the joint. Load-carrying ball joints may be compression loaded or tension loaded. If the control arm is mounted above the lower end of the knuckle and rests on the knuckle, the ball joint is **compression loaded**. In this type of ball joint, the vehicle weight is pushing downward on the control arm. This weight is supported on the tire and

Shop Manual
Chapter 6, page 230

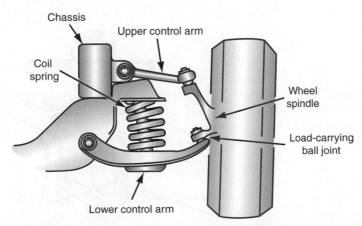

Figure 6-13 Load-carrying ball joint mounted on the control arm on which the spring is seated.

Figure 6-14 Compression-loaded ball joint.

Figure 6-15 Compression-loaded ball joint mounting.

wheel, which are attached to the steering knuckle. Since the ball joint is mounted between the control arm and the steering knuckle, the vehicle weight squeezes the ball joint together (**Figure 6-14**). In this type of ball-joint mounting, the ball joint is mounted in the lower control arm and the ball joint stud faces downward (**Figure 6-15**).

When the lower control arm is positioned below the steering knuckle, the vehicle weight is pulling the ball joint away from the knuckle (**Figure 6-16**). This type of ball joint mounting is referred to as **tension loaded**. This type of ball joint is mounted in the lower control arm with the ball joint stud facing upward into the knuckle (**Figure 6-17**).

Since the load-carrying ball joint supports the vehicle weight, this ball joint wears faster compared with a non-load-carrying ball joint. Many load-carrying ball joints have built-in **wear indicators**. These ball joints have an indicator on the grease nipple surface that recedes into the housing as the joint wears. If the ball joint is in good condition, the grease nipple shoulder extends a specified distance out of the housing. If the grease nipple shoulder is even with or inside the ball joint housing, the ball joint is worn, and replacement is necessary (**Figure 6-18**).

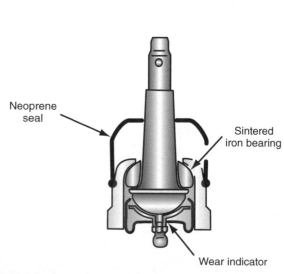

Figure 6-16 Tension-loaded ball joint.

Figure 6-17 Tension-loaded ball joint mounting.

Figure 6-18 Ball joint wear indicator.

Most states have safety inspection procedures for testing ball joints and other safety-related components. These safety inspection procedures include specifications. Always follow the guidelines when performing safety inspections.

Non-Load-Carrying Ball Joint

A **non-load-carrying ball joint** may be referred to as a stabilizing or follower ball joint. A non-load-carrying ball joint is designed with a preload, which provides damping action (**Figure 6-19**). This ball joint preload provides improved steering quality and vehicle stability.

Low-Friction Ball Joints

Low-friction ball joints are standard equipment on many vehicles. Low-friction ball joints provide precise low-friction movement of the ball socket in the ball joint. Compared to conventional ball joints, two-thirds of the internal friction is eliminated in a low-friction ball joint. The smooth ball socket movement in a low-friction ball joint provides improved steering performance, better steering wheel return, and longer ball joint life. Low-friction ball joints have a highly polished ball socket surface surrounded by a high-strength polymer bearing (**Figure 6-20**).

A **non-load-carrying ball joint** maintains knuckle position, but this ball joint does not support the vehicle weight.

Figure 6-19 Non-load-carrying ball joint.

Figure 6-20 Low-friction ball joint.

Figure 6-21 Strut, or radius, rod and bushings.

Strut Rod

On some front suspension systems, a strut rod is connected from the lower control arm to the chassis. The strut rod is bolted to the control arm, and a large rubber bushing surrounds the strut rod in the chassis opening. The outer end of the strut rod is threaded, and steel washers are positioned on each side of the strut rod bushing. Two nuts tighten the strut rod into the bushing (**Figure 6-21**). The strut rod prevents fore-and-aft movement of the lower control arm. In some suspension systems, the position of the strut rod nuts provides proper front wheel adjustment.

Some strut rods are presently manufactured from tubular steel to reduce strut rod and unsprung weight.

Shop Manual
Chapter 6, page 249

AUTHOR'S NOTE It has been my experience that the most important item when diagnosing automotive problems is your knowledge of automotive systems and components. You must understand the operation of automotive systems and the individual components within these systems. You must also be familiar with the problems that certain worn, defective, or misadjusted components can cause. When diagnosing an automotive problem, you must identify the exact problem, and then think of all the causes of the problem. When you are familiar with all the causes of the problem, you will know the system and component where diagnosis should begin. The cause of an automotive problem may not be in the system where the problem seems to appear. For example, a problem of steering pull during braking would usually be caused by a defect in the brake system. However, a worn front strut rod bushing may allow the front suspension on one side to move rearward during braking, which changes the front suspension alignment angles and causes steering pull when braking. Therefore, if you know that a worn radius arm bushing may cause steering pull when braking, you will inspect this component in the diagnosis as well.

SHORT-AND-LONG ARM FRONT SUSPENSION SYSTEMS

Upper and Lower Control Arms

Many years ago, trucks and cars were equipped with I-beam front suspension systems designed with kingpins and longitudinally mounted leaf springs. A few trucks still used this type of suspension until the late 1980s. The kingpins were retained in a vertical

Figure 6-22 I-beam front suspension with kingpins.

Shop Manual
Chapter 6, page 231

opening in the ends of the axle with a lock bolt. Upper and lower bushings in the steering knuckle pivoted on the upper and lower ends of the kingpins (**Figure 6-22**). In this type of front suspension, if one front wheel moves upward or downward, some movement is transferred to the other end of the axle. This action tends to result in reduced ride quality, steering instability, and tire wear.

As automotive technology evolved, the control arm front suspension system replaced the I-beam front suspension system. This type of front suspension system has coil springs with upper and lower control arms. Since wheel jounce or rebound movement of one front wheel does not directly affect the opposite front wheel, the control arm suspension is an independent system. Many rear-wheel-drive cars have control arm front suspension systems. Early front suspension systems had equal-length upper and lower control arms. On these early suspension systems, the bottom of the tire moved in and out with wheel jounce and rebound travel. This action constantly changed the tire tread width and caused tire scuffing and wear problems (**Figure 6-23**).

In later **short-and-long arm front suspension systems**, the upper control arm is shorter than the lower control arm. During wheel jounce and rebound travel in this suspension system, the upper control arm moves in a shorter arc than the lower control arm. This action moves the top of the tire in and out slightly, but the bottom of the tire remains in a more constant position (**Figure 6-24**). This short-and-long arm front suspension

An independent suspension system is one in which wheel jounce or rebound travel of one wheel does not directly affect the movement of the opposite wheel.

Figure 6-23 Early front suspension system with equal length upper and lower control arms.

Figure 6-24 Short-and-long arm front suspension system.

Bushings

Ball joint
(base)

An upper control arm. Upper control arm shaft
and bushings are visible at top.

Figure 6-25 The inner end of the control arm
contains large rubber insulating bushings, and the ball
joint is attached to the outer end of the control arm.

Inner sleeve

Rubber   Outer sleeve

Cutaway of a control arm bushing

Figure 6-26 The control arm bushings have a metal inner sleeve and
outer shell with rubber sandwiched in between.

system provides reduced tire tread wear, improved ride quality, and better directional
stability compared to I-beam suspension systems and suspension systems with equal-
length upper and lower control arms.

The inner end of the control arm contains large rubber insulating bushings, and the
ball joint is attached to the outer end of the control arm that attaches to the steering
knuckle (**Figure 6-25**). A shaft or bolt passes through the eye of the bushing to connect
it to the frame of the vehicle. The rubber bushings are used to separate and insulate parts
while providing up-and-down pivot points for the suspension. The rubber cushions
reduce noise and vibration from being transmitted to the vehicle body. The control arm
bushings have a metal inner sleeve and outer shell with rubber sandwiched in between
(**Figure 6-26**). The bushings are usually pressed into the control arms. The inner sleeve
is bolted to the frame and remains stationary, whereas the outer shell of the bushing moves
with the control arm. Thus, the control arm moves as the rubber is twisted and flexed
allowing for up and down movement and some shock absorption. As was seen in other
areas of the suspension system, bushings are found on shock absorbers, strut rods, leaf
springs, and sway bars, among other places.

The lower control arm is bolted to the front crossmember, and the attaching bolts are
positioned in the center of the lower control arm bushings (**Figure 6-27**). The ball joint
may be riveted, bolted, pressed, or threaded into the control arm (**Figure 6-28**). A spring

Upper
ball joint

Upper control arm
bushings

Lower control arm
bushings

Lower ball joint   Lower control arm

Figure 6-27 Ball joints and complete short-and-long arm front suspension system.

Bolt/rivet   Pressed fit   Threaded

Ball joint retention methods

Control arm

Ball joint

Steering knuckle

Taper

Cotter pin

Castle nut

Ball joint taper

Figure 6-28 The ball joint may be riveted, bolted, pressed, or threaded into the control arm.

seat is located in the lower control arm. An upper control arm shaft is bolted to the frame, and rubber insulators are located between this shaft and the control arm.

On some suspension systems that utilize a lightweight lower control arm that has a single anchor point at the frame instead of two, a strut rod is used to provide the second anchor point (**Figure 6-29**). The strut rod provides stability to the control arm. This second anchor point is required to keep the control arm from moving backward (aft) or

Frame

Strut rod

Lower control arm

Figure 6-29 A strut rod is used with a control arm that uses only one anchor point at the frame to provide the second anchor point.

forward (fore) as a result of acceleration and braking forces. The strut rod is bolted to the lower control arm and the frame. At least one end of the strut rod is insulated with a bushing. The other end may be rigidly attached to either the control arm or the frame or both ends may contain a bushing. On some platforms the strut rod provides adjustment for wheel alignments settings.

Some current vehicles have hydraulic control arm bushings on the front suspension. Hydraulic control arm bushings do a superior job of preventing road shocks and vibrations supplied to the control arms from reaching the body. This action improves ride quality.

Many current vehicles are equipped with aluminum upper and lower control arms and/or steering knuckles to reduce unsprung weight and improve ride quality. Any significant weight reduction also helps to improve fuel economy and vehicle performance, as well as reduces CO_2 emissions. When servicing aluminum control arms, the use of the wrong tools may damage these components. Always use the tools specified by the vehicle manufacturer.

On some short-and-long arm front suspension systems, the coil spring is positioned between the upper control arm and the chassis (**Figure 6-30**). In these suspension systems, the upper ball joint is compression loaded.

Shop Manual
Chapter 6, page 230

Steering Knuckle

The upper and lower ball joint studs extend through openings in the steering knuckle. Nuts are threaded onto the ball joint studs to retain the ball joints in the knuckle, and the nuts are secured with cotter pins. The wheel hub and bearings are positioned on the steering knuckle extension, and the wheel assembly is bolted to the wheel hub. When the steering wheel is turned, the steering gear and linkage turn the steering knuckle. During this turning action, the steering knuckle pivots on the upper and lower ball joints. The upper and lower control arms must be positioned properly to provide correct tracking and wheelbase between the front and rear wheels. The control arm bushings must be in satisfactory condition to position the control arms properly.

Coil Spring and Shock Absorber

The coil spring is positioned between the lower control arm and the spring seat in the frame. A spring seat is located in the lower control arm, and an insulator is positioned between the top of the coil spring and the spring seat in the frame. The shock absorber is mounted in the center of the coil spring, and the lower shock absorber bushing is bolted to the lower control arm. The top of the shock absorber extends through an opening in the frame above the upper spring seat. Washers, grommets, and a nut retain the top of the shock absorber to the frame. Side roll of the front suspension is controlled by a steel

Figure 6-30 Short-and-long arm front suspension with the coil spring between the upper control arm and the chassis.

Figure 6-31 Short-and-long arm front suspension system with coil spring mounted on the shock absorber.

ITEM	DESCRIPTION
1	Steering knuckle
2	Nut, shock absorber upper mount
3	Upper control arm
4	Coil spring
5	Coil spring seat
6	Lower control arm
7	Tie rod end

Figure 6-32 Component identification in short-and-long arm suspension with the coil spring mounted on the shock absorber.

stabilizer bar, which is mounted to the lower control arms and the frame with rubber bushings.

On some later model short-and-long arm front suspension systems, the lower end of the coil spring is mounted on a seat attached to the shock absorber. A rubber bushing containing a metal bar is installed in the lower end of the shock absorber. This bushing is bolted to the lower control arm (**Figure 6-31** and **Figure 6-32**). The upper end of the coil spring is seated on an upper shock absorber mount. The rod in the center of the shock absorber extends through this mount, and a nut on the threaded end of the shock absorber rod retains the mount on the shock absorber. Bolts in the top of the upper shock absorber mount extend through the upper control arm.

The upper control arm is mounted high in the suspension system, and the upper end of the knuckle has a gooseneck shape. The lower control arm is made from stamped steel to reduce weight. The rear lower control arm bushing is mounted vertically and carries only fore-and-aft loads. This mounting allows the use of a softer rear bushing in the lower control arm. The horizontal front lower control arm bushing and the lower shock absorber mounting are aligned with the wheel center. This provides a direct path for lateral cornering loads. This design allows the use of a hard front lower control arm bushing.

When servicing this suspension system, a special spring compressing tool must be used to compress all the spring tension before loosening the nut on top of the shock absorber rod (**Figure 6-33** and **Figure 6-34**). After this nut is removed, the compressing tool is operated to gradually release the spring tension, and then the upper mount may be removed. (Refer to Chapter 5 in the Shop Manual for strut and coil spring service.)

Figure 6-33 Lower end of spring compressing tool.

Figure 6-34 Upper end of spring compressing tool.

Figure 6-35 Short-and-long arm front suspension with yoke-type shock absorber mounting.

Figure 6-36 Shock absorber with yoke attachment to lower control arm.

⚠ **Caution**

Never loosen the nut on top of the shock absorber rod until a coil spring compressing tool is used to compress all the spring tension. Failure to follow this procedure may cause the spring tension to suddenly release, resulting in personal injury.

In the short-and-long arm front suspension system on the Trailblazer SUV, the coil spring is mounted over the shock absorber; the coil spring tension is applied against the upper mount and the lower spring seat on the shock absorber (**Figure 6-36**). A special yoke attaches to the lower end of the shock absorber, and the lower end of this yoke pivots on a bolt in the lower control arm (**Figure 6-37**). This type of suspension system provides a more positive shock absorber and spring mounting because the rubber bushing-type mounting between the shock absorber and the lower control arm is not required. The steering knuckle pivots on the upper and lower ball joints in the control arms. Therefore, the shock absorber and coil spring do not have to turn when the knuckle turns, which allow the use of a more rigid upper spring mount. This suspension design provides improved vehicle handling and ride characteristics.

Improved Designs in Short-and-Long Arm Front Suspension Systems

Some vehicles now have a reverse-L front suspension system. In these suspension systems, the rear attachment point on the lower control arm extends rearward and attaches to the engine cradle at a point farther toward the rear of the vehicle (Figure 6-37). The lower control arm has an L-shaped design. A firm bushing is installed at the location where the

Figure 6-37 Reverse-L front suspension system with firm, front lower control arm bushing and more compliant rear, lower control arm bushing.

shorter forward leg of the lower control arm attaches to the cradle to control lateral control arm movement and quicken steering response. A more compliant bushing is installed between the longer rear leg and the rear attachment point to absorb the impact of longitudinal forces caused by road irregularities. This type of suspension separates the control of suspension loads into fore-and-aft control and side-to-side control to improve ride and steering quality.

Other modern short-and-long arm front suspension systems have high upper control arms that place the upper ball joints above the tires to provide suspension articulation that helps to keep the tire perpendicular to the road surface while cornering (**Figure 6-38**). This type of front suspension system has a lateral link and a tension strut to position the lower end of the knuckle and the lower ball joint rather than a lower control arm. The lateral link extends slightly forward from the lower ball joint to an attachment point on the chassis. Bushings are mounted in both ends of the lateral link. The tension strut extends rearward from the lower ball joint to the attachment location on the chassis. This type of front suspension also separates the control of suspension loads to improve ride quality and steering control. This type of front suspension may be called a four-link suspension system.

Sway Bar

The sway bar prevents excessive body roll or lean by resisting centrifugal forces. During a turn the inside of the suspension rises, and the outside of the suspension drops. The sway bar (stabilizer bar) is made of spring steel much like a torsion bar and is designed to absorb and transfer energy as it is twisted during suspension jounce and rebound. As the bar twists during a cornering maneuver the inside of the suspension is forced down and the outside of the suspension is forced up as energy is transmitted from one side to the other. In this manner, body roll is reduced, and the vehicle maintains a flatter profile with improved vehicle control and maneuverability during turns. If a sway bar were to fail or if one was not present, the inside tire of the vehicle would have less weight on it during a turn as the weight was transferred to the outside of the vehicle. This would reduce the ability of the driver to control the vehicle during the turn, thereby reducing the stability of the vehicle. A sway bar may be found on both the front

Figure 6-38 Front suspension system with lateral link and tension strut in place of lower control arm.

Figure 6-39 The sway bar (stabilizer bar) reduces body roll and is attached to the vehicle with mounting bushings and link pin bushings.

and rear suspension system of many vehicles. The front sway bar on a short-and-long arm suspension system connects both lower control arms to one another by sway bar links (**Figure 6-39**). Generally, each sway bar link has four bushings and a link pin and is a frequent area of wear and failure especially on higher-mileage vehicles. The sway bar is connected to the frame or unibody of the vehicle with one-piece sway bar mounting bushings and brackets. Some vehicles may exhibit a squeak or squawking noise over bumps, especially on cold mornings, as a result of faulty sway bar mounting bushings (**Figure 6-40**). Shock absorbers and struts also help in reducing body roll, so it is important to inspect all suspension components when excessive body role is suspected.

AUTHOR'S NOTE Driving over speed bumps is very useful to diagnose the presence of suspension noises. It is also helpful to drive the vehicle first thing in the morning when suspension components are cold and the suspension has settled from sitting for a long period, especially for intermittent noises.

 A BIT OF HISTORY

Rear-wheel-drive cars usually have short-and-long arm or torsion bar front suspension systems. Since space and weight are important factors on today's smaller, more efficient front-wheel-drive cars, the lighter, more compact MacPherson strut front suspension is used on most of these cars.

Migrating out of bracket

Large cracks and splits
in link bushings

Missing bushing

Large cracks, chunks
missing, wallowed out

Sway bar bushing and link kit inspection

Figure 6-40 The sway bar (stabilizer bar) mounting bushings and link pin bushings are common areas of wear and damage on higher-mileage vehicles.

MACPHERSON STRUT FRONT SUSPENSION SYSTEM DESIGN

Lower Control Arms and Support

When smaller front-wheel-drive cars became popular, most of these cars had **MacPherson strut front suspension systems**. In these suspension systems, the lower end of the strut is bolted to the top of the steering knuckle, and the lower end of the knuckle is attached to the ball joint in the lower control arm (**Figure 6-41**). An upper strut mount connects the top of the strut to the chassis. An upper control arm is not required in this type of suspension system, because the strut supports the top of the steering knuckle. Since the upper control arm is not required in these suspension systems, they are more compact and therefore very suitable for smaller cars.

On some MacPherson strut front suspension systems, a steel support is positioned longitudinally on each side of the front suspension. These supports are bolted to the unitized body. The inner ends of the lower control arms contain large insulating bushings with a bolt opening in the bushing center. The control arm retaining bolts extend through the center of these bushings and openings in the support (**Figure 6-42**).

Road irregularities cause the tire and wheel to move up and down vertically, and the lower control arm bushings pivot on the mounting bolts during this movement. When the vehicle is driven over road irregularities, vibration and noise are applied to the tire and wheel. The control arm bushings help prevent the transfer of this noise and vibration to the support, the unitized body, and the passenger compartment. Proper location of the support and lower control arm is important to provide correct vehicle tracking. The supports also carry the engine and transaxle weight. Large rubber mounts are positioned between the supports and the engine and transaxle. These mounts absorb engine vibration.

A stabilizer bar may be called a sway bar.

Shop Manual
Chapter 6, page 232

Stabilizer Bar

The **stabilizer bar** is attached to the chassis and interconnects the lower control arms. Rubber insulating bushings are used at all stabilizer bar mounting positions. Some stabilizer bars are attached to the underside of the front crossmember, and the outer ends of the stabilizer bar are connected to the lower side of the front control arms (**Figure 6-43**). This type of torsion bar may be called direct contact, because the outer ends of the bar

MacPherson strut suspension

Figure 6-41 The MacPherson strut suspension system replaces the upper control arm and combines the shock absorber, spring, and upper pivot point into one assembly.

Figure 6-42 Lower control arm and support.

Figure 6-43 Stabilizer bar with mounting bushings.

are in direct contact with the lower control arms. Other stabilizer bars are attached to the upper side of the front longitudinal supports, and links are connected between the outer ends of the bar and the upper side of the front control arms (**Figure 6-44**). On some MacPherson strut front suspension systems, the stabilizer bars are attached to the upper side of the front subframe and the outer ends of the bar are linked to the front struts (**Figure 6-45**). Stabilizer bars with links between the outer ends of the bar and the suspension components may be called indirect contact. Large rubber bushings with steel mounting caps attach the stabilizer bar to the chassis. The linkages at the outer ends of the

Figure 6-44 Stabilizer bar connected between the two lower control arms.

Figure 6-45 Stabilizer bar connected to the front struts.

stabilizer bar are connected to the control arms or struts with retaining bolts, small rubber bushings, steel washers, and sleeves.

The outer ends of the stabilizer bar move up and down with the control arm movement. When jounce or rebound wheel movement occurs, the stabilizer bar transmits part of this movement to the opposite front wheel to reduce and stabilize body roll. The rubber stabilizer bar mounting and linkage bushings prevent noise.

Some current vehicles have front and rear stabilizer bars with flat areas on the bars in the bushing contact areas. The bushings used with these bars have oval openings in the center of the bushing to match the flat areas on the bars. This stabilizer bar design reduces body roll while cornering or driving on irregular road surfaces, because more force is required to twist the stabilizer bar. Some vehicles are now equipped with aluminum stabilizer bars to reduce unsprung weight. Hollow stabilizer bars are used on some current vehicles for weight reduction.

An aftermarket electronically controlled stabilizer bar is available for off-road vehicle operation (**Figure 6-46**). In this type of stabilizer bar system, the driver may press a switch in the instrument panel, and the electronic control disconnects one side of the stabilizer bar from the opposite side, thus making this component ineffective. When the stabilizer bar is disconnected electronically, full vertical wheel travel is allowed on rough terrain. The driver can press the control switch again to return the stabilizer bar to normal operation for stabilized control of body roll during on-road operation.

Lower Ball Joint

The lower ball joint is attached to the outer end of the lower control arm. Methods used to attach the ball joint to the control arm include bolting, riveting, pressing, and threading. A threaded stud extends from the top of the lower ball joint. This stud fits snugly into a hole in the bottom of the steering knuckle. When the ball joint stud is installed in the steering knuckle opening, a nut and cotter pin retain the ball joint (**Figure 6-47**).

Steering Knuckle and Bearing Assembly

The front wheel bearing assembly is bolted to the outer end of the steering knuckle, and the brake rotor and wheel rim are retained on the studs in the wheel bearing assembly.

Electronic controller

Right independent stabilizer

Left independent stabilizer

Figure 6-46 Electronically controlled stabilizer bar.

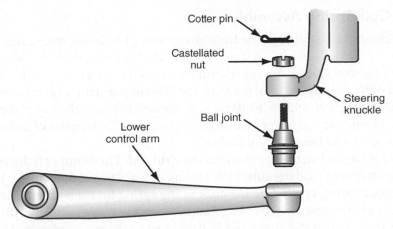

Figure 6-47 Lower ball joint.

This front wheel bearing assembly is a complete, non-serviceable, sealed unit. The front drive shaft is splined into the center of the wheel bearing hub. Thus, drive axle torque is applied to the front wheel. A tie rod end connects the steering linkage from the steering gear to the steering knuckle. The top end of the steering knuckle is bolted to the lower end of the strut (**Figure 6-48**). Many steering knuckles are manufactured from metal, but some current vehicles have aluminum steering knuckles to reduce vehicle weight and unsprung weight. Reducing unsprung weight improves ride quality and vehicle performance and reduces fuel consumption and CO_2 emissions.

Figure 6-48 The front wheel bearing assembly is bolted to the outer end of the steering knuckle and the strut assembly is attached to the upper portion of the steering knuckle where the upper ball joint would have been located on a short-and-long-arm suspension system.

Strut and Coil Spring Assembly

The strut is the shock absorber in the front suspension. The lower spring seat is attached near the center area of the strut. An insulator is located between the lower spring seat and the bottom of the coil spring. An upper strut mount is retained on top of the strut with a nut threaded onto the upper end of the strut rod. The upper strut mount contains a bearing and upper spring seat, and an insulator is positioned between the top of the coil spring and the seat. The upper and lower insulators help prevent the transfer of noise and vibration from the spring to the strut and body.

A bumper is located on the upper end of the strut rod. This bumper reduces harshness while driving on severe road irregularities. During upward wheel movement, the bumper strikes the upper spring seat before the coils in the spring hit each other. Therefore, this bumper reduces harshness when the wheel and suspension move fully upward. The spring tension is applied against the upper and lower spring seats and insulators. However, the nut on top of the upper mount holds the spring in the compressed position between the upper and lower spring seats. When the steering wheel is turned, the steering linkage turns the steering knuckles to the right or left. During this front wheel turning action, the strut-and-spring assembly pivots on the lower ball joint and the upper strut mount bearing.

All the suspension-to-chassis mounting devices such as the lower control arm bushings and the upper strut mount must be positioned properly and be in satisfactory condition to provide correct vehicle tracking and the same wheelbase on both sides of the vehicle.

The purpose of the main components in a MacPherson strut front suspension system may be summarized as follows:

1. Lower control arm—controls lateral (side-to-side) movement of each front wheel.
2. Stabilizer bar—reduces body roll when a front wheel strikes a road irregularity.
3. Coil springs—allow proper setting of suspension ride heights and control suspension travel during driving maneuvers.
4. Shock absorber struts—provide necessary suspension damping and limit downward wheel movement with an internal rebound stop and upward wheel movement with an external jounce bouncer.
5. Strut upper mount—insulates the strut and spring from the body and provides a bearing pivot for the strut-and-spring assembly.
6. Ball joint—connects the outer end of the lower control arm to the steering knuckle and acts as a pivot for the strut, spring, and knuckle assembly.

MacPherson strut front suspension systems are all similar in design, but some vehicle manufacturers provide unique differences in their suspension systems. For example, the MacPherson strut front suspension system on the new Jaguar X-type car has two significant differences: an arm extends several inches inward from the top of the steering knuckle, and the lower end of the strut is attached to the inner end of this arm (**Figure 6-49**). The upper strut mount contains a specially designed bearing that allows the strut-and-spring assembly to rotate freely regardless of the forces supplied to the front suspension. This upper strut mount design reduces friction within the struts and provides very smooth steering action.

MODIFIED MACPHERSON STRUT SUSPENSION

A modified MacPherson strut front suspension is used on some older vehicles. This type of suspension has MacPherson struts with coil springs positioned between the lower control arms and the frame (**Figure 6-50**). The struts in these systems may be gas-filled or oil-filled.

Figure 6-49 MacPherson strut front suspension system on Jaguar X-type car.

Figure 6-50 Modified MacPherson strut front suspension system.

HIGH-PERFORMANCE FRONT SUSPENSION SYSTEMS

Multilink Front Suspension System

High-performance suspension systems are usually installed on sports cars. In these cars, the driver expects improved steering quality, especially when driving and cornering at higher speeds.

In a **multilink front suspension**, a short upper link is attached to the chassis with a bracket, and the outer end of this upper link is connected to a third link. Large rubber insulating bushings are mounted in each end of the upper link. The lower end of the third link is connected through a heavy pivot bearing to the steering knuckle (**Figure 6-51**).

> The **multilink front suspension** may be referred to as a double wishbone suspension because of the link design. It has upper and lower links, and a third link connects the upper link to the top of the knuckle through a bearing.

Figure 6-51 Multilink front suspension system.

The lower link is similar to the conventional lower control arm. A rubber insulating bushing connects the inner end of the lower link to the front crossmember, and a ball joint is connected from the outer end of the lower link to the steering knuckle. In the multilink suspension system, the ball joint axis extends through the lower ball joint and upper pivot bearing, but the ball joint axis is independent from the upper and third links. The extra links in a multilink suspension system maintain precise wheel position during cornering to provide excellent directional stability and steering control while minimizing tire wear.

The shock absorbers are connected from the lower end of the third link to the fender reinforcement. A coil spring seat is attached to the lower end of the shock absorber, and the upper spring seat is located on the upper shock absorber mounting insulator (**Figure 6-52**). Since the steering knuckle pivots on the lower ball joint and the upper pivot bearing, the coil spring and shock absorber do not rotate with the knuckle as they do in a MacPherson strut suspension. Tension or strut rods are connected from the lower links to tension rod brackets attached to the chassis. A stabilizer bar is mounted on rubber insulating bushings in the tension rod brackets, and the outer ends of this bar are attached to the third link.

Double Wishbone Front Suspension System

Double wishbone suspension systems provide increased suspension rigidity and maintain precise wheel position under all driving conditions to supply improved directional stability and steering control. In the double wishbone front suspension system, the upper and lower control arms are manufactured from lightweight, high-strength aluminum alloys designed for maximum strength and rigidity. These lighter control arms decrease the unsprung weight of the vehicle, which improves traction and ride quality. Since the upper and lower control arms have a wishbone shape, the term *double wishbone* is used for this type of suspension. Suspension rigidity is also increased by positioning the ball joints and steering knuckle inside the wheel profile (**Figure 6-53**). On each side of the car, the front suspension is attached to the chassis by a cast aluminum subframe. This design also reduces vehicle weight. The double wishbones are attached to the chassis at the most efficient locations to maintain precise wheel position and provide improved ride quality.

Figure 6-52 Complete multilink suspension system.

Figure 6-53 Double wishbone front suspension.

The front ends of the upper and lower control arms are attached to a compliance pivot assembly. Bushings are mounted in the front ends of the upper and lower control arms, and these bushings are bolted to the compliance pivot (**Figure 6-54**). When one of the front wheels is subjected to rearward force by hard braking or a road irregularity, the coil spring is compressed and the ride height is lowered. This rearward force on the front wheel twists the compliance pivot, allowing both control arms to pivot slightly. Under this condition, the upper and lower control arm movement allows the front wheel to move rearward a small amount, and this wheel movement absorbs energy to significantly improve ride quality (**Figure 6-55**). During this upper and lower control arm movement,

Figure 6-54 Compliance pivot on double wishbone front suspension.

Compliance
pivot

Figure 6-55 Compliance pivot action.

track width and wheel geometry changes are minimal and do not affect steering control. While cornering, the compliance pivot does not move, and lateral suspension stiffness is maintained to supply excellent steering control.

Multilink Front Suspension with Compression and Lateral Lower Arms

Some lateral multilink front suspensions have compression and lateral lower arms. The lateral arm prevents front wheel movement, and the compression arm prevents fore-and-aft front wheel movement. A rubber insulating bushing in the inner end of the lateral arm is bolted to the chassis. A second rubber insulating bushing near the outer end of the lateral arm is bolted to the damper fork. The upper end of the damper fork is bolted to the front strut (**Figure 6-56** and **Figure 6-57**). A ball joint in the outer end of the lateral arm is bolted into the steering knuckle. A rubber insulating bushing in the inner end of the compression arm is bolted to the chassis, and a ball joint in the outer end of this arm is bolted into the steering knuckle.

The upper control arm is mounted higher, so it is above the front tire. The higher upper control arm and the lateral and compression lower arms provide excellent suspension stability and steering control, especially during high-speed cornering or when driving

Compression
arm

Chassis

Steering
knuckle

Lateral
arm

Figure 6-56 Multilink front suspension with lateral and compression lower arms.

Figure 6-57 Multilink front suspension damper fork.

Figure 6-58 Multilink front suspension upper control arm.

on irregular road surfaces. A ball joint in the outer end of the upper control arm is attached to the top of the knuckle, and two shafts in the inner end of this arm are bolted into the strut tower (**Figure 6-58**). There are no provisions for camber or caster adjustments on this multilink front suspension.

Shop Manual
Chapter 6, page 231

Twin I-Beam Suspension Systems

Some Ford trucks and Ford two-wheel-drive Super Duty trucks are equipped with twin I-beam front suspension systems. In this type of suspension system, each front wheel is connected to a separate I-beam. The outer ends of the I-beam are connected to the spindles, and the inner ends of the beams are connected through a rubber pivot bushing to the chassis (**Figure 6-59**). Coil springs are positioned between the I-beams and the chassis to support the vehicle weight. Radius arms are connected rearward from each I-beam to the chassis to prevent longitudinal wheel movement. Since each front wheel can move independently in a twin I-beam suspension system, the problems associated with straight I-beam systems are greatly reduced.

In some older I-beam front suspension systems, kingpins were used to attach the I-beams to the spindles. In other I-beam suspension systems, ball joints connect the I-beams to the spindles.

Some Ford four-wheel-drive trucks used twin I-beam front suspension systems with a coil spring twin I-beam front suspension (**Figure 6-60**). The lower ends of the coil springs are seated on the twin I-beams, and the upper spring seat is positioned on the chassis. Heavy radius arms are bolted to the twin I-beams, and the other ends of the radius

Figure 6-59 Twin I-beam front suspension.

Figure 6-60 Four-wheel-drive twin I-beam front suspension with coil springs.

Figure 6-61 Radius arms on a twin I-beam front suspension system with coil springs.

arms are mounted in a bushing and frame bracket (**Figure 6-61**). The radius arms prevent axle movement. The bushings in the inner ends of the twin I-beams are mounted on pivots attached to the front crossmember. Since each front wheel can move independently, twin I-beam suspensions are independent suspension systems. Upper and lower ball joints are pressed in the ends of the steering knuckle. The studs on these ball joints extend through openings in the outer ends of the twin I-beams. The ball joint studs are

Leaf springs

Axle pivot
attachment bushing

Figure 6-62   Four-wheel-drive twin I-beam front suspension system with leaf springs.

retained in the twin I-beams with nuts and cotter pins. Universal joints in the outer ends of the drive axles allow simultaneous wheel rotation and wheel turning to the right or left. The right-side drive axle also has an inner universal joint near the differential. Newer design Ford Super Duty four-wheel-drive trucks use a coil spring solid front axle assembly with heavy radius arms.

Older Ford F-250 and F-350 four-wheel-drive trucks used a leaf-spring twin I-beam front suspension system (**Figure 6-62**). U-bolts retain the leaf springs to the twin I-beams. A bushing in the front of the leaf-spring eye is bolted into a frame bracket. The bushing in the rear spring eye is connected to the frame through a conventional spring shackle (**Figure 6-63**). Since the leaf springs maintain the axle position, the radius arms are not required.

Some light-duty four-wheel-drive trucks have a straight front drive axle housing and a coil spring suspension system (**Figure 6-64**). Upper and lower radius rods control fore-and-aft drive axle movement, and a track bar controls lateral axle movement. Ball joints connect the steering knuckles to the outer ends of the drive axle housing.

Ford four-wheel drive Super Duty trucks use a similar design but with a single piece radius arm (**Figure 6-65**) to secure the front axle assembly with improved handling and stability.

Twin I-beam front suspension systems have recently been replaced with short-and-long arm or torsion bar suspension systems, because these systems are light weight and provide improved ride quality.

Frame

Shackle

Frame
bracket

Leaf
spring

Figure 6-63   Leaf spring mounting, four-wheel-drive twin I-beam front
suspension.

Figure 6-64 Four-wheel-drive front suspension with straight drive axle and coil springs.

Figure 6-65 Ford Super Duty front radius arm.

Figure 6-66 Ride height determines the amount of vehicle chassis ground clearance.

Ride Height

Vehicle **ride height** determines the amount of ground clearance from the bottom of the tire touching the ground and the lowest point of the vehicle's structure (**Figure 6-66**). Ride height determines the amount of vehicle clearance from the ground. Proper vehicle ride height is a compromise between practicality, handling, and stability without sacrificing safety. Regular inspection and proper maintenance of suspension systems are extremely important to maintaining vehicle safety. The curb riding height is determined mainly by spring condition.

Trucks and SUVs have a higher ride height than passenger vehicles to allow them to be driven on uneven road surfaces and over small obstacles. The negatives of increased ride height are that it raises the center of mass, reducing handling and causing a decrease in aerodynamics, and increases the chance of rollover under some conditions. Stability control became mandatory on passenger vehicles and light trucks for the 2012 model year, which has improved vehicle safety regardless of vehicle ride height. Sports cars have a very low ride height, improving handling characteristics, aerodynamics, and causing the vehicle to have a lower center of gravity, which reduces the risk of rollover in some conditions. But ground clearance is sacrificed, making these vehicles poorly suited for uneven and rough roads. Low ground clearance vehicles require that they be driven at very low speeds on very rough and uneven surfaces, and generally not at all on off-road conditions.

A vehicle's ride height should be maintained with manufacturer specifications. Most manufacturers allow up to 1 in. lower than specified ride height. If ride height is below manufacturer specifications it is often an indication of weak springs, but could also be caused by faulty suspension components, structural damage, or, on torsion bar suspensions, the need for height adjustment. Sagged springs cause insufficient curb riding height; therefore, the distance is reduced between the rebound bumper and its stop. This distance reduction causes the bumper to hit the stop frequently with resulting harsh ride quality.

Other suspension components such as worn control arm bushings will affect curb riding height. Since incorrect curb riding height affects most of the other suspension angles, this measurement is critical. Sagged springs change the normal operating arc of the lower ball joint. This action causes excessive lateral movement of the tire during wheel jounce and rebound with resulting tire wear (**Figure 6-67**).

Manufacturers specify vehicle ride height in their service information and the point at which ride height is measured which is critical (**Figure 6-68**). The curb riding height must be measured at the vehicle manufacturer's specified location, which varies depending on the type of suspension system. When the vehicle is on a level floor or an alignment rack, measure the curb riding height from the floor to the manufacturer's specified location on the chassis.

The measurement reference point may be on the fender, frame, body rocker panel, bumper, or suspension control arm. On some applications, such as light trucks, the measurement may be made between two points on the vehicle's suspension system.

Always verify where the measurement is supposed to be made (**Figure 6-69**). For most conventional ride height measurements, all that is required is a tape measure or ruler and possibly a level. Verify that the vehicle is equipped with the manufacturer-specified

Figure 6-67 Sagged springs change the normal operating arc of the lower ball joint and cause excessive lateral tire movement during wheel jounce and rebound with resulting tire wear.

Figure 6-68 Manufacturers specify specific location where ride height is measured and compared to vehicle specifications.

Where ride height may be specified & measured

Figure 6-69 Ride height measurements are specified at different locations depending on the year, make, and model of vehicle.

tire and rim size. On vehicles where the tire and wheel assemblies have been replaced and that differ in overall assembly height (shorter or taller) than that specified by the manufacturer, ride height measurements will be affected. Shorter assemblies will decrease ride height, and taller assemblies will increase ride height if the measurement is being made from a vehicle reference point to the ground.

The vehicle's ride height should be measured at all four corners of the vehicle. Comparing side-to-side measurements on both the front and the rear of the vehicle will aid in identifying weak springs, frame damage, and body misalignment issues. A ride height difference of more than 1 in. could cause the vehicle to pull to one side.

Errors in ride height can cause caster measurements to be out of specifications. The caster angle is the number of degrees between the true vertical centerline of the tire and wheel, and an imaginary line through the center of the upper strut mount and lower ball joint (**Figure 6-70**). If the front of the vehicle is lower than specified, the caster measurement will be more negative (**Figure 6-71**). If the rear ride height is lower than specified, the caster measurement will be more positive (**Figure 6-72**).

If both the front and rear ride height are equally low or high, there will be no net change.

Incorrect vehicle ride height can cause camber measurements to be out of specifications (**Figure 6-73**); this is especially true if there is a side-to-side ride height difference in the front. Side-to-side ride height difference in the rear can affect rear camber readings on vehicles with independent rear suspension systems. The amount of camber change will depend on the specific vehicle geometry and the amount of ride height difference. If a spring is sagged on one side of the front suspension, the camber angle is negative on that

Caster angle is the tilt of an imaginary line through the center of the upper strut mount and ball joint in relation to the true vertical line through the center of the wheel and spindle as viewed from the side.

Figure 6-70 Positive and negative caster.

Figure 6-71 Effects of front spring sag on caster angle.

Figure 6-72 Effects of rear spring sag and incorrect curb riding height on caster angle.

Figure 6-73 Lower than specified ride height can cause camber to become more negative.

side. Excessive negative camber results in rapid wear on the inside edge of the tire tread and a decrease in directional stability. *Therefore, riding height is extremely important in maintaining correct wheel alignment angles, normal tire wear, and satisfactory directional stability.*

Ride height changes can increase tire wear due to alterations in vehicle geometry. This is particularly evident on suspension system where toe angle changes occur with a change in ride height.

Vehicles with lower than specified ride height may generate customer complaints related to suspension bottoming, nose diving while braking, instability and wandering when hauling heavy loads or towing, or excessive body roll when cornering. Lower than specified vehicle ride height as a result of sagging springs causes less total suspension travel increasing the likelihood of bottoming out on bumps or while towing or carrying heavy loads. This may result in accelerated suspension component wear and damage to suspension stops, struts, or shock absorbers. In some cases, the coils on coil spring suspensions may even contact one another, showing wear marks.

Checking vehicle ride height is one of the easiest methods for diagnosing weak and sagging springs.

Camber angle is the tilt of a line through the center of the wheel in relation to the vertical centerline viewed from the front of the wheel.

SUMMARY

- Coil springs may be classified as linear rate or variable rate. A linear-rate coil spring has a constant spring rate regardless of spring load. Variable-rate coil springs have unequally spaced coils. The coils in these springs may be referred to as inactive, transitional, or active.

- Heavy-duty coil springs have larger-diameter wire than regular-duty coil springs. Most vehicle manufacturers recommend that springs be replaced in pairs. In a torsion bar suspension system, the torsion bars replace the coil springs. Torsion bars are directional, and they are marked right or left. In a leaf spring, zinc and plastic spacers between the leaves reduce noise and friction problems.

- Leaf springs may be mono-leaf or multiple-leaf design. Spring leaves may be manufactured from steel or fiberglass. Although many leaf springs are mounted longitudinally, some of these springs are mounted transversely.

- Ball joints may be classified as load carrying or non-load carrying. A load-carrying ball joint may be compression loaded or tension loaded. The

load-carrying ball joint wears faster than the non-load-carrying ball joint.

- Many ball joints have wear indicators that provide visual ball joint wear inspection. A stabilizer bar reduces body roll or sway when one front wheel strikes a road irregularity. A strut rod prevents fore-and-aft lower control arm movement. In many MacPherson strut front suspension systems, the lower end of the strut is bolted to the steering knuckle, and the upper end of the strut is connected through an upper mount to the fender reinforcement.

- Torsion bars may be transversely or longitudinally mounted in a front suspension system. In a short-and-long arm suspension system, the coil springs may be mounted between the lower control arm and the frame or between the upper control arm and the chassis.

- In a twin I-beam suspension system, the spindles may be connected to the I-beams with kingpins or ball joints.

- Curb riding height is extremely important for maintaining normal tire wear and proper suspension alignment angles.

REVIEW QUESTIONS

Short Answer Essays

1. Explain the meaning of constant spring rate.

2. Describe the main difference in the design of heavy-duty coil springs compared with regular-duty coil springs, and compare the free diameter of each type of spring.

3. Explain the type of load condition that requires variable-rate coil springs, and describe the load-carrying advantage of these springs.

4. Describe the design of taper-wire closed-end springs.

5. Explain torsion bar action during wheel jounce and rebound.

6. Describe the action of a leaf spring as it compresses.

7. Explain the position of the lower control arm when the lower ball joint is compression loaded.

8. Describe the position of the wear indicator when a ball joint is worn.

9. Explain the basic purpose of the lower control arms.

10. Describe the effect of sagged front springs on caster angle and directional stability.

Fill-in-the-Blanks

1. Heavy-duty coil springs are designed to carry _____ to _____ percent greater loads than regular-duty coil springs.

2. A variable-rate coil spring provides automatic _____ adjustment.

3. In a variable-rate spring, the _____ coils operate during the complete range of spring loading.

4. The load-carrying ball joint is attached to the control arm on which the _____ is seated.

5. If the lower control arm is mounted above the steering knuckle and rests on the knuckle, the lower ball joint is_____.

6. When the grease nipple shoulder is extended from the ball joint housing, the ball joint is_____.

7. When one front wheel strikes a road irregularity, the stabilizer bar reduces_____.

8. The strut rod prevents _____ and _____ lower control arm movement.

9. When the front springs are sagged, the positive caster is_____.

10. If a front spring is sagged, the camber angle moves toward a _____ position.

Multiple Choice

1. A 400-lb load deflects a linear-rate coil spring 1 in. An 800-lb load will deflect this coil spring:
   A. 1.5 in.
   B. 2 in.
   C. 3 in.
   D. 4 in.

2. Heavy-duty coil springs have:
   A. The same wire diameter as a regular-duty coil spring.
   B. A shorter free height compared to a regular-duty coil spring.
   C. A 10 percent increase in load-carrying capacity.
   D. A high aluminum content in the spring material.

3. A car's rear coil springs keep breaking. This problem is likely because of:
   A. Continual driving on rough road surfaces.
   B. Continual driving on curved roads.
   C. Excessive air pressure in the rear tires.
   D. Constant overloading of the rear suspension.

4. Aluminum control arms have all of these advantages EXCEPT:
   A. Reduce unsprung weight.
   B. Provide improved road feel.
   C. Provide improved ride quality.
   D. Contribute to improved fuel economy.

5. While discussing a torsion bar front suspension system:
   A. One end of the torsion bar is attached to the lower control arm.
   B. A suspension height adjustment is positioned on the end of the torsion bar connected to the control arm.
   C. Torsion bars may be used with coil springs in a front suspension system.
   D. Torsion bars eliminate the need for shock absorbers in the front suspension system.

6. While discussing leaf springs:
   A. A fiberglass mono-leaf spring is heavier than a steel mono-leaf spring.
   B. Some mono-leaf springs are mounted transversely.
   C. In a multiple-leaf spring, the lower leaf is called the main leaf.
   D. A multiple-leaf spring has the same stiffness regardless of how much it is compressed.

7. All these statements about ball joints are true EXCEPT:
   A. A load-carrying ball joint wears faster than a non-load-carrying ball joint.
   B. A non-load-carrying ball joint is designed with a preload.
   C. Wear indicators may be positioned in the side of the ball-joint housing.
   D. Low-friction ball joints have a highly polished ball socket surrounded by a high-strength polymer bearing.

8. While discussing MacPherson strut suspension systems:
   A. During a turn, the strut, spring, and knuckle rotate on the upper strut mount.
   B. The lower end of the coil spring is seated on the lower control arm.
   C. The strut is welded into the knuckle and these components are replaced as an assembly.
   D. The inner ends of the lower control arms are bolted to the strut.

9. While discussing short-and-long arm front suspension systems:

   A. The coil springs are positioned between the upper and lower control arms.

   B. Compared with a suspension system with equal length control arms, the short-and-long design reduces track width change and tire wear.

   C. The outer ends of the stabilizer bar may be attached to the steering knuckles.

   D. The coil spring is retained in the lower control arm with a clamp and bolt.

10. While discussing sagged springs:

   A. Sagged front springs cause harsh riding.

   B. Sagged front springs cause excessive positive caster on the front wheels and increased directional stability.

   C. Sagged rear springs cause excessive negative caster on the front wheels.

   D. Sagged rear springs cause excessive positive camber on the front wheels.

CHAPTER 7
REAR SUSPENSION SYSTEMS

Upon completion and review of this chapter, you should be able to understand and describe:

- A live axle rear suspension system.
- The advantages and disadvantages of a live axle leaf-spring rear suspension system.
- The movement of the rear axle housing during vehicle acceleration.
- How the differential torque is absorbed in a live axle coil spring rear suspension system.
- The purpose of a tracking bar in a live axle coil spring rear suspension system.
- The difference between a semi-independent and an independent rear suspension system.
- How individual rear wheel movement is provided in a semi-independent rear suspension system.

- The difference between a MacPherson strut and a modified MacPherson strut rear suspension.
- The advantage of attaching the differential housing to the chassis in an independent rear suspension system.
- How differential and suspension vibration, noise, and shock are insulated from the chassis in a multilink independent rear suspension system.
- How the top of the knuckle is supported in a multilink independent rear suspension system.
- The advantages of hydraulic suspension mounts.
- The effect of sagged rear springs on caster angle and steering.

Terms To Know

Asymmetrical leaf spring	Five-link suspension	Semielliptical springs
Axle tramp	Independent rear suspension	Semi-independent rear suspension
Axle windup	Live axle rear suspension systems	Solid axle beam
Braking and deceleration torque	Multilink independent rear suspension system	Sprung weight
Caster angle	Rear suspension system	Symmetrical leaf spring
Double wishbone rear suspension system	Rear wheel toe	Track bar
		Unsprung weight

INTRODUCTION

A **rear suspension system** with two longitudinal leaf springs and a one-piece rear axle housing may be called a Hotchkiss drive.

The **rear suspension system** plays a very important part in ride quality and in the control of suspension and differential noise, vibration, and shock. Although the front wheels actually steer the vehicle, the rear suspension is also vital to steering control. The rear suspension must also provide adequate tire life and maintain tire traction on the road surface. Rear suspension systems described in this chapter include live axle, semi-independent, and independent. **Live axle rear suspension systems** are found on rear-wheel-drive trucks and vans, a few rear-wheel-drive cars, and some four-wheel-drive cars. Most front-wheel-drive vehicles have semi-independent or independent rear suspensions. Independent rear suspensions are also found on rear-wheel-drive cars and four-wheel-drive cars.

> A **live axle rear suspension system** may be defined as one in which the differential axle housing, wheel bearings, and brakes act as a unit.

LIVE AXLE REAR SUSPENSION SYSTEMS

Leaf-Spring Rear Suspension

A leaf spring is mounted longitudinally on each side of the rear suspension on some rear-wheel-drive cars and trucks (**Figure 7-1**). These relatively flat springs provide excellent lateral stability and reduce side sway, which contribute to a well-controlled ride with very good handling characteristics. However, leaf-spring rear suspension systems have a lot of **unsprung weight**, and leaf springs require a considerable amount of space.

> **Unsprung weight** refers to the weight that is not supported by the springs, which includes the weight of the suspension system.

  A BIT OF HISTORY

The leaf-spring rear suspension was one of the first widely used rear suspension systems. Because this type of rear suspension system has weight and ride quality disadvantages, it has been replaced on many vehicles with independent or semi-independent rear suspension systems.

The **semielliptical springs** have steel leaves and zinc or plastic interleaves to reduce corrosion, friction, and noise. A large rubber bushing is installed in the front eye of the main spring leaf, and a bolt retains this bushing to the front spring hanger (**Figure 7-2**). The rear spring shackle is bolted to a rubber bushing in the rear main leaf eye, and the upper shackle bolt extends through a similar rubber bushing in the rear spring hanger (**Figure 7-3**). Some rear spring shackles contain threaded steel bushings or a slipper mount in which the end of the spring slides through the shackle. Shackle insulating bushings help prevent the transfer of noise and road shock from the suspension to the chassis and vehicle interior. When a rear wheel strikes a road irregularity, the spring is compressed and the spring length changes. The rear shackle provides fore-and-aft movement with variations in spring length.

> **Semielliptical springs** have individual leaves stacked with the shortest leaf at the bottom and the longest leaf at the top.

> **Shop Manual**
> Chapter 7, page 288

Figure 7-1 Rear suspension system with long torque arm and track bar.

Figure 7-2 Rear leaf-spring eye bushing.

Figure 7-3 Rear leaf-spring shackle.

Because the differential axle housing is a one-piece unit, jounce and rebound travel of one rear wheel affects the position of the other rear wheel. This action increases tire wear and decreases ride quality and traction.

The differential axle housing is mounted above the springs, and a spring plate with an insulating clamp and U-bolts retains the springs to the rear axle housing (**Figure 7-4**). The shock absorbers are mounted between the spring plates and the frame.

The vehicle **sprung weight** is supported by the springs through the rear axle housing and wheels. When the vehicle accelerates, the rear wheels turn counterclockwise when viewed from the left vehicle side. One of Newton's laws of motion states that for every action there is an equal and opposite reaction. Therefore, when the wheels turn counterclockwise (when viewed from the left), the rear axle housing tries to rotate clockwise. This rear axle torque action is absorbed by the rear springs and the chassis moves downward (**Figure 7-5**). Engine torque supplied through the driveshaft to the differential tends to twist the differential housing and the springs. This twisting action may be referred to as

> **Sprung weight** refers to the weight carried by the springs, which includes the chassis and all components attached to the chassis.

Figure 7-4 Individual leaf-spring suspension components.

Braking torque
reaction

Acceleration torque
reaction

Figure 7-5 Rear axle torque action during acceleration and deceleration.

axle windup. Many leaf springs have a shorter distance from the center bolt to the front of the spring compared to the distance from the center bolt to the rear of the spring. This type of leaf spring is referred to as an **asymmetrical leaf spring**, and the shorter distance from the center bolt to the front of the spring resists axle windup. A **symmetrical leaf spring** has the same distance from the center bolt to the front and rear of the spring.

When braking and decelerating, the rear axle housing tries to turn counterclockwise. This rear axle torque action applied to the springs lifts the chassis. This action may be called **braking and deceleration torque**.

During hard acceleration, the entire power train twists in the opposite direction to engine crankshaft and drive shaft rotation. The engine and transmission mounts absorb this torque. However, the twisting action of the drive shaft and differential pinion shaft tends to lift the rear wheel on the passenger's side of the vehicle. Extremely hard acceleration may cause the rear wheel on the passenger's side to lift off the road surface. Once this rear wheel slips on the road surface, engine torque is reduced, and the leaf spring forces the wheel downward. When this rear tire contacts the road surface, engine torque increases and the cycle repeats. This repeated lifting of the differential housing is called **axle tramp**, and this action occurs on live axle rear suspension systems. Axle tramp is more noticeable on live axle leaf-spring rear suspension systems in which the springs have to absorb all the differential torque. For this reason, only engines with moderate horse-power were used with this type of rear suspension. Rear suspension and axle components such as spring mounts, shock absorbers, and wheel bearings may be damaged by axle

AUTHOR'S NOTE Leaf-spring rear suspension systems are still used on many light-duty trucks because of their load-carrying capability. However, today's design engineers have improved the ride quality of these suspension systems compared with past models. Ride quality in these leaf-spring suspension systems has been improved by installing longer leaf springs and using larger, improved rubber insulating bushings in the spring eye and shackle. Ride quality has also been improved by maximizing the shock absorber mounting location and matching the shock absorber design more closely to the leaf-spring jounce and rebound action. Optimizing the rear axle mounting position on the leaf springs also improves ride quality.

Figure 7-6 Rear suspension system with long torque arm and track bar.

tramp. Mounting one rear shock absorber in front of the rear axle and the other rear shock behind the rear axle helps reduce axle tramp.

In some cars with higher torque engines, a long torque arm is bolted to the rear axle housing (**Figure 7-6**). This torque arm helps prevent differential rotation during hard acceleration and braking. The front of this torque arm is mounted in a rubber insulator and bracket that is bolted to the back of the transmission housing. This long torque arm helps prevent differential rotation when high torque is delivered from the engine to the differential. The center bearing assembly on the drive shaft is bolted to the long torque arm. This rear suspension system has a track bar (tie rod) connected from the left side of the rear axle housing to the chassis to help prevent lateral rear axle movement. A track bar brace is connected from the chassis end of the track bar to the other side of the chassis to provide extra rigidity.

Coil Spring Rear Suspension

⚠WARNING Compressed coil springs contain a large amount of energy. Never disconnect any suspension component that suddenly releases the coil spring tension. This action may result in personal injury and vehicle damage.

⚠WARNING Always follow the vehicle manufacturer's recommended rear suspension service procedures in the service manual to avoid personal injury.

Some rear-wheel-drive cars have a coil spring rear suspension. Upper and lower suspension arms with insulating bushings are connected between the differential housing and the frame (**Figure 7-7**). The upper arms control lateral movement, and the lower trailing control arms absorb differential torque. In some rear suspension systems, the upper arms are replaced with strut rods. The front of the upper and lower arms contains large rubber bushings. When strut rods are used in place of the upper arms, both ends of these rods contain large rubber bushings to prevent noise and vibration transfer from the suspension to the chassis. The coil springs are usually mounted between the lower suspension arms and the frame, whereas the shock absorbers are mounted between the back of the suspension arms and the frame.

Figure 7-7 Live axle coil spring rear suspension.

Figure 7-8 Rear suspension system with coil springs and shock absorbers mounted separately.

Some rear suspension systems have a **track bar** connected from one side of the differential housing to the chassis to prevent lateral chassis movement. Large rubber insulating bushings are positioned in each end of the track bar.

Some late-model sport utility vehicles (SUVs) have a rear suspension system with coil springs and shock absorbers mounted separately from the springs (**Figure 7-8**). Notice that a control rod is mounted in rubber insulating bushings on the bottom of the rear axle housing, and the outer end of this bar is connected through links to the chassis. This rear suspension has an upper and a lower control rod on each side of the suspension. Each lower control rod is connected from a bracket on the lower side of the axle housing to a frame bracket. Rubber insulating bushings are located in both ends of the lower control rods. The frame bracket is located ahead of the rear axle. The upper control rod is connected from a bracket on top of the axle housing to a frame bracket near the lower control rod frame bracket (**Figure 7-9**). Both ends of the upper control rod contain rubber

A **track bar** may be referred to as a Panhard rod or Watts rod.

Figure 7-9 Rear suspension system with upper and lower control rods.

insulating bushings. The upper and lower control rods prevent rear axle windup during hard acceleration.

This SUV rear suspension also has a track bar connected from a bracket on the rear of the axle housing to a frame bracket on the opposite side of the vehicle. The track bar prevents lateral rear axle movement. A rear axle brace is also connected from the track bar frame bracket to another frame bracket on the opposite side of the vehicle (**Figure 7-10**). The rear axle brace prevents any movement of the track bar frame bracket during off-road or severe driving conditions. The dual rear axle rods on each side of this suspension with the track bar and brace provide a very stable rear axle position during hard acceleration and severe driving conditions. This rear axle stability improves tire life, ride comfort, and steering control.

SEMI-INDEPENDENT REAR SUSPENSION SYSTEMS

A **semi-independent rear suspension** allows some individual rear wheel movement when one rear wheel strikes a bump.

Shop Manual
Chapter 7, page 289

Many front-wheel-drive vehicles have a **semi-independent rear suspension** that has a **solid axle beam** connected between the rear trailing arms (**Figure 7-11**). A solid axle beam is usually a transverse inverted U-section channel connected between the rear wheels in a semi-independent rear suspension system. When one rear wheel strikes a bump, this beam twists to allow some independent wheel movement. Some of these rear axle beams are fabricated from a transverse inverted U-section channel.

In some rear suspension systems, the inverted U-section channel contains an integral tubular stabilizer bar. When one rear wheel strikes a road irregularity and the wheel moves upward, the inverted U-section channel twists, which allows some independent rear wheel movement. The trailing arms are connected to chassis brackets through rubber insulating bushings. In some semi-independent rear suspension systems, the coil springs are mounted on the rear struts, the lower spring seat is located on the strut, and the upper spring seat is positioned on the upper strut mount.

In other semi-independent rear suspension systems, the coil springs are mounted separately from the shock absorbers. Coil spring seats are located on the trailing arms, and the shock absorbers are connected from the trailing arms to the chassis. A crossmember connected between the trailing arms provides a twisting action and some independent rear wheel movement (**Figure 7-12**).

Figure 7-10 Track bar and brace.

Figure 7-11 Semi-independent rear suspension system.

Figure 7-12 Semi-independent rear suspension system with coil springs and shock absorbers mounted separately.

Figure 7-13 Semi-independent rear suspension with track bar and brace.

Some semi-independent rear suspension systems have a track bar connected from a rear axle bracket to a chassis bracket. In some applications, an extra brace is connected from this chassis bracket to the rear upper crossmember (**Figure 7-13**). The track bar and the brace prevent lateral rear axle movement.

INDEPENDENT REAR SUSPENSION SYSTEMS

MacPherson Strut Independent Rear Suspension System

In an **independent rear suspension** system, each rear wheel can move independently from the opposite rear wheel. Independent rear suspension systems may be found on front-wheel-drive and rear-wheel-drive vehicles. When rear wheel movement is independent, ride quality, tire life, steering control, and traction are improved. In a MacPherson strut rear suspension system, the coil springs are mounted on the rear struts. A lower spring seat is located on the strut, and the upper spring seat is positioned on the upper strut mount. This upper strut mount is bolted into the inner fender reinforcement. Dual lower control arms on each side of the suspension are connected from the chassis to the lower end of the spindle (**Figure 7-14**).

The lower end of each strut is bolted to the spindle. Two strut rods are connected forward from the spindles to the chassis. Rubber insulating bushings are located in both

In an **independent rear suspension**, vertical movement of one rear wheel does not affect the opposite rear wheel.

Shop Manual
Chapter 7, page 290

Figure 7-14 MacPherson strut independent rear suspension system.

Figure 7-15 Upper and lower spring insulators, MacPherson strut independent rear suspension.

ends of the strut rods. A stabilizer bar is mounted in rubber bushings connected to the chassis, and the ends of this bar are linked to the struts.

Insulators are mounted between the lower end of the coil spring and the lower spring seat, and the top of the coil spring and the upper spring support (**Figure 7-15**). These insulators help prevent the transfer of spring noise and vibration to the chassis and passenger compartment.

Modified MacPherson Strut Independent Rear Suspension System

Some front-wheel-drive vehicles have a modified MacPherson strut independent rear suspension. Each side of the rear suspension has a shock strut, lower control arm, tie rod, forged spindle, and coil spring mounted between the lower control arm and the crossmember side rail (**Figure 7-16**).

The shock absorber strut has a rubber isolated top mount with a one-piece jounce bouncer dust shield. This top mount is attached to the body side panel and the lower end of the strut is bolted to the spindle. The stamped lower control arms are bolted to the crossmember and the spindle. A tie rod is connected from the spindle to the underbody. The purpose of each rear suspension component may be summarized as follows:

1. Stamped lower control arm—controls the lateral (side-to-side) wheel movement and contains the lower spring seat.
2. Tie rod—controls fore-and-aft wheel movement and positions the spindle properly.
3. Shock absorber strut—reacts to braking forces and provides suspension damping. A strut internal rebound stop provides rebound control, and an external jounce bumper supplies jounce control.
4. Coil spring—controls suspension travel, provides ride height control, and acts as a metal-to-metal jounce stop.
5. Forged spindle—supports the wheel bearings and attaches to the lower control arms, tie rod, brake assembly, and strut.

Figure 7-16 Modified MacPherson strut independent rear suspension.

6. Suspension bushings—insulate the chassis and passenger compartment from road noise and vibration.
7. Suspension fasteners—connect components such as the spindle and strut. These fasteners must always be replaced with equivalent quality parts, and each fastener must be tightened to the specified torque.

Independent Rear Suspension with Lower Control Arm and Ball Joint

Some front-wheel-drive cars have an independent rear suspension system with a ball joint pressed into the outer end of the lower control arm. The ball joint contains a conventional wear indicator. The upper end of the ball joint stud is bolted into the knuckle (**Figure 7-17**). The inner end of the lower control arm is connected to the chassis through two rubber insulating bushings (**Figure 7-18**).

The lower end of the strut is bolted to the knuckle, and the upper strut mount is bolted to the inner fender reinforcement (**Figure 7-19**). A stabilizer bar is mounted in bushings attached to the chassis, and the outer ends of this bar are linked to brackets connected to the strut (**Figure 7-20**).

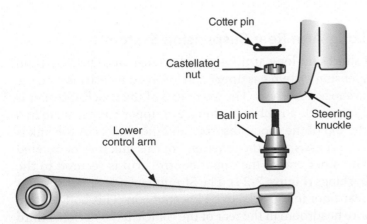

Figure 7-17 Lower control arm and ball joint.

Figure 7-18 Lower control arm bushings.

Figure 7-19 Upper and lower strut mounting.

Figure 7-20 Stabilizer bar mounting.

Figure 7-21 Suspension adjustment link.

Figure 7-22 Coil spring mounting.

Rear wheel toe is the variation between the distance measured between the tires at the front edge and the rear edge. If the distance between the front edge of the tires is less than the distance between the rear edge of the tires, the rear wheels have toe-in.

A suspension adjustment link is connected from the lower control arm to the knuckle (**Figure 7-21**). This link provides a **rear wheel toe** adjustment. A coil spring seat is located in the lower control arm and the coil spring is mounted between this seat and an upper seat in the chassis (**Figure 7-22**). Upper and lower insulators are mounted between the coil spring and the seats.

Independent Short-and-Long Arm Rear Suspension System

Some late-model SUVs have a short-and-long arm independent rear suspension system (**Figure 7-23**). This type of suspension system has upper and lower control arms, and the coil springs are mounted on the shock absorbers. The lower end of the shock absorber is mounted to the lower control arm with a rubber bushing. An upper rubber-insulated mount is positioned between the top of the shock absorber and the chassis. A toe link is connected from the spindle to the chassis, and the inner end of this link may be rotated to adjust the rear wheel toe. The inner end of the upper control arm is secured to the chassis with rubber insulating bushings (**Figure 7-24**). The SLA independent rear suspension system design allows the rear floor in the SUV to be lowered 5 in. This body design allows extra cargo space and more headroom in the rear of the vehicle if the optional third seat is installed.

Figure 7-23 Short-and-long arm independent rear suspension system.

Figure 7-24 Upper control arm, short-and-long arm independent rear suspension system.

Independent Rear Suspension System with Rear Axle Carrier

Some import rear-wheel-drive cars have an independent rear suspension with a large rear axle carrier extending across the width of the chassis. A large extension on the rear axle carrier is bolted to the top of the differential, and the outer ends of this carrier are connected through heavy insulating bushings to chassis brackets (**Figure 7-25**).

A large final drive mount extends from the rear of the differential housing, and this mount is bolted to the chassis. When the differential is attached to the chassis, rather than being connected to the suspension, the unsprung weight is reduced. Two rubber insulating bushings connect the large trailing arms to the rear axle carrier. The rear axle carrier provides very stable differential and wheel position, which improves steering control, ride quality, and tire life. Lower coil spring seats are located on the trailing arms, and upper coil spring seats are positioned in the chassis.

Shock absorbers are connected from the back of the trailing arms to the chassis. The rear wheel bearings are mounted in the outer ends of the trailing arms. Drive axles with inner and outer drive joints are connected from the differential to the rear wheels. When the drive axles have inner and outer joints, rear wheel camber change is minimized during wheel jounce and rebound. On some early independent rear suspension systems in rear-wheel-drive cars, the drive axles had only inner joints. With this type of rear suspension

Figure 7-25 Independent rear suspension with rear axle carrier.

and drive axle, camber change was excessive during wheel jounce and rebound, and this action caused wear on the tire edges.

A stabilizer bar is bolted to the rear axle carrier through rubber insulating bushings. The outer ends of this bar are linked to the trailing arms.

Multilink Independent Rear Suspension System

Many vehicles have a **multilink independent rear suspension system**. A typical multilink rear suspension system has upper and lower control arms and additional links connected from the rear knuckles to the chassis to stabilize the rear wheel position. The suspension components are attached to the rear frame. The lower control arm is connected from the rear frame to the lower end of the knuckle. A large bushing is mounted in the inner end of the control arm at the frame attachment location (**Figure 7-26**). The coil springs are mounted between the lower control arms and the rear chassis, and insulators are positioned on the upper and lower ends of the springs. The upper control arms are connected from the rear frame to the top of the knuckle. The inner end of each upper control arm has two attachment points to frame, and large insulating bushings are installed at these attachment locations (**Figure 7-27**). A ball joint in the outer end of each upper control arm connects the control arm to the top end of the knuckle. Many current rear suspension systems have aluminum components such as upper and lower control arms. Aluminum components reduce weight and improve fuel mileage, which in turn reduces emissions.

Figure 7-26 Lower control arm with upper and lower spring insulators.

Figure 7-27 The upper control arm is connected from the frame to the top of the knuckle.

Figure 7-28 The trailing arm is connected from the lower end of the knuckle to the frame.

A trailing arm is connected from the lower end of the knuckle to the frame. Large insulating bushings are mounted at each trailing arm attachment location (**Figure 7-28**). The trailing arm prevents fore-and-aft wheel movement. An adjustment link is also connected from the lower end of the knuckle to the frame through insulating bushings (**Figure 7-29**). The adjustment link and lower control arm prevent lateral wheel movement, and this link also provides a method of rear wheel toe adjustment. This type of rear suspension may be called a **five-link suspension** because there are five attachments points between each knuckle and the frame. The lower control arm, trailing arm, and adjustment link have one attachment location, and the upper control arm has two attachment points. A five-link rear suspension provides a very stable rear wheel position, which improves steering control and rear tire tread life.

The shock absorbers are connected from the rear knuckles to the chassis (**Figure 7-30**). Two bolts retain the upper insulating shock absorber mount to the chassis. In this rear-wheel-drive car, the top of the differential is bolted to the upper part of the rear frame. Two large insulating bushings are mounted between the differential and the frame (**Figure 7-31**). Mounting the differential to the frame reduces the unsprung weight, which improves ride quality, because the unsprung weight forces the wheels downward during wheel rebound, which contributes to harsh ride quality.

A **five-link suspension** has five attachment locations between each side of the suspension and the frame.

Figure 7-29 Adjustment link.

Figure 7-30 Shock absorber mounting.

Figure 7-31 Differential to frame mounting bushings.

In some current multilink rear suspension systems on rear-wheel-drive vehicles, four large rubber mounts are positioned between the rear suspension member or frame and the chassis. These rubber mounts are designed such that they are soft in fore-and-aft movement but stiff in lateral movement. These rubber mounts reduce the transfer of road vibrations and shocks from the rear suspension system to the body. The stiff lateral movement in these bushings maintains lateral rear wheel position to reduce rear tire wear and improve vehicle stability.

Some multilink rear suspension systems have four hydraulic mounts mounted between the rear suspension member or frame and the chassis. The differential is bolted to the rear frame. These hydraulic mounts contain silicone oil that helps prevent noise, vibration, and shock transfer from the differential and suspension to the chassis and vehicle interior (**Figure 7-32**).

The rear suspension member is connected to the outer shell of the hydraulic mounts, and the inner shell is attached to the chassis. Silicone oil fills the area between the inner and outer shells in each mount (**Figure 7-33**). Noise, vibration, and shock are transferred from the differential and suspension to the rear suspension member, but the silicone oil in the hydraulic mounts prevents the transfer of these undesirable forces to the chassis and vehicle interior. These hydraulic mounts have superior noise and vibration dampening characteristics compared to rubber bushings.

Figure 7-32 Multilink independent rear suspension system with hydraulic mounts.

Figure 7-33 Hydraulic mount design.

Hydraulic bushing and mounts are gaining in popularity and may also be used as subframe mounts. Hydraulic bushings are tuned to the suspension system and are designed to work with the manufacturer tire and wheel combination to isolate noise frequencies and vibrations from the vehicle. Advancements in hydraulic bushings are in response to customer demands for quieter, better handling and smoother rides from today's vehicles. While inspecting the suspension system, these bushings should be inspected for leaks.

A rear upper link is connected from each side of the rear suspension member to the top of the knuckles. Both ends of this rear upper link contain rubber insulating bushings. The lower end of the shock absorber strut extends through a circular opening in the rear upper link. A front upper link is also connected from the rear suspension member to the knuckle. The top of the knuckle is supported by the front and rear upper links, rather than being supported by the shock absorber strut. The coil springs are mounted on the shock absorber struts, and the lower spring seat is attached to the strut. An upper spring seat is attached to the top of the shock absorber strut, and this upper seat is bolted into the inner fender reinforcement. The lower end of the shock absorber strut is connected to the back of the lower control arm. Lower control arms are connected from the rear suspension member to the lower end of the knuckles.

Some multilink rear suspension systems have a lower control arm and a toe control arm parallel to the lower control arm (**Figure 7-34**). A bolt in the outer end of the lower control arm extends through a bushing in the lower end of the steering knuckle. A ball joint in the outer end of the toe control arm is bolted into the lower end of the knuckle.

Figure 7-34 Multilink rear suspension with parallel lower control arm and toe control arm.

Rubber insulating bushings are pressed into the inner ends of the toe control arm and lower control arm. Bolts extend through these bushings and openings in the chassis. The rear end of the trailing arm is bolted to a bushing in the knuckle, and a bushing in the front end of this arm is bolted to the chassis. The trailing arm, lower control arm, and toe control arm prevent lateral and fore-and-aft wheel movement and provide excellent suspension rigidity when cornering or driving on irregular road surfaces. An eccentric cam bolt on the inner end of the toe control arm provides a rear wheel toe adjustment (**Figure 7-35**).

A bushing in the upper end of the knuckle is bolted to the outer end of the upper control arm (**Figure 7-36**). Front and rear bushings in the inner ends of the upper control arm are bolted to brackets that are in turn bolted to the chassis.

The one-piece rear wheel bearing assemblies are bolted to the knuckles (**Figure 7-37**). If the vehicle has an antilock brake system, the toothed ring is attached to the inner end of each rear wheel bearing and stub axle shaft, and a wheel speed sensor is mounted in the knuckle. The rear wheel bearings assemblies are non-serviceable.

Double Wishbone Rear Suspension

In the **double wishbone rear suspension system**, the upper and lower control arms are manufactured from lightweight, high-strength aluminum alloys designed for maximum strength and rigidity. These lighter control arms decrease the unsprung weight, which helps improve traction and ride quality. Suspension rigidity is also increased by positioning the ball joints and steering knuckle inside the wheel profile (**Figure 7-38**). Since the upper and lower control arms have a wishbone shape, the term *double wishbone* suspension system is used for this type of rear suspension. On each side of the car, the rear suspension is attached to the chassis by a cast aluminum subframe (**Figure 7-39**). This design also helps reduce vehicle weight and transmits suspension loads to the chassis at the most efficient locations.

Figure 7-35 Eccentric cam bolt on inner end of toe control arm.

Figure 7-36 Upper control arm and strut in multilink rear suspension.

Figure 7-37 One-piece rear wheel bearing assembly.

Aluminum alloy
lower control arm

Wishbone

Figure 7-38 Double wishbone rear suspension.

Subframe

Aluminum alloy
lower control arm

Wishbone

Figure 7-39 Double wishbone rear suspension and subframe.

Some independent rear suspension systems experience undesirable toe changes during wheel jounce and rebound. These toe changes cause vehicle instability during cornering and acceleration. In the double wishbone rear suspension system, the control arm design and the pivot locations on the rear toe control arm provide minimal change in toe-in during wheel jounce (**Figure 7-40**). This action results in extremely stable steering while cornering, accelerating, or driving on irregular road surfaces. The toe control arm may be lengthened or shortened to adjust rear wheel toe. An eccentric cam on the rear upper control arm bolt provides a rear wheel camber adjustment (**Figure 7-41** and **Figure 7-42**).

Independent Rear Suspension with Transverse Leaf Spring

Some cars have an independent rear suspension system with a transverse mono-leaf fiberglass spring. This type of spring is compact, lightweight, and corrosion free. Dual trailing arms are connected rearward from the chassis to the knuckle on each side of the suspension, and spindle support rods are attached from the center of the suspension to the bottom of the knuckle. Tie rods are connected from the rear of the knuckle to the center of the suspension (**Figure 7-43**).

Figure 7-40 Double wishbone rear suspension and toe control arm.

Figure 7-41 Rear suspension camber and toe adjustments.

Figure 7-42 Eccentric cam for camber adjustment on rear upper control arm bolt.

Figure 7-43 Independent rear suspension with mono-leaf transverse fiberglass spring.

Other independent rear suspension systems have a multiple-leaf transversely mounted rear spring. In these suspension systems, heavy control arms extend rearward from the chassis to the knuckles, and strut rods are connected from the bottom of the knuckles to the center of the suspension. The shock absorbers are connected from the lower end of the knuckles to the chassis. A suspension member connects the differential housing to the chassis. All suspension component mounting locations are insulated with rubber bushings.

CURB RIDING HEIGHT

Regular inspection and proper maintenance of suspension systems are extremely important to maintaining vehicle safety. *The curb riding height is determined mainly by spring condition.* Other suspension components such as control arm bushings will affect curb riding height if they are worn. Since incorrect curb riding height affects most of the other suspension angles, this measurement is critical. The curb riding height must be measured at the vehicle manufacturer's specified location, which varies depending on the type of suspension system. When the vehicle is on a level floor or an alignment rack, measure the curb riding height from the floor to the manufacturer's specified location (**Figure 7-44**).

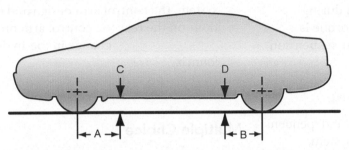

Figure 7-44 Front and rear curb ride height measurement locations.

SUMMARY

- A live axle rear suspension is one in which the differential housing, wheel bearings, and brakes act as a unit.

- A live axle rear suspension system provides good lateral stability, sway control, and steering characteristics, but this type of rear suspension causes increased tire wear with decreased ride quality and traction compared with other types of rear suspensions.

- Compared with other rear suspensions, the live axle leaf-spring suspension is more subject to axle tramp problems.

- In a live axle leaf-spring rear suspension system, the leaf springs absorb differential torque and provide lateral control.

- In a live axle coil spring rear suspension system, the lower control arms absorb differential torque and the upper arms control lateral movement.

- Live axle coil spring rear suspension systems may have a tracking bar to control lateral movement.

- A semi-independent rear suspension has a limited amount of individual rear wheel movement provided by a steel U-section channel or crossmember.

- In an independent rear suspension, each rear wheel can move individually without affecting the opposite rear wheel.

- Compared with a live axle rear suspension system, an independent rear suspension provides improved ride quality, steering control, tire life, and traction.

- In a MacPherson strut independent rear suspension, the coil springs are mounted on the struts.

- In a modified MacPherson strut independent rear suspension, the coil springs are mounted separately from the struts.

- In some independent rear suspension systems, the knuckle is positioned by a ball joint on the lower end and the strut on the upper end, and the coil spring is positioned between the lower control arm and the chassis.

- In a multilink independent rear suspension, the top of the knuckle is positioned by a rear upper link and a front upper link, and the lower end of the knuckle is positioned by a lower arm.

- Some rear-wheel-drive vehicles have an independent rear suspension with the differential connected through insulating mounts to the chassis.

- Some front-wheel-drive and rear-wheel-drive vehicles have an independent rear suspension with a mono-leaf or multiple-leaf transversely mounted leaf spring.

REVIEW QUESTIONS

Short Answer Essays

1. Describe a live axle rear suspension system.

2. Explain the disadvantages of a live axle leaf-spring rear suspension.

3. Describe the purpose of a leaf-spring shackle.

4. Explain the differential torque action during acceleration, and describe how this torque is absorbed in a live axle coil spring rear suspension.

5. Define axle tramp.

6. Describe the purpose of a tracking bar.

7. Explain the difference between a semi-independent and an independent rear suspension system.

8. Explain the advantages of an independent rear suspension compared with a live axle rear suspension.

9. Describe the advantage of mounting the differential to the chassis in a rear-wheel-drive car with an independent rear suspension.

10. Describe the components that position the upper end of the knuckle in a multilink rear suspension.

Fill-in-the-Blanks

1. A live axle leaf-spring rear suspension provides excellent _____ stability.

2. During braking and deceleration, the front of the differential is twisted_____.

3. Axle tramp occurs during hard_____.

4. In many live axle coil spring rear suspensions, the differential torque is absorbed by the trailing lower_____ _____.

5. In a modified MacPherson strut independent rear suspension, the coil spring seat is located on the_____ _____.

6. In a multilink independent rear suspension, the rear suspension member supports the suspension and the_____.

7. In some multilink rear suspension systems, the rubber mounting bushings between the suspension member and the chassis are _____ in the fore and aft direction and _____ in the lateral direction.

8. A fiberglass mono-leaf rear spring is compact, lightweight, and _____ free.

9. When the rear springs are sagged, the caster angle on the front suspension becomes more_____.

10. In the double _____rear suspension system, the control arm design and the pivot locations on the rear toe control arm provide _____ change in toe-in during wheel jounce.

Multiple Choice

1. Rear axle tramp is caused by:
   A. Irregular road surfaces.
   B. A bent rear control arm.
   C. Engine torque transmitted through the drive shaft.
   D. Improper rear wheel alignment.

2. All of these statements about a live axle leaf-spring rear suspension are true EXCEPT:
   A. During acceleration the front of the differential twists upward.
   B. The differential torque in a live axle rear suspension is applied to the springs.
   C. While decelerating and braking, the front of the differential twists downward.
   D. This type of rear suspension has a small amount of unsprung weight.

3. Rear axle tramp occurs during:
   A. Hard acceleration.
   B. Deceleration.
   C. High-speed driving.
   D. Cornering at high speed.

4. In a semi-independent rear suspension system, some individual wheel movement is provided by:
   A. The U-section channel and integral stabilizer bar.
   B. The struts.
   C. The trailing arms.
   D. The track bar and brace.

5. In a semi-independent rear suspension system:

   A. The track bar and brace absorb differential torque.

   B. The trailing arms prevent lateral wheel movement.

   C. The lower coil spring seats may be on the trailing arms.

   D. The lower end of the shock absorbers may be attached to the track bar.

6. Advancements in hydraulic bushings are in response to customer demands for:

   A. Quieter suspension.

   B. Lower ride height.

   C. Smoother ride quality.

   D. Improved performance.

7. All of these statements about modified MacPherson strut suspension systems are true EXCEPT:

   A. The lower control arms prevent lateral wheel movement.

   B. The tie rods control fore-and-aft wheel movement.

   C. The strut is bolted to the lower end of the spindle.

   D. The upper strut mount is retained on top of the strut.

8. In an independent rear suspension with a lower control arm and ball joint:

   A. The ball joint is pressed into the tie rod.

   B. The strut is bolted to the top of the knuckle.

   C. The lower coil spring seat is mounted on the strut.

   D. The suspension adjustment link is connected from the lower control arm to the chassis.

9. Sagged rear springs may cause:

   A. Slow steering wheel return after turning a corner.

   B. Decreased steering effort.

   C. Decreased positive caster on the front wheels.

   D. Harsh ride quality.

10. The adjustment link in a multilink rear suspension:

   A. Provides a rear wheel caster adjustment.

   B. Provides a rear wheel toe adjustment.

   C. Reduces fore-and-aft rear wheel movement.

   D. Absorbs engine torque transmitted through the drive shaft.

CHAPTER 8
STEERING COLUMNS AND STEERING LINKAGE MECHANISMS

Upon completion and review of this chapter, you should be able to understand and describe:

- How the steering column provides driver safety during a frontal collision.
- Two methods of steering column movement to protect the driver in a frontal collision.
- The purpose of a clock spring electrical connector.
- The mechanism that locks the steering wheel and gear shift when the ignition switch is in the lock position.
- A parallelogram steering linkage and explain the advantage of this type of linkage.

- The components in a parallelogram steering linkage.
- The purposes of the pitman arm and the idler arm.
- Two possible rack and pinion steering gear mountings.
- The rack and pinion steering linkage, and explain the advantages of this type of linkage.
- The design and operation of an active steering column.
- The design and operation of a driver protection module.

Terms To Know

Air bag deployment module	Idler arm	Steering linkage
Center link	Parallelogram steering linkage	Silencer
Clock spring electrical connector	Pitman arm	Spherical bearing
Energy-absorbing lower bracket	Pre-safe systems	Spiral cable
	Pyrotechnic	Tie rod
	Rack and pinion	Toe plate

INTRODUCTION

Steering columns play a significant part in steering control, safety, and driver convenience. The steering column connects the steering wheel to the steering gear. The column components must be in satisfactory condition to minimize free play and to provide adequate steering control. If steering column components such as universal joints are worn, column free play is excessive and steering control is reduced.

Most steering columns provide some method of column collapse during a collision. Some vehicles have plastic pins that shear off in the column jacket, gearshift tube, and steering shaft if the driver hits the steering wheel in a frontal collision. This shearing action of the plastic pins allows the column to collapse away from the driver. In other vehicles, the column-to-instrument panel mounting is designed to allow column movement if the driver hits the steering wheel in a collision. This action helps prevent driver injury.

All air bag-equipped vehicles have a driver's side air bag located in the upper side of the steering wheel, and most vehicles also have a passenger's side air bag mounted in the instrument panel. Some vehicles now have seat belt pretensioners that tighten the seat belts and hold the driver or passengers back against the seat if the vehicle is involved in a collision. Many vehicles now have side air bags mounted in the outer edge of the seat back, or an air bag curtain mounted just above the front and rear door openings. The side air bags have a separate module and sensors, and deploy only when the vehicle is involved in a side collision.

A recent development on some vehicles is the installation of **pre-safe systems** that react during the few milliseconds before a collision occurs to increase driver and passenger safety. The input signals to the pre-safe system include vehicle speed, braking torque, brake pedal application speed, wheel slip, vehicle acceleration around the vertical axis, spring compression and rebound travel, steering wheel rotational speed, and tire pressure. The pre-safe system recognizes and acts in three crucial situations:

1. Sideways skidding
2. Avoidance maneuvers
3. Emergency braking beyond ABS operation

The pre-safe system can differentiate between a drama and a crisis. For example, if the vehicle is driven over a patch of ice and the brakes are applied lightly without any vehicle skidding, the ABS system can operate to prevent wheel lockup, but the pre-safe system recognizes this condition as a drama and remains inoperative. The pre-safe system also remains inoperative during mild avoidance maneuvers.

When the inputs indicate an impending crisis driving situation, the pre-safe system performs these functions:

1. Pretensions the driver and front passenger seat belts so the driver and front passenger are optimally restrained in their seats during emergency braking or a skid prior to a collision. When a collision occurs, pyrotechnic seat belt tensioners provide increased seat belt tension. If a collision does not occur, the seat belts return to normal tension.
2. If the front passenger seat is moved too far forward, the pre-safe system moves the seat rearward before a collision occurs. If appropriate, the front passenger seat cushion and backrest are moved to a position that provides increased passenger protection if a collision occurs. For example, a steeply inclined seat backrest or a flat seat cushion may impair passenger restraint during a collision.
3. The electronically adjustable individual rear seat cushions are moved to the best possible angle to improve rear passenger restraint.
4. The sliding sunroof is closed to reduce the risk of occupant injury if a rollover occurs.

The brake assist system (BAS) also plays a crucial role in the pre-safe system operation. The BAS system operates beyond the ABS and electronic stability program (ESP) parameters. If the BAS detects an unstable driving condition such as severe vehicle skidding, it uses brake intervention and engine power reduction to correct the situation.

When the vehicle is involved in a frontal collision above a specific speed, the air bag sensor signals deploy the air bag or bags in a few milliseconds. As the driver is thrust forward during the frontal collision, the inflated air bag prevents the driver from striking the steering wheel, windshield, or instrument panel. The air bag deflates very quickly so it does not block the driver's vision in case the driver is still attempting to steer the vehicle. Since the driver's side air bag and some connecting components are mounted in the steering column, this column is a very important part of vehicle safety equipment.

Shop Manual
Chapter 8, page 310

60 mm (2.36 in.)

20 mm (0.79 in.)

Figure 8-1 Tilt steering column.

Steering columns may be classified as non-tilting, tilting, and tilting or telescoping. Tilt steering columns facilitate driver entry to and exit from the front seat. These columns also allow the driver to position the steering wheel to suit individual comfort requirements (**Figure 8-1**). A tilting or telescoping steering column allows the driver to tilt and extend or retract the steering wheel. In this type of steering column, the driver has more steering wheel position choices.

NON-TILT STEERING COLUMN

Design

The **toe plate** surrounds the steering column and covers the opening where the column extends through the vehicle floor.

The **silencer** is mounted with the toe plate and helps to reduce the transmission of engine noise to the passenger compartment.

Many steering columns contain a two-piece steering shaft connected by two universal joints. A jacket and shroud surround the steering shaft. The upper shaft is supported by two bearings in the jacket. A **toe plate**, seal, and **silencer** surround the lower steering shaft and cover the opening where the shaft extends through the floor (**Figure 8-2**). The lower steering shaft is surrounded by a shield underneath the toe plate.

The lower universal joint couples the lower shaft to the stub shaft in the steering gear. In some steering columns, a flexible coupling is used in place of the lower universal joint (**Figure 8-3**).

Studs and nuts retain the steering column bracket to the instrument panel support bracket. The steering column is designed to protect the driver if the vehicle is involved in

Toe plate

Silencer

Coupling assembly

Steering column assembly

Pin

Steering gear

Seal

Shield

Figure 8-2 Toe plate and silencer.

Figure 8-3 Flexible coupling.

a frontal collision. An **energy-absorbing lower bracket** and lower plastic adapter are used to connect the steering column to the instrument panel mounting bracket. This bracket allows the column to slide down if the driver is thrown forward into the wheel in a frontal collision. The mounting bracket is also designed to prevent rearward movement toward the driver in a collision.

Shop Manual
Chapter 8, page 315

In some steering columns, the outer column jacket is a two-piece unit retained with plastic pins (**Figure 8-4**). In this type of column, the lower steering shaft is a two-piece sliding unit retained with plastic pins (**Figure 8-5**). When the driver is thrown against the steering wheel in a frontal collision, the plastic pins shear off in the lower steering shaft and the outer column jacket. The shearing action of the plastic pins allows the steering column to collapse away from the driver, which reduces the impact as the driver hits the steering wheel.

A few current vehicles have steering columns with magnesium jackets or housings to reduce vehicle weight. Vehicle weight reduction improves fuel economy and reduces emissions.

An adaptive steering column has recently been introduced to the automotive market on some new models. The adaptive steering column collapses at two different speeds based on information received about the driver. Many vehicles have sensors in the lower

Figure 8-4 Injection plastic in collapsible outer steering column jacket.

Figure 8-5 Injection plastic in collapsible lower steering shaft.

part of a front seat to determine the weight of the driver and the weight and the presence of a passenger. The driver's weight signal, and other input signals are used by the restraints control module (RCM) to vary the air bag deployment force. The driver's weight input signal also determines how quickly the steering column should collapse. The adaptive steering column tailors the steering column collapse load to the driver's mass, seat belt use, and seat track position.

An adaptive steering column contains an energy-absorbing steel that buckles between the upper and lower portions of the steering column. The RCM receives input signals from various crash sensors indicating the severity of the crash, and this module determines the speed of steering column collapse. If input signals indicate a faster steering column collapse is required, the RCM fires a **pyrotechnic** device in the steering column that pulls a pin in the column and allows the energy-absorbing steel to buckle and provide faster column collapse. The result is a softer impact between the driver and the steering wheel. The adaptive steering column is designed to operate with the driver's air bag.

When the driver is thrown against the steering wheel during a collision, many steering wheels are designed to deform away from the driver to reduce the force on the driver's body.

⚠ **WARNING** Small amounts of sodium hydroxide are a by-product of an air bag deployment. Sodium hydroxide is a caustic chemical that causes skin irritation and eye damage. Always wear eye protection and gloves when servicing and handling a deployed air bag.

On many cars, the **air bag deployment module** is mounted in the top of the steering wheel (**Figure 8-6**). A **clock spring electrical connector**, or **spiral cable**, is mounted under the steering wheel. This component contains a ribbon-type conductor that maintains constant electrical contact between the air bag module and the air bag electrical system during steering wheel rotation.

The steering wheel splines fit on matching splines on the top of the upper steering shaft, and a nut retains the wheel on the shaft. Most steering wheels and shafts have matching alignment marks that must be aligned when the steering wheel is installed.

An ignition switch cylinder is usually mounted in the upper right side of the column housing, and the ignition switch is bolted on the lower side of the housing (**Figure 8-7**). An operating rod connects the ignition switch cylinder to the ignition switch. Ignition switches are integral with the lock cylinder in some steering columns.

A **pyrotechnic** device contains an explosive and an ignition source.

An **air bag deployment module** contains the air bag and the inflator device.

The clock spring maintains positive electrical contact between the air bag module and the air bag electrical system regardless of steering wheel position. A clock spring electrical connector may be called a coil, **spiral cable**, or cable reel, depending on the manufacturer.

Figure 8-6 Air bag inflator module mounted in the steering wheel and clock spring electrical connector located under the steering wheel.

Figure 8-7 Ignition switch and ignition switch cylinder mounted in steering column.

Figure 8-8 Turn signal switch, hazard warning switch, dimmer switch, and wipe or wash switch mounted in steering column.

Figure 8-9 Upper steering column with locking plate and lever.

The turn signal switch and hazard warning switch are mounted on top of the steering column under the steering wheel. Lugs on the bottom of the steering wheel are used to cancel the signal lights after a turn is completed. On many vehicles, the signal light lever also operates the wipe or wash switch and the dimmer switch (**Figure 8-8**).

If the gear shift is mounted in the steering column, a tube extends from the gear shift housing to the shift lever at the lower end of the steering column. This shift lever is connected through a linkage to the transaxle or transmission shift lever. A lock plate is attached to the upper steering shaft, and a lever engages the slots in this plate to lock the steering wheel and gear shift when the gear shift is in Park and the ignition switch is in the Lock position (**Figure 8-9**).

TILT STEERING COLUMN

Design

Tilt steering columns have a short upper steering shaft connected to the steering wheel in the usual manner. This upper steering shaft is connected through a universal joint to the lower steering main shaft. An upper column tube surrounds the upper steering shaft, and the lower steering main shaft is supported on bushings in the lower column tube (**Figure 8-10**).

The combination turn signal, hazard warning, wipe or wash, and dimmer switch are sometimes called a smart switch.

Multiaxial movement refers to movement in any direction around a pivot point.

⚠ **Caution**

When servicing a vehicle equipped with an air bag or bags, follow all service precautions in the vehicle manufacturer's service manual. Failure to follow these precautions may result in an expensive accidental air bag deployment.

Shop Manual
Chapter 8, page 317

Figure 8-10 Tilt steering column components.

When the steering wheel is tilted, the wheel and upper steering column tube pivot on two bolts connected between the upper and lower column tubes. During the steering wheel tilting motion, the upper steering shaft pivots on the universal joint connected between the upper steering shaft and the lower steering main shaft. A release lever on the side of the steering column must be activated to allow the wheel and upper column to tilt. This steering wheel action allows the driver to position the steering wheel for greater comfort and easier movement in and out of the driver's seat.

In some steering columns, a **spherical bearing** acts as a pivot between the upper steering shaft and the lower steering main shaft (**Figure 8-11**).

Four bolts attach the lower column support tube to the instrument panel. The ignition lock cylinder and switch are mounted in an upper bracket that is clamped to the lower column tube. A universal joint is connected between the lower steering main shaft and the intermediate shaft assembly, which is attached to the stub shaft on the steering gear (**Figure 8-12**).

Figure 8-11 Spherical bearing acts as a pivot between the upper steering shaft and the lower steering main shaft in some tilt steering columns.

Figure 8-12 Tilt steering column with universal joint, intermediate shaft, combination switch, and steering wheel.

Figure 8-13 Clock spring electrical connector, or spiral cable, maintains electrical contact between the air bag inflator module and the air bag electrical system.

A steering wheel pad containing the air bag deployment module is mounted in the top of the steering wheel. Electrical contact between the air bag deployment module and the air bag electrical system is maintained by a clock spring electrical connector, or spiral cable, mounted directly under the steering wheel (**Figure 8-13**). A combination switch is mounted directly under the clock spring electrical connector. This switch contains the signal light switch, hazard light switch, dimmer switch, and wipe or wash switch. The upper side of the steering wheel also contains the horn switch contacts.

ELECTRONIC TILT AND TELESCOPING STEERING COLUMN

Some vehicles are equipped with an electronically controlled tilt and telescoping steering column. A driver-operated switch mounted in the steering column below the signal light lever controls the tilt and telescoping functions (**Figure 8-14**). A tilt and telescoping motor is mounted in the steering column power assembly. A potentiometer is mounted

Figure 8-14 Tilt and telescoping steering column switch.

Telescoping motor

Steering column position sensor

Figure 8-15 Tilt and telescoping motor and steering column position sensor.

Tilt cable

Figure 8-16 Tilt cable for tilt and telescoping steering column.

on the side of the tilt and telescoping motor, and this potentiometer is a steering column position sensor (**Figure 8-15**). This potentiometer sends a voltage signal to the driver position module (DPM) in relation to the motor and steering wheel position. The steering column position sensor signal is sent to the DPM, and the DPM uses this signal when storing and recalling steering column memory settings. A short tilt cable assembly is connected between the tilt and telescoping motor and the tilt mechanism in the steering column (**Figure 8-16**).

When the tilt and telescoping switch is operated by the driver, it sends voltage input signals to the DPM. If the driver operates the tilt and telescoping switch to request the telescoping function, the DPM operates the tilt and telescoping motor to move the steering wheel toward or away from the driver depending on the switch request. When the driver operates the tilt and telescoping switch to request the tilt function, the DPM operates the tilt and telescoping motor to tilt the wheel upward or downward depending on the switch request. The DPM is interconnected via data links to some of the other onboard computers. Therefore, the DPM can receive input signals from other computers such as the driver door switch (DDS) module via the data links. When the ignition switch is turned off and the driver pushes the unlock button on the driver's door, the DPM tilts the steering column to a position that allows easier exit from the driver's seat. When the driver enters the vehicle and turns on the ignition switch, the DPM moves the steering column back to the previous setting.

ACTIVE STEERING COLUMN

The active steering column in some vehicles actively adjusts the energy-absorbing capability of the steering column using a pyrotechnic actuator. This type of steering column has a section of energy-absorbing steel that holds the upper and lower halves of the steering column together (**Figure 8-17**). The energy-absorbing steel section is designed to control the collapse of the steering column during impacts.

A RCM operates a pyrotechnic actuator in the steering column. Input sensors inform the RCM regarding driver weight, front seat position, driver seat belt usage, and crash severity. When the RCM receives input signals indicating softer steering column collapse is necessary to protect the driver, the RCM fires the pyrotechnic device in the steering column. This action pulls a pin out of the column, and reduces the columns' resistance to collapse. The steering column collapse is designed to operate with the air

Figure 8-17 Active steering column.

bag level that is being activated. The ease of steering column collapse is different for a belted or unbelted driver. The active steering column helps the vehicle manufacturers to meet enhanced federal safety regulations, which require crash protection for 5th percentile, smaller female drivers, and 50th percentile, large male drivers with and without seat belts.

DRIVER PROTECTION MODULE

In the driver protection module, the steering column, knee bolster, and pedals are mounted in a module that allows these components to move away from the driver in a controlled manner during a vehicle crash (**Figure 8-18**). The driver protection module contains steel tubes supported by aluminum extrusions. This design allows the steering column, knee bolster, and pedals to move along the trajectory of the driver during a severe vehicle crash. This action helps to maintain air bag position. During a vehicle crash, the driver protection module movement may be controlled actively by a pyrotechnic device operated by an electronic module. The driver protection module will provide adequate crash protection for drivers from the 5th percentile to the 95th percentile. It is expected that this level of protection will be required by federal legislation in the future.

Figure 8-18 Driver protection module.

Rack and Pinion Steering

Figure 8-19 The rack and pinion steering linkage is used with rack and pinion steering gears.

STEERING LINKAGE MECHANISMS

Rack and Pinion Steering Linkage

The **rack and pinion steering linkage** is used with rack and pinion steering gears (**Figure 8-19**). In this type of steering gear, the rack is a rod with teeth on one side. This rack slides horizontally on bushings inside the gear housing. The rack teeth are meshed with teeth on a pinion gear, and this pinion gear is connected to the steering column. When the steering wheel is turned, the pinion rotation moves the rack sideways. Tie rods are connected directly from the ends of the rack to the steering arms. An inner tie rod end connects each tie rod to the rack, and bellows boots are clamped to the gear housing. The bellows boots keep dirt out of these joints (**Figure 8-20**). The inner tie rod end contains a spring-loaded ball socket. The outer tie rod ends connected to the steering arms are basically the same as those in parallelogram steering linkages (**Figure 8-21**). Some inner tie rod ends contain a bolt and a bushing. These tie rod ends are threaded onto the rack (**Figure 8-22**). Since the rack is connected directly to the tie rods, the rack replaces the center link in a parallelogram steering linkage.

Shop Manual
Chapter 8, page 327

Figure 8-20 Rack and pinion steering gear.

Figure 8-21 Inner tie rod and outer tie rod end, rack and pinion steering.

Figure 8-22 The inner tie rod connects directly to the rack and pinion steering gear and is protected by a bellows boot.

Some inner tie rod ends have a mirror-finished ball and a high-strength polymer bearing to ensure low torque, minimal friction, and extended life (**Figure 8-23**). A hardened alloy steel rod extends from the ball to the outer tie rod end and provides maximum strength and durability.

The rack and pinion steering gear may be mounted on the front subframe (**Figure 8-24**) or attached to the firewall behind the engine (**Figure 8-25**). Rubber

Figure 8-23 Inner tie rod design.

Figure 8-24 Rack and pinion steering gear mounting on front crossmember.

Figure 8-25 Rack and pinion steering gear mounted on cowl.

bushings insulate the steering gear, and these bushings are attached to the subframe or firewall. The rack and pinion steering gear is mounted at the proper height to position the tie rods and lower control arms parallel to each other. The number of friction points is reduced in a rack and pinion steering system, and this system is light and compact. Most of today's cars, sport utility vehicles (SUVs) and light trucks use rack and pinion steering. Since the rack and pinion is linked directly to the steering arms, this type of steering linkage provides good road feel and responsiveness, but does not provide as much road isolation as parallelogram steering.

Parallelogram Steering Linkage

A **parallelogram steering linkage** may be defined as one in which the tie rods are mounted parallel to the lower control arms.

Steering linkage mechanisms are used to connect the steering gear to the front wheels. A **parallelogram steering linkage** may be mounted behind the front suspension (**Figure 8-26**) or in front of the front suspension (**Figure 8-27**). The parallelogram steering linkage must not interfere with the engine oil pan or chassis components.

⚠ **WARNING Always remember that customers' lives may depend on the condition of the steering linkages on their vehicles. State safety inspections play a very important role in maintaining suspension, steering, and other vehicle systems in safe driving condition and saving lives. During undercar service, always make a quick check of the steering linkage condition.**

Figure 8-26 Parallelogram steering linkage behind the front suspension.

Figure 8-27 Parallelogram steering linkage in front of the front suspension.

Regardless of the parallelogram steering linkage mounting position, this type of steering linkage contains the same components. The main components in this steering linkage mechanism are:

1. Pitman arm
2. Center link
3. Idler arm assembly
4. Tie rods, inner
5. Tie rods, outer

Parallelogram steering linkages are found on independent front suspension systems (**Figure 8-28**). In a parallelogram steering linkage, the tie rods are connected parallel to the lower control arms. Road vibration and shock are transmitted from the tires and wheels to the steering linkage, and these forces tend to wear the linkages and cause steering looseness. If the steering linkage components are worn, steering control is reduced. Since loose steering linkage components cause intermittent toe changes, this problem increases tire wear. The wear points in a parallelogram steering linkage are the tie rod sockets and ends, idler arm, and center link end.

Tie Rods

The **tie rod** assemblies connect the center link to the steering arms, which are bolted to the front steering knuckles. In some front suspensions, the steering arms are part of the

Figure 8-28 In a parallelogram steering linkage, the tie rods are connected parallel to the lower control arms.

steering knuckle; in other front suspension systems, the steering arms are bolted to the knuckle. On current parallelogram steering systems, the inner tie rods (**Figure 8-29**) and the outer tie rods are similar to those found on rack and pinion steering systems. The inner tie rod is threaded into the center link assembly and the outer tie rod is threaded onto the inner tie rod, there is not a separate adjuster sleeve (**Figure 8-30**). On older design inner tie rods, a ball socket is mounted on the inner end to each tie rod, and a tapered stud on this socket is mounted in a center link opening. A lock nut or castellated nut and cotter pin retain the tie rods to the center link. A threaded sleeve is mounted on the outer end of each tic rod, and a tie rod end is threaded into the outer end of this sleeve (**Figure 8-31**).

Some outer tie rod ends have a ball stud that is surrounded by an upper hardened steel bearing and a high-strength polymer lower bearing seat (**Figure 8-32**). The hardened steel

Figure 8-29 Inner tie rod design, which threads into the center link assembly, and outer tie rod threads onto inner tie rod, similar to a rack and pinion design.

Figure 8-30 On many current design parallelogram steering systems, the inner tie rod is threaded into the center link assembly and the outer tie rod is threaded onto the inner tie rod; there is no separate adjuster sleeve.

Outer tie rod

Steering arm

Castellated nut and cotter pin

Center link

Inner tie rod

Adjusting sleeve

Outer tie rod

Figure 8-31 Tie rod design.

Exclusionary seal

Hardened steel bearing provides strength and durability under any load condition

Smooth "full ball" stud

High-strength polymer lower bearing seat for smooth movement

Self-adjusting wear take-up internal spring ensures constant tension.

Figure 8-32 Outer tie rod end with hardened steel upper bearing and high-strength polymer lower bearing.

upper bearing provides strength and durability, and the polymer lower bearing seat provides smooth rotation of the ball stud in the tie rod end. An internal spring between the polymer lower bearing seat supplies self-adjusting action and constant tension on this seat. A seal in the upper part of the ball joint housing seals the ball stud to prevent contaminants from entering the tie rod end. These tie rod ends are installed on some original

Smooth "full ball"
stud

Rubber
bushing

Figure 8-33 Outer tie rod end with rubber-encapsulated ball stud.

equipment manufacturers' vehicles, and they are available as replacement tie rod ends on most vehicles.

Some cars have rubber-encapsulated outer tie rod ends in which a rubber bushing surrounds the lower end of the ball stud (**Figure 8-33**). Special service procedures required on these tie rod ends are explained in the Shop Manual. Similar outer tie rod ends are used on parallelogram steering linkages and rack and pinion steering linkages.

Each tie rod sleeve contains a left-hand and a right-hand thread where it is threaded onto the tie rod end and the tie rod. Therefore, sleeve rotation changes the tie rod length and provides a toe adjustment. Clamps are used to tighten the tie rod sleeves. The clamp opening must be positioned away from the slot in the tie rod sleeve. The design of the steering linkage mechanism allows multiaxial movement, since the front suspension moves vertically and horizontally. Ball-and-socket-type pivots are used on the tie rod assemblies and center link.

If the front wheels hit a bump, the wheels move up and down and the control arms move through their respective arcs. Since the tie rods are connected to the steering arms, these rods must move upward with the wheel. Under this condition, the inner end of the tie rod acts as a pivot, and the tie rod also moves through an arc. This arc is almost the same as the lower control arm arc because the tie rod is parallel to the lower control arm. Maintaining the same arc between the lower control arm and the tie rod minimizes toe change on the front wheels during upward and downward wheel movement. This action improves the directional stability of the vehicle and reduces tread wear on the front tires.

> Steering wheel free play is the amount of steering wheel movement before the front wheels start to move to the right or left.

AUTHOR'S NOTE It has been my experience that premature wear on outer tie rod ends and other pivot points in a steering linkage is most commonly caused by lack of lubrication or contamination that enters through broken seals. If the tie rod ends and other pivot points have grease fittings, lubrication at the manufacturer's recommended interval is important to maintain component life. However, you should not over-lubricate the steering linkage pivot points with a high-pressure grease gun, because this may rupture the seals. The steering linkage components are exposed to a large amount of water and dirt contamination. If the seals are leaking on any of the linkage pivot points, the pivots will soon be contaminated with moisture and dirt that acts as an abrasive to shorten component life.

Pitman Arm

The **pitman arm** connects the steering gear to the center link. This arm also supports the left side of the center link. Motion from the steering wheel and steering gear is transmitted to the pitman arm, and this arm transfers the movement to the steering linkage. This pitman arm movement forces the steering linkage to move to the right or left, and the linkage moves the front wheels in the desired direction. The pitman arm also positions the center link at the proper height to maintain the parallel relationship between the tie rods and the lower control arms.

Wear-type pitman arms have ball sockets and studs at the outer end, and this stud fits into the center link opening (**Figure 8-34**). The ball stud and socket are subject to wear, and pitman arm replacement is necessary if the ball stud is loose in the pitman arm. A non-wear pitman arm has a tapered opening in the outer end. A ball stud in the center link fits into this opening. The non-wear pitman arm only needs replacing if the arm is damaged or bent in a collision. The opening in the inner end of both types of pitman arms has serrations that fit over matching serrations on the steering gear shaft. A nut and lock washer retain the pitman arm to the steering gear shaft.

Idler Arm

An idler arm support is bolted to the frame or chassis on the opposite end of the center link from the pitman arm. The **idler arm** is connected from the support bracket to the center link. Two bolts retain the idler arm bracket to the frame or chassis. In some idler arms, a ball stud on the outer end of the arm fits into a tapered opening in the center link (**Figure 8-35**), whereas in other idler arms, a ball stud in the center link fits into a tapered opening in the idler arm.

🖋📖 A BIT OF HISTORY

For many years, rear-wheel-drive vehicles were equipped with parallelogram steering linkages. The smaller, lighter front-wheel-drive cars introduced in large numbers in the late 1970s and 1980s required a more compact, lighter steering gear. The rack and pinion steering gear is used almost exclusively on front-wheel-drive cars because of its reduced weight and space requirements.

Figure 8-34 Pitman arm design.

Figure 8-35 Idler arm design.

The idler arm supports the right side of the center link and helps maintain the parallel relationship between the tie rods and the lower control arms. The outer end of the idler arm is designed to swivel on the idler arm bracket, and this swivel is subject to wear. A worn idler arm swivel causes excessive vertical steering linkage movement and erratic toe. This action results in excessive steering wheel free play with reduced steering control and front tire wear.

Some idler arms contain an upper and lower gusher bearing that surrounds the bearing surface on the lower end of the bracket. A newly designed idler arm has a one-piece powdered metal gusher bearing that extends the full length of the friction surface on the lower end of the bracket (**Figure 8-36**). Compared with previous designs, this new design has 50 percent more bearing surface, which provides extended service life. A Belleville washer installed below the gusher bearing maintains a precision preload on the thrust washer, bracket, and gusher bearing.

Other idler arms have a conical machined surface on the lower end of the bracket. This conical surface is seated on a matching conical surface on the bearing (**Figure 8-37**). A coil spring between the lower end of the bracket and the housing maintains constant upward pressure on the bracket to maintain minimal endplay and constant turning torque as the bearing wears. Since this type of idler arm maintains minimal end play, more precise front suspension alignments are possible.

Figure 8-36 Idler arm internal design.

Figure 8-37 Idler arm with conical bearing.

Center Links

The **center link** controls the sideways steering linkage and wheel movement. The center link together with the pitman arm and idler arm provides the proper height for the tie rods, which is very important to minimizing toe change on road irregularities. Some center links have tapered openings in each end, and the studs on the pitman arm and idler arm fit into these openings. This type of center link may be called a taper end, or non-wear, link. Other wear-type center links have ball sockets in each end with tapered studs extending from the sockets (**Figure 8-38**). These tapered studs fit into openings in the pitman arm and idler arm, and they are retained with castellated nuts and cotter pins.

A **center link** may be referred to as a drag link, a steering link, or an intermediate link.

Haltenberger

A Haltenberger steering system is similar to the parallelogram and is used with a steering gear box. It was once popular on some light trucks, but over the last several decades this style steering system has primarily been used on just a few Jeep platforms. When the steering wheel is turned, the pitman arm will rotate. The pitman arm is connected to the drag link. The drag link is either connected to the tie rod assembly or directly to an arm on the left steering knuckle (**Figure 8-39**) depending on design. The drag link movement causes the tie rod assembly to move both the left and right steering arms thus turning the wheels. Some advantages to this system include its high strength, and the fact that it is less expensive than parallelogram linkage, as it has fewer parts. A major disadvantage is its less precise geometry.

STEERING DAMPER

A steering damper, or stabilizer, may be found on some parallelogram steering linkages. The steering damper is similar to a shock absorber. This component is connected from one of the steering links to the chassis or frame (**Figure 8-40**). When a front wheel strikes a road irregularity, a shock is transferred from the front wheel to the steering linkage, steering gear, and steering wheel. The steering damper helps absorb this road shock and

Figure 8-38 Center link design.

Figure 8-39 On a Haltenberger style steering system, the drag link is either connected to the tie rod assembly or directly to an arm on the left steering knuckle. The figure shows three variants of the design.

Figure 8-40 Steering gear damper and linkage.

prevent it from reaching the steering wheel. Heavy-duty steering dampers are available for severe road conditions such as those sometimes encountered by four-wheel-drive vehicles.

SUMMARY

- Steering columns help provide steering control, driver convenience, and driver safety. Many steering columns provide some method of energy absorption to protect the driver during a frontal collision.

- Steering wheels and columns now contain an air bag deployment module to protect the driver in a frontal collision.

- Tilt steering columns increase driver comfort and ease while driving or getting in or out of the driver's seat.

- A clock spring electrical connector supplies positive electrical contact between the air bag module in the steering wheel and the air bag electrical system.

- The ignition switch, dimmer switch, signal light switch, hazard switch, and wipe or wash switch may be mounted in the steering column.

- When the ignition switch is in the Lock position, a locking plate and lever in the upper steering column locks the steering wheel and the gear shift.

- In some tilt steering columns, the upper column housing pivots on two bolts, and the upper steering shaft pivots on a universal joint.

- In a parallelogram steering linkage, the tie rods are parallel to the lower control arms. The parallelogram steering linkage minimizes toe change as the control arms move up and down on road irregularities.

- A rack and pinion steering linkage has reduced friction points; it is lightweight and compact compared with a parallelogram steering linkage.

REVIEW QUESTIONS

Short Answer Essays

1. Explain how a collapsible steering column protects the driver in a frontal collision.

2. Explain how the driver's side air bag protects the driver in a frontal collision.

3. Describe the purpose of a clock spring.

4. List the switches commonly found in a steering column.

5. Describe the type of mechanism used to lock the steering wheel and gear shift when the ignition is in the Lock position.

6. Describe the pivot points in the upper shaft and upper column tube in a tilt steering wheel.

7. List the wear points in a parallelogram steering linkage.

8. List the five main components in a parallelogram steering linkage, and explain the purpose of each component.

9. Describe the basic design of a rack and pinion steering linkage.

10. Explain the advantages of a rack and pinion steering linkage compared with a parallelogram steering linkage.

Fill-in-the-Blanks

1. In some collapsible steering columns, _____ in the outer column jacket and steering shaft shear off if the driver is thrown against the steering wheel in a frontal collision.

2. The driver's side air bag protects the driver if the vehicle is involved in a _____ collision above a specific speed.

3. In an air bag system, a _____ in the steering column maintains positive electrical contact between the air bag module and the air bag electrical system as the steering wheel is rotated.

4. In a parallelogram steering linkage, the center link connects the pitman arm to the _____.

5. In a parallelogram steering linkage, the pitman arm and the _____ position the center link and tie rods at the correct height.

6. In a wear-type pitman arm, a _____ is positioned in the outer end of the pitman arm.

7. In a parallelogram steering linkage, the tie rods are parallel to the _____.

8. The clamp opening in a tie rod sleeve must be positioned away from the _____ in the tie rod sleeve.

9. In a rack and pinion steering system, the rack is connected directly to the _____.

10. Compared with a parallelogram steering linkage, a rack and pinion steering linkage has a reduced number of _____ points.

Multiple Choice

1. A typical air bag deployment time is:
   A. 1.5 minutes.
   B. 1 minute.
   C. 30 seconds.
   D. 40 milliseconds.

2. Many collapsible steering columns have:
   A. Plastic pins in the two-piece outer jacket.
   B. A collapsible bellows in the two-piece outer jacket.
   C. Steel pins in the two-piece lower steering shaft.
   D. A rubber spacer in the two-piece lower steering shaft.

3. The clock spring electrical connector:
   A. Maintains electrical contact between the air bag inflator module and the air bag electrical system.
   B. Is mounted above the steering wheel.
   C. Contains three spring-loaded copper contacts.
   D. Provides electrical contact between the signal light switch and the signal lights.

4. An active steering column has all of these components EXCEPT:
   A. A pyrotechnic actuator.
   B. A section of energy-absorbing steel.
   C. A telescoping cylinder.
   D. A pull-out pin that allows easier column collapse.

5. All of these statements about rack and pinion steering linkages are true EXCEPT:
   A. The tie rods are parallel to the lower control arms.
   B. The tie rod position depends on the steering gear mounting.
   C. The tie rods are connected to the pinion in the steering gear.
   D. The outer tie rod ends connect the tie rods to the steering arms.

6. All these statements about parallelogram steering linkages are true EXCEPT:

   A. Tie rod sleeves have the same type of thread in both ends of the sleeve.

   B. Loose steering linkages may cause excessive tire tread wear.

   C. Loose steering linkage causes excessive steering wheel free play.

   D. The pitman arm helps maintain the proper center link and tie rod height.

7. While discussing idler arms:

   A. Some idler arms contain a tapered roller bearing.

   B. The idler arm bracket is bolted to the upper control arm.

   C. A worn idler arm has no effect on front wheel toe.

   D. A partially seized idler arm bearing increases steering effort.

8. In a parallelogram steering linkage, the tie rods are parallel to the lower control arms to:

   A. Improve ride quality.

   B. Provide longer steering linkage life.

   C. Extend shock absorber and spring life.

   D. Reduce toe change during upward and downward front wheel movement.

9. A rack and pinion steering gear:

   A. Has tie rods that connect the rack directly to the steering arms.

   B. May be bolted to the vehicle frame.

   C. Has inner tie rod ends that are pressed onto the rack.

   D. Has more friction points compared with a parallelogram steering linkage.

10. While discussing steering linkages and dampers:

   A. In a rack and pinion steering gear, the rack positions the tie rods parallel to the lower control arms.

   B. A rack and pinion steering gear is used on most rear-wheel-drive cars.

   C. A steering damper is used on many front-wheel-drive vehicles.

   D. A defective steering damper may cause excessive steering effort.

CHAPTER 9

FOUR-WHEEL ALIGNMENT, PART 1 PRIMARY ANGLES

Upon completion and review of this chapter, you should be able to understand and describe:

- The effect of excessive positive camber on front suspension systems.
- Camber changes during front wheel jounce and rebound travel.
- The relationship between camber and vehicle directional stability.
- How front suspension camber and caster may be adjusted to compensate for road crown.
- The effects of positive and negative caster on directional control and steering effort.
- Positive and negative caster as they relate to ride quality.
- How higher- or lower-than-specified front or rear suspension height affects front suspension caster.
- Toe-in and toe-out on front suspension systems.
- The toe-in settings required on front-wheel-drive and rear-wheel-drive vehicles.
- Tire tread wear caused by excessive toe-in.
- The customer complaints that may arise from incorrect rear wheel toe and thrust line adjustments.
- Suspension and chassis defects that may cause improper rear wheel toe, thrust line, or camber.

- The variables that affect wheel alignment.
- Wheel alignment, and explain five reasons for performing a wheel alignment.
- Why four-wheel alignment is essential on front-wheel-drive unitized body cars.
- Thrust line and geometric centerline and describe the effect of an improper thrust line on vehicle steering.
- The result of improper rear wheel toe and its effect on the thrust line.
- The causes of improper rear wheel toe.
- The causes of rear axle offset.
- Rear axle sideset, setback, and dog tracking.
- Front and rear tire tread wear caused by inaccurate rear wheel toe and thrust line settings.
- Geometric centerline alignment and explain the problem with this type of alignment.
- Thrust line alignment and describe the shortcoming of this type of alignment.
- Total four-wheel alignment and explain the advantage of this type of alignment.
- The safety hazards created by incorrect wheel alignment or worn suspension and steering components.

Terms To Know

Bump steer
Directional stability
Four-wheel alignment
Geometric centerline

Geometric centerline alignment
High-frequency transmitter
Jounce

Memory steer
Motorist Assurance Program (MAP)
Negative camber

Negative caster	Road variables	Toe-out
Positive camber	Steering drift	Torque steer
Positive caster	Steering pull	Tracking
Rear axle offset	Steering wander	Wheel alignment
Rear axle sideset	Thrust angle	Wheel sensors
Rebound	Thrust line	Wheel setback
Receiver	Thrust line alignment	Wheel shimmy
Road crown	Toe-in	

INTRODUCTION

Automotive engineers design suspension and steering systems that provide satisfactory vehicle control with acceptable driver effort and road feel. The vehicle should have a tendency to go straight ahead without being steered. This tendency is called **directional stability**. A vehicle must have predictable directional control, which means the steering must provide a feeling that the vehicle will turn in the direction steered. The wheels must be reasonably easy to turn and tire wear should be minimized. These steering qualities and tire conditions are achieved when front and rear **wheel alignment** angles are within the vehicle manufacturer's specifications.

The condition of suspension system components and wheel alignment is extremely important to maintaining driving safety and normal tire wear. Worn suspension components such as tie rod ends, ball joints, and control arms can suddenly fall apart and cause complete loss of steering. This disastrous event may result in not only some very expensive property damage, but also the loss of human life. When alignment angles are incorrect, an uncontrolled vehicle swerve or skid may occur during hard braking, resulting in a serious accident. Severe misalignment may reduce tire life to one-third of the normal expected tire life with correct alignment. After suspension components such as ball joints or control arms are replaced, wheel alignment is essential.

Technicians must be familiar with the symptoms that indicate incorrect rear wheel alignment. The rear wheel alignment procedures required to correct improper rear wheel alignment must be understood.

CAMBER FUNDAMENTALS

Camber

Camber is defined as the inward or outward tilt of the tire-and-wheel assembly as viewed from the front of the vehicle (**Figure 9-1**). A reference line begins at the base (ground) center of the tire and wheel and extends upward at a 90-degree angle to the road. Put another way, camber refers to the tilt of a line through the tire and wheel centerline in relation to the true vertical centerline of the tire and wheel. **Positive camber** is obtained when the top of the tire and wheel is tilted outward, away from the true vertical line of the wheel assembly. **Negative camber** occurs when the tire and wheel centerline tilts inward in relation to the wheel assembly true vertical centerline. The camber angle is measured and referred to the same for both the front and rear tires, meaning negative camber is the inward tilt of the tire at the top whether it is in the front or the rear of the vehicle.

Camber angle is displayed in degrees on an alignment computer. Camber is an angle that can affect both tire wear and directional stability, such as a pull or drift. If the side-to-side difference in front camber is greater than 0.5 to 0.75 degree, the vehicle may exhibit a pull or drift to the most positive side (**Figure 9-2**) as a general rule. This left

---

Wheel alignment may be defined as an adjustment and refitting of suspension parts to original specifications that ensure design performance.

Shop Manual
Chapter 9, page 373

Figure 9-1 Negative and positive camber angles on a MacPherson strut front suspension system.

Figure 9-2 Cross camber is the side-to-side difference in camber measurements. This vehicle will pull right toward the most positive side.

Left camber   **Right camber**

1.0° | | −1.0°   −1.0° | | 1.0°
0.6°   −0.7° | | 0.7°   **−0.4°**
1.0°

Cross camber

Figure 9-3 The left and right camber preferred specification is 0 degree and the tolerance is + or − 1.0 degree in this example. The actual left camber reading is positive 0.6 degree and the actual right camber reading is −0.4 degree. The maximum allowable cross camber is 0.7 degree with an actual cross camber of 1.0 degree.

Figure 9-4 Tire shoulder wear because of excessive camber.

side to right side difference is referred to as **cross camber** and is shown on the alignment computer's readings screen (**Figure 9-3**). Some manufactures specify a maximum allowable amount of cross camber. Other manufactures do not specify an acceptable amount and the general assumption as stated above is that a vehicle may exhibit a drift if the cross camber difference is between 0.5 and 0.75 degrees and a pull if the cross camber difference is greater than 0.75 degree toward the side with the higher camber reading. Rear camber differences do not cause vehicles to pull or drift, but do cause tire wear just as errors in the front.

Camber can also cause smooth, tapered tire wear to either the outer or inner edge of the tire tread around the entire tire's circumference, and is typically isolated to the tread shoulder area of the tire (**Figure 9-4**). Too much positive camber will cause outside edge wear to the tire shoulder (**Figure 9-5**). Too much negative camber will cause inside edge wear to the tire shoulder.

A camber specification and tolerance is specified by the vehicle manufacturer for both the front and rear (**Figure 9-6**). The specification is the preferred angle and the tolerance

Figure 9-5 Tire shoulder wear because of excessive positive camber.

Vehicle Specifications and Tolerances		
Front	**Spec.**	**Tol.**
Left Camber	0.00°	1.00°
Right Camber	0.00°	1.00°
Cross Camber		0.70°
Left Ca		00°
Right C		00°
Cross C		70°
Total To	1.0° −0.0° −1.0°	25°
Left SA		°
Right S		°
Rear		
Camber	−0.40°	0.75°
Total Toe	0.20°	0.30°
Thrust Angle		0.20°

Figure 9-6 Typical vehicle specification and tolerance data sheet with an alignment screen bar graph superimposed for camber.

Figure 9-7 Negative and positive camber angles on a short-and-long arm front suspension system.

is the permitted amount that can be increased (+) or decreased (−) from the preferred angle. The tolerance is a means provided by the manufacturer to fine tune the suspension system to specific driving habits. It also allows for some change and wear that may naturally occur between alignments without a noticeable impact on directional stability or excessive tire wear.

During an alignment, camber is considered a live measurement because the computer is continually updating the sensor information on the computer measurement screen. Some alignment machines will allow you to adjust camber with the weight on or off the wheels using a program designed to allow the vehicle to be lifted. Hunter Engineering calls this program "Jack-Up Selected Axle."

The camber may be adjusted with the cam on the upper strut-to-steering knuckle bolt on some MacPherson strut front suspension systems. Negative and positive camber angles are the same on a short-and-long arm front suspension system as they are on a MacPherson strut suspension system (**Figure 9-7**).

Front cradle position is extremely important to provide proper front wheel camber on a front-wheel-drive vehicle. For example, if the vehicle is involved in a side collision and most of the collision force is supplied to the front wheel and chassis, the cradle may be pushed sideways. On the side of the vehicle impacted during the collision, the front suspension and cradle may be pushed inward, and the other side of the cradle may be forced outward. When the front suspension and cradle are moved inward, the camber on that side of the vehicle becomes more positive. When the cradle, suspension, and bottom of the wheel on the opposite side are moved outward, the camber is moved toward a negative position. Often this is noted as an equal and opposite difference in side-to-side camber.

DRIVING CONDITIONS AFFECTING CAMBER

Jounce and Rebound

Upward wheel movements are referred to as **jounce**, whereas downward wheel movements are termed **rebound**. On most modern suspension systems during wheel jounce, the top of the wheel moves outward and creates a more positive camber angle (**Figure 9-8**). During wheel rebound, it is desirable to have very little camber change to minimize tire tread wear. A short-and-long arm front suspension system will have a slightly negative camber during wheel rebound travel (**Figure 9-9**).

Figure 9-8 Camber change during wheel jounce.

Figure 9-9 Camber change during wheel rebound.

Cornering Forces

During hard cornering at high speeds, centrifugal force attempts to move the vehicle sideways to the outside of the turn. Under this condition, more vehicle weight is transferred to the side of the vehicle on the outside of the curve. Therefore, the front suspension on the outside of the curve is forced downward, whereas the front suspension on the inside of the curve is lifted upward. When this action occurs, the camber on the inside wheel becomes less positive and the inside edge of the tire grips the road surface to help prevent sideways skidding (**Figure 9-10**).

Simultaneously, the camber on the outside wheel moves to a slightly more positive position. This action on the outside wheel causes the outside edge of the tire to grip the road surface, which helps prevent sideways skidding (**Figure 9-11**). Frequent, hard, high-speed cornering causes wear on the edges of the tire treads.

Tire Tread Wear

The camber angle may be referred to as one of the tire wear alignment angles. When the front wheels are adjusted to the manufacturer's specified camber setting, the front wheels will remain at, or very close to, the 0-degree camber position during average driving conditions. Therefore, maximum tire tread life and directional stability are maintained.

If a front wheel has excessive positive camber, the wheel is tilted outward and the vehicle weight is concentrated on the outside edge of the tire. Under this condition, the

Inside wheel less positive camber

Figure 9-10 While cornering at high speed, the camber on the inside front wheel becomes less positive and the inside edge of the tire grips the road surface to resist lateral skidding.

Outside wheel positive camber

Figure 9-11 While cornering at high speed, the camber on the outside front wheel moves to a slightly more positive position and the outside edge of the tire grips the road surface to resist lateral kidding.

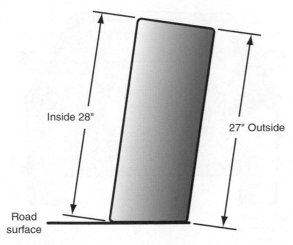

Figure 9-12 Excessive positive camber causes a smaller diameter on the outside edge of the tire compared to the inside edge of the tire.

Figure 9-13 Tire tread wear caused by incorrect camber adjustment.

outside edge of the tire has a smaller diameter than the inside tire edge. Therefore, the outside tire edge has to complete more revolutions to travel the same distance as the inside tire edge. Because both edges are on the same tire, the outside edge must slip and scuff on the road surface as the tire and wheel revolve (**Figure 9-12**).

Excessive negative camber tilts the wheel inward and concentrates the vehicle weight on the inside edge of the tire. This condition causes wear and scuffing on the inside edge of the tire tread (**Figure 9-13**). Therefore, correct camber adjustment is extremely important to providing normal tire tread life.

Shop Manual
Chapter 9, page 359

⚠ **WARNING** **Improper camber adjustment may result in rapid tire tread wear and steering pull, which is a vehicle safety hazard.**

Road Crown

A wheel that is tilted tends to steer in the direction it is tilted (**Figure 9-14**). For example, a bicycle rider tilts the bicycle in the direction he or she wishes to turn, making the turning process easier.

When camber angles are equal on both front wheels, the camber steering forces are equal, and the vehicle tends to maintain a straight-line position. If the camber on the front wheels is significantly unequal, the vehicle will drift to the side with the greatest degree of positive camber.

Figure 9-14 A wheel turns in the direction of the tilt.

Figure 9-15 A crowned road surface causes the vehicle steering to pull to the right.

Right front wheel
camber = 1/8°+
caster = 1°+

Left front wheel
camber = 3/8°+
caster = 1°+

Figure 9-16 Camber setting to offset road crown.

Crowned highway design prevents water buildup on the driving surface. When a vehicle is driven on a crowned road, it is actually driven on a slight slope, which causes the vehicle steering to pull toward the right (**Figure 9-15**). Some car manufacturers use the pulling effect of camber to offset this pull to the right caused by road crown. In these vehicles, the left front wheel may have 0.25 to 0.5 degree more positive camber than the right front wheel.

When the camber on the front wheels is adjusted to the manufacturer's specifications, vehicle directional stability is maintained. If the camber adjustment on the left front wheel is used to offset road crown, the caster on both front wheels must be the same (**Figure 9-16**).

CASTER FUNDAMENTALS

Caster Definition

When viewed from the side, caster is the angle formed when a line is drawn through the upper and lower pivot points in reference to true vertical. On a MacPherson strut suspension system, the caster line intersects the center of the upper strut mount and lower ball joint (**Figure 9-17**). While on a short-and-long arm suspension system, the caster line intersects the center of the upper and lower ball joints (**Figure 9-18**). **Positive caster** occurs when the caster line is tilted backward toward the rear of the vehicle in relation to

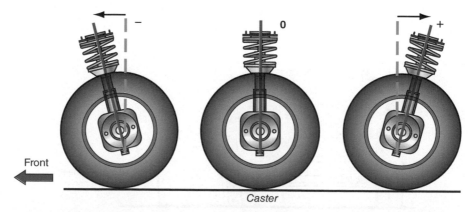

Front

Caster

Figure 9-17 On a vehicle with MacPherson strut front suspension, the caster line intersects the upper strut mount and lower ball joint.

Figure 9-18 On a vehicle with upper and lower ball joints, the caster line intersects the upper and lower ball joints.

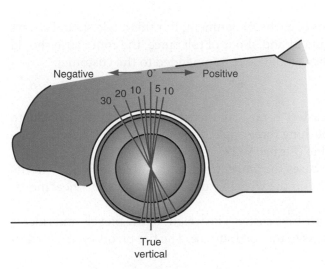

Figure 9-19 Positive and negative caster.

Figure 9-20 Cross caster is the side-to-side difference in caster and may cause a drift or pull to the side with the least positive caster.

the vertical centerline of the spindle and wheel viewed from the side (**Figure 9-19**). **Negative caster** occurs when the caster centerline is tilted toward the front of the vehicle in relation to the spindle and wheel vertical centerline.

Caster is not considered a direct tire wear angle, but is a directional stability angle and is used to improve cornering and steering wheel returnability. Caster is measured in degrees, and positive caster is specified on consumer production vehicles. Cross caster is the difference in caster between the right and left side of the vehicle (**Figure 9-20**). If one front wheel has more positive caster than the other front wheel, the steering pulls toward the side with the least amount of positive caster. In general, a vehicle may exhibit a drift if the cross caster difference is between 0.5 and 1.0 degrees and a pull if the cross caster difference is greater than 1.0 degree toward the side with the least positive caster. Rear caster is not measured.

The caster specification and tolerance are specified by the vehicle manufacturer for the front wheels (**Figure 9-21**). The specification is the preferred angle and the tolerance is the permitted amount that can be increased (+) or decreased (−) from the preferred angle. Like camber, the caster tolerance is a means provided by the manufacture to fine-tune the suspension system to specific driving habits. Many vehicles today do not allow

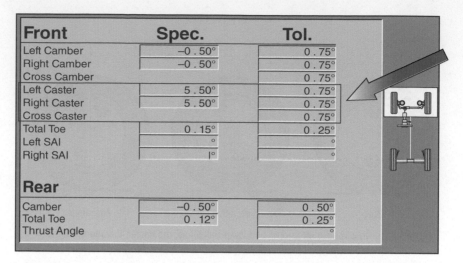

Front	Spec.	Tol.
Left Camber	−0.50°	0.75°
Right Camber	−0.50°	0.75°
Cross Camber		0.75°
Left Caster	5.50°	0.75°
Right Caster	5.50°	0.75°
Cross Caster		0.75°
Total Toe	0.15°	0.25°
Left SAI	°	°
Right SAI	|°	°
Rear		
Camber	−0.50°	0.50°
Total Toe	0.12°	0.25°
Thrust Angle		°

Figure 9-21 Typical alignment specification sheet for a vehicle. Actual specifications vary between year, make, and model of vehicle.

for a change in the caster angle as part of a wheel alignment. But this angle should always be measured as part of a wheel alignment and, if out of tolerance, the root cause should be determined and appropriate repairs should be recommended to the customer.

Effects of Positive Caster

If a piece of furniture mounted on casters is pushed, the casters turn on their pivots to bring the wheels into line with the pushing force applied to the furniture. Therefore, the furniture moves easily in a straight line (**Figure 9-22**).

Any force exerted on the pivot causes the wheel to turn until it is lined up with the force on the pivot, because the weight on the wheel results in resistance to wheel movement (**Figure 9-23**).

Most bicycles are designed with positive caster. The weight of the bicycle and rider is projected through the bicycle front forks to the road surface. The tire pivots on the vertical

Figure 9-22 Caster action on furniture.

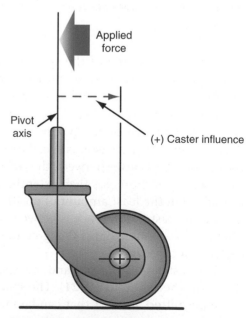

Figure 9-23 Furniture caster wheel aligned with the force on the pivot.

centerline of the spindle and wheel when the handlebars and front wheel are turned. Notice that the caster line through the center of the front forks is tilted rearward in relation to the vertical centerline of the spindle and wheel as viewed from the side (**Figure 9-24**).

Because the pivot point is behind the caster line where the bicycle weight is projected against the road surface, the front wheel tends to return to the straight-ahead position after a turn. The wheel also tends to remain in the straight-ahead position as the bicycle is driven. Therefore, the caster angle on a bicycle front wheel provides the same action as the caster angle on a piece of furniture.

Positive caster projects the vehicle weight ahead of the wheel centerline, whereas negative caster projects the vehicle weight behind the wheel centerline. Because positive caster causes a larger tire contact area behind the caster pivot point, this large contact area tends to follow the pivot point. This action tends to return the wheels to a straight-ahead position after a turn. It also helps maintain the straight-ahead position. Positive caster increases steering effort because the tendency of the tires to remain in the straight-ahead position must be overcome during a turn. The returning force to the straight-ahead position is proportional to the amount of positive caster. Positive caster helps maintain vehicle directional stability. Excessive positive caster is undesirable because it increases steering effort and creates a rapid steering wheel return.

Caster causes camber to change as the wheels are turned left or right. This effect is referred to as camber roll and is the effect that positive caster has on camber as the wheels are turned (**Figure 9-25**). If the caster angle were 0 degree, the front spindles would rotate horizontally in relation to the road surface. However, a positive caster angle causes the left front spindle to tilt toward the road surface during a left turn (**Figure 9-26**). Caster is the angle causing camber to increase or decrease when the wheels are turned, but camber is the angle causing the tire wear. This effect means that caster can be an indirect tire wear angle if the caster angle is excessively out of specifications. If caster is severely out of specifications, the tires may develop wear on both the inner and the outer

Wheel centerline → ← Pivot or caster centerline

Figure 9-24 Caster line on a bicycle.

Figure 9-25 Camber roll is the effect that positive caster has on camber during a turn.

During a left turn the spindle moves downward.

Figure 9-26 Left front spindle movement during a left turn.

Pivot axis ℄

(+) Caster

Road shock

Figure 9-27 Harsh riding quality caused by excessive positive caster.

edges. Because the front wheels are the drive wheels on a front-wheel-drive vehicle, camber roll wear does occur on the front tires as part of the normal tire wear pattern on the front tires but the wear is minimal. Inner and outer edge wear becomes very noticeable on the front tires of a front-wheel-drive car if they are not rotated regularly. The severity of front tire wear because of camber roll is directly proportional to the amount of positive caster present.

This downward spindle movement tends to drive the tire into the road surface. Because this action cannot take place, the left side of the suspension and chassis is lifted. When the driver begins to return the wheel to the centered position, gravity forces the vehicle weight to its lowest position, which helps return the steering wheel to the straight-ahead position. Excessive positive caster increases the left front spindle downward tilt during a left turn, which increases the suspension and chassis lift. Therefore, excessive positive caster increases steering effort. The same action occurs at the right front spindle, but this spindle tilts downward during a right turn.

Harsh riding may be caused by excessive positive caster because the caster line is actually aimed at some road irregularities (**Figure 9-27**). Excessive positive caster may cause the front wheels to shimmy from side to side at low speeds.

When the caster line is aimed directly at the road irregularity, road shock is transmitted through the ball joint and upper strut mount to the suspension and chassis. A small degree of positive caster or a negative caster setting allows the front wheel to roll into a road depression without the caster line being aimed at the hole in the road. Therefore, this type of caster line improves ride quality.

Improper front wheel caster may be caused by an improperly positioned front cradle on a front-wheel-drive car. For example, if one or both sides of the cradle are driven rearward in a front-end collision, the front wheel caster is moved from a positive position toward a negative position. This condition may cause steering wander, reduced directional control, and slower steering wheel return.

Some car manufacturers recommend less positive caster on the left front wheel than on the right front wheel to offset the steering pull to the right caused by the road crown. If the caster adjustment is used to compensate for road crown, the camber on both front wheels should be the same. The most important facts about positive caster are the following:

1. Positive caster helps the front wheels return to the straight-ahead position after a turn.
2. Correct positive caster provides improved directional stability of a vehicle.

Shop Manual
Chapter 9, page 387

3. Excessive positive caster produces harsh riding quality.
4. Excessive positive caster promotes sideways front wheel shimmy.
5. The left front wheel may be adjusted with less positive caster than the right front wheel to compensate for road crown.

Effects of Negative Caster

Negative caster moves the centerline of the upper strut mount and lower ball joint behind the vertical centerline of the spindle and wheel at the road surface. If this condition is present, the friction of the tire causes the tire to pivot around the point where the centerline of the upper strut mount and ball joint meets the road surface. When this occurs, the wheel is pulled away from the straight-ahead position, which decreases directional stability.

Negative caster reduces steering effort. Because excessive positive caster increases road shock transmitted to the suspension and chassis, negative caster reduces this shock and improves ride quality. This improvement occurs because the front wheel rolls into a road depression without the caster line being aimed at the hole in the road. But, if caster is below manufacturers' specifications, poor steering wheel return may be experienced as well as a wondering feeling when driving straight ahead because of reduced directional stability that may result from a low or negative caster setting. Caster is never specified as a negative number. This is why we say that caster may pull to the side with the "least" positive caster because it is always specified to be a positive setting.

Effects of Suspension Height on Caster

When the rear springs become sagged or overloaded, the caster on the front wheels becomes more positive. This action explains why front wheel shimmy may occur when a trunk is severely overloaded.

If the rear suspension height is above the vehicle manufacturer's specification, the caster on the front wheels becomes less positive. Under this condition, the front wheel caster may change from positive to negative. This explains why a vehicle may have reduced directional stability and control when the rear suspension height is raised.

On many short-and-long arm front suspension systems, the inner front end of the upper control arm is higher than the inner rear side of this arm where it is attached to the frame. This design causes the front wheel caster to become more positive if the front suspension height is lowered. When the front suspension height is above the vehicle manufacturer's specification, the front wheel caster becomes less positive.

Most front-wheel-drive cars have a MacPherson strut front suspension system with a slightly positive caster setting. Therefore, the top strut mount is tilted rearward. If the front suspension height is lowered on this type of front suspension system, the front wheel caster becomes less positive. Conversely, if the suspension height is raised, positive caster increases. The most important facts about negative caster include the following:

1. Negative caster does not help return the front wheels to the straight-ahead position after a turn.
2. Negative caster contributes to directional instability and reduced directional control.
3. Negative caster does not contribute to front wheel shimmy.
4. Negative caster reduces road shock transmitted to the suspension and chassis.

SAFETY FACTORS AND CASTER

Directional Control

As explained earlier in this chapter, positive caster provides increased directional stability and control, whereas negative caster reduces directional stability. Therefore, front wheel caster must be adjusted to the manufacturer's specifications to maintain vehicle directional control and safe handling characteristics.

Suspension Height

We have already explained how incorrect front or rear suspension height results in changes in front wheel caster. Therefore, abnormal suspension heights may contribute to reduced directional control and unsafe steering characteristics.

⚠ **WARNING** **Improper caster adjustment may result in decreased directional stability of the vehicle and reduced driving safety.**

TOE DEFINITION

Total toe is the measurement difference between the distances between the fronts of the tires compared to the distance between the rears of the same tires (**Figure 9-28**). When the distance between the rear inside tire edges is greater than the distance between the front inside tire edges (**Figure 9-29**), the front wheels have a **toe-in** setting and is more commonly referred to as positive toe and is displayed on an alignment computer in degrees (**Figure 9-30**). **Toe-out** occurs when the distance between the inside front tire edges exceeds the distance between the inside rear tire edges and is more commonly referred to as negative toe and is displayed on an alignment computer as a negative number in degrees (**Figure 9-31**).

A total toe specification and tolerance is specified by the vehicle manufacturer for both the front and the rear (**Figure 9-32**). The specification is the preferred angle and the tolerance is the permitted amount that can be increased (+) or decreased (−) from the preferred angle. Manufacturers generally publish the toe specifications as total toe. On most vehicles individual right and left toe is adjustable independently by changing the length of right and left tie rod assemblies. The individual toe specification is one-half the total toe specification and may be measured in either degrees or inches. Individual toe is the difference between the front of one tire and a reference line and the rear of the tire and the same reference line. The reference line is either the centerline or the thrust line of the vehicle, depending on the type of alignment being performed (**Figure 9-33**). Because the front tires are the steering wheels, front individual toe will attempt to equalize while driving. If one front tire's toe specification is out of adjustment, tire wear will occur

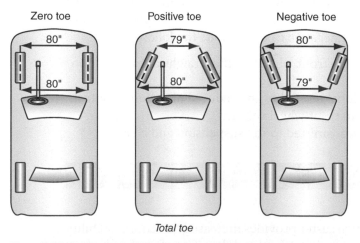

Figure 9-28 Total toe is the difference in distance measurements between the front measurement compared to the rear measurement.

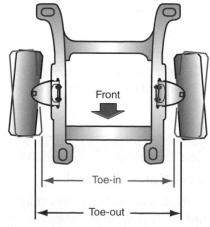

Figure 9-29 Toe-in and toe-out.

Toe-in

Toe-out

Figure 9-30 If distance between the rear inside tire edges is greater than the distance between the front inside tire edges, the front wheels have a positive (toe-in) setting.

Figure 9-31 If distance between the inside front tire edges exceeds the distance between the inside rear tire edges, the front wheels have a negative (toe-out) setting.

Figure 9-32 A total toe specification and tolerance is specified by the vehicle manufacturer for both the front and the rear. The individual toe specification is half (1/2) the total toe specification and may be measured in either degrees or inches.

Figure 9-33 Individual toe is the difference between the front of one tire and a reference line and the rear of the tire and the same reference line.

on both front tires, and the steering wheel will not be centered when driving straight ahead. Changes in front individual toe can occur as a result of road impacts such as curbs, potholes, and severally rough roads or from loose or damaged components. Rear toe, unlike front toe, cannot equalize between the two rear tires because they are not steering wheels and the linkage is independent of one another. A rear toe error will cause a dog track condition, which is discussed in further detail in the thrust angle alignment section, and tire wear to the effected wheel. On the rear tires, a toe angle error will cause the

Figure 9-34 When rear tire's toe angle error is extreme, it can cause a diagonal cupping often referred to as diagonal wipe on the tread surface.

effected tire to squirm as it rolls down the road and when extreme can cause a diagonal cupping, often referred to as diagonal wipe on the tread surface (**Figure 9-34**).

Toe is considered a major tire wear angle and incorrect total toe will wear tires extremely quickly. Toe angles that are outside the allowed tolerance will cause tire wear. Positive toe will wear the outside edge of the tire and negative toe will wear the inside edge of the tire. The wear will begin on the shoulder of the tire and taper toward the center of the tire.

TOE SETTING FOR FRONT-WHEEL-DRIVE AND REAR-WHEEL-DRIVE VEHICLES

On a front-wheel-drive vehicle, the front drive axle torque forces the front wheels toward an increased positive (toe-in) position. Therefore, car manufacturers usually specify a slight toe-out position for the front wheels on these vehicles.

On rear-wheel-drive vehicles, front tire friction on the road surface moves the wheels toward the toe-out position when the car is driven. On this type of vehicle, manufacturers usually specify a slight toe-in on the front wheels. The front wheels are adjusted to a slight toe-in or toe-out with the vehicle at rest, so the wheels will be parallel to each other when the vehicle is driven on the road. A slight amount of lateral movement always exists in steering linkages. Forces acting on the front wheels try to compress or stretch the steering linkages when the vehicle is driven. Whether a compressing or stretching action occurs on the steering linkage depends on whether the steering linkages are located at the rear edge or front edge of the front wheels.

Shop Manual
Chapter 9, page 390

TOE ADJUSTMENT AND TIRE WEAR

When the front wheel toe is not adjusted to the manufacturer's specifications, front tire tread wear is excessive. If one of the tie rods is not parallel to the lower control arms, toe change on the front wheels is not equal during front wheel jounce and rebound. This action may result in bump steer when a front wheel strikes a road irregularity. While checking front wheel toe, improper toe change during suspension jounce and rebound should be checked to be sure the tie rods are parallel to the lower control arms. Technicians must understand front wheel toe and improper toe changes.

Figure 9-35 Feathered tire tread wear caused by incorrect toe adjustment.

Ford Motor Company has calculated that a toe-in error of 1/8 in. (3.17 mm) is equivalent to dragging the tires crosswise for 11 ft (3.3 m) for each mile the vehicle is driven. This crosswise movement causes severe feathered tire tread wear (**Figure 9-35**). Improper toe adjustment is the most common cause of rapid tire tread wear.

Excessive toe-out causes wear on the inside of the tire tread ribs and a sharp feathered edge on the outside of the tread ribs. If excessive toe-in is present, the tire tread wear is reversed.

Worn steering linkage components such as tie rod ends cause incorrect and erratic toe-in settings. If the front springs become weak, the front suspension height is lowered. When this occurs, the pitman arm and the idler arm move downward with the chassis. This action moves the tie rods to a more horizontal position that tends to push outward on the steering arms and increases front wheel toe-in. The toe-in change just described occurs when the steering linkage is located at the rear of the front wheels.

 Caution
Improper toe adjustment causes rapid tire tread wear, which may result in tire failure.

⚡ **WARNING** **Worn steering components may suddenly become disconnected, causing complete loss of steering control, collision damage, and personal injury.**

STEERING TERMINOLOGY

In the automotive service industry, certain terms are used for specific steering problems. Some of these problems are related to camber and caster, others are caused by various suspension or steering defects. Technicians must be familiar with both the steering terminology and the causes of the problems.

A front-wheel-drive vehicle with unequal-length drive axles produces some **torque steer** on hard acceleration. Some front-wheel-drive vehicles, especially those with higher horsepower, have equal-length front drive axles to reduce torque steer. On a front-wheel-drive vehicle, torque steer is aggravated by different tire tread designs on the front tires or uneven wear on the front tires.

Bump steer occurs if the tie rods are not the same height, meaning one of the tie rods is not parallel to the lower control arm. This condition may be caused by a bent or worn idler arm or pitman arm on a parallelogram steering linkage. On a rack and pinion steering system, worn steering gear mounting bushings may cause this unparallel condition between one of the tie rods and the lower control arm.

Memory steer may be caused by a binding condition in the steering column or in the steering shaft universal joints. A binding upper strut mount may result in memory steer. Negative caster or reduced positive caster also causes memory steer.

Steering pull is the tendency of the steering to gradually pull to the right or left when the vehicle is driven straight ahead on a level road. Steering pull or drift may be caused by

Torque steer may be defined as the tendency of the steering to pull to one side during hard acceleration.

Bump steer is the tendency of the steering to veer suddenly in one direction when one or both of the front wheels strikes a bump.

When **memory steer** occurs, the vehicle doesn't want to steer straight ahead after a turn because the steering does not return to the straight-ahead position.

Shop Manual
Chapter 9, page 354

improper caster or camber adjustments. **Steering drift** is the tendency of the steering to slowly drift to the right or left when driving straight ahead on a level road.

When **steering wander** occurs, the vehicle tends to steer in either direction rather than straight ahead on a level road surface. Steering wander may be caused by improper caster adjustment.

WHEEL ALIGNMENT THEORY

Road Variables

Vehicles are subjected to many **road variables** that affect wheel alignment. These variables must be counteracted by the suspension design and alignment, or steering would be very difficult. Some of the variables that affect wheel alignment and suspension design follow:

Road variables are differences in road surface conditions, vehicle loads, and weather conditions.

Road crown refers to the high portion in the center of the road with a gradual slope to each side.

1. **Road crown** (the curvature of the road surface)
2. Bumps and holes
3. Natural crosswinds or crosswinds created by other vehicles
4. Heavy loads or unequal weight distribution
5. Road surface friction and conditions such as ice, snow, and water
6. Tire traction and pressure
7. Side forces while cornering
8. Drive axle forces in front-wheel-drive vehicles
9. Relationship between suspension parts, as the front wheels turn and move vertically when road bumps and holes are encountered

A desirable plan to reduce tire wear would be to place the front wheels and tires so they are perfectly vertical. The tires would then be flat on the road. However, if the wheels and tires are perfectly vertical, such variables as the driver entering the car, turning a corner, or adding weight to the luggage compartment would move the tire from its true vertical position. Therefore, tire wear and steering operation would be adversely affected.

Rather than allowing the variables to adversely affect tire wear and steering operation, the suspension and steering are designed with characteristics to provide directional stability, predictable directional control, and minimum tire wear. Wheel alignment angles are designed to provide these desired requirements despite road variables. Wheel alignment angles also control the **tracking** of the rear wheels in relation to the front wheels.

Vehicle **tracking** is the straightness of the rear wheels in relation to the front wheels.

A customer may enter your shop and request a wheel alignment for many reasons. The three most common reasons for customers requesting a wheel alignment are to straighten a steering wheel that is not centered (crooked), to prevent tire wear, or to address a vehicle handling issue such as a pull while driving straight ahead. In addition, a customer may request a wheel alignment for a steering wheel that is shaking. A shaking of the vehicle is not generally caused by a wheel alignment angle error. A tire or wheel balance or runout condition as well as bearing and axle faults are more likely causes of the steering wheel shake condition. Tire rotation and balancing are additional services that should be considered when a wheel alignment is performed, because customers are often expecting a smooth, vibration-free ride after an alignment is performed.

After an alignment is performed, the benefits your customer can expect are as follows:

- The steering wheel position is level and pointing straight down the road while driving the vehicle straight ahead.
- The steering wheel automatically returns to the straight-ahead position after making a right or left turning maneuver.

- The vehicle exhibits directional stability without pulling or wandering while driving straight ahead reducing driver fatigue.
- Improved tire tread life and fuel economy.
- Proper vehicle tracking whereby the front and rear wheels follow each other.
- Improved cornering performance, directional control, and feedback while maintaining optimum road isolation.

TYPES OF WHEEL ALIGNMENT

Geometric Centerline Alignment

Toe refers to the angle between the plane of a front or rear wheel and a reference line. Either the geometric centerline or the thrust line may be used as a toe reference line. When **geometric centerline alignment** is used as a reference for front wheel toe, the toe on each front wheel is adjusted to specifications using the geometric centerline as a reference (**Figure 9-36**). This type of wheel alignment has been used for many years. It may provide a satisfactory wheel alignment if the rear wheels are properly positioned and the thrust line is at the vehicle centerline. However, if the rear wheels are not positioned properly, the thrust line is not at the geometric centerline and steering problems will occur. Therefore, geometric centerline front wheel reference ignores rear wheel misalignment.

> In a **geometric centerline alignment**, the front wheel toe is adjusted using the geometric centerline as a reference.

Thrust Line Alignment

If **thrust line alignment** is used, the thrust line created by the rear wheels is used as a reference for front wheel toe adjustment (**Figure 9-37**). When the front wheel toe is adjusted with a thrust line reference, and this line is not at the geometric centerline, neither the front nor rear wheels are parallel to the geometric centerline. Under this

Figure 9-36 Front wheel alignment with the front wheel toe referenced to the geometric centerline.

Figure 9-37 Thrust line alignment with the front wheel toe referenced to the thrust line.

Figure 9-38 Positive and negative thrust line referenced to the vehicle centerline as viewed from above. A positive thrust line is to the right of the centerline and a negative thrust line is to the left of the centerline.

condition, none of the four wheels is facing straight ahead when the vehicle is driven straight ahead. This action results in excessive wear on all four tire treads. This type of alignment ensures a centered steering wheel when the vehicle is driven straight ahead.

A positive thrust line (angle) is to the right of the vehicle centerline when viewed from above and a negative thrust line (angle) is to the left of the vehicle centerline when viewed from above (**Figure 9-38**). If thrust angle error is excessive a visible dog track condition will be present (**Figure 9-39**). Dog tracking occurs when the front wheels follow the direction that the rear wheels are pointing. If excessive, the vehicle will appear to be going down the road sideways. The next time you are driving on the highway in a straight-ahead path pay attention to the vehicles in front of you and look at the rear and front tires on the same side of the vehicle. Look for a dog track condition; if you can see both the rear and front tire when viewed from behind, the vehicle is dog tracking. When viewing the actual alignment readings for a vehicle on a computer-based alignment information screen, pay attention to whether the thrust angle is a positive or negative number. This information can aid your diagnosis when dealing with frame and structural damage. The maximum allowable thrust angle can vary between manufacturers but is generally not more than 0.125 degree.

Figure 9-39 If a severe thrust angle condition exists a vehicle will dog track.

Four-Wheel Alignment

When a **four-wheel alignment** is completed, the thrust line position is measured in relation to the geometric centerline. Individual rear wheel toe-in is measured and adjusted to manufacturer's specifications. This adjustment moves the thrust line to the geometric centerline. Thrust angle is measured to be sure the thrust line is positioned at the geometric centerline. Front wheel toe is measured using the common geometric centerline and thrust line as a reference (**Figure 9-40**). When a four-wheel alignment is completed, all four wheels are parallel to the geometric centerline and the steering wheel is centered as the vehicle is driven straight ahead. Wheel aligners have different alignment capabilities. Modern computer wheel alignment systems have four-wheel alignment capabilities, and this type of wheel alignment must be performed on today's vehicles to ensure proper steering control and vehicle safety.

When performing a wheel alignment, the sequence that the angles are adjusted is critical, as a change in one angle may affect another angle. For example, a change in rear toe will change the position of the vehicles thrust angle, which is the reference line for front toe. So, a change in rear toe will also change the reading for front toe.

The sequence the wheel alignment angles must be adjusted in (if an angle adjustment is necessary) is:

1. Adjust rear camber (if a means of adjustment is available).
2. Adjust rear individual toe (if a means of adjustment is available).
3. Adjust front caster and camber (if a means of adjustment is available).
4. Adjust front individual toe.

The advantages of four-wheel alignment are the following:

1. *Improved fuel mileage.* After a four-wheel alignment, all four wheels are parallel, and this condition combined with proper tire inflation decreases rolling resistance, which improves fuel mileage.
2. *Longer tire life.* When all four wheels are aligned properly, tire tread wear is minimized.
3. *Improved vehicle handling.* When all four wheels are properly aligned and all steering and suspension components are in satisfactory condition, steering pulls, vibrations, and abnormal steering conditions are eliminated to ensure improved vehicle handling.
4. *Safer driving.* Proper alignment of all four wheels, plus inspection and replacement of all worn or defective steering and suspension components improves vehicle handling, and this reduces the possibility of a collision and provides safer driving.

Figure 9-40 Four-wheel alignment with the thrust line adjusted so it is at the geometric centerline and the front wheel toe is referenced to the geometric centerline.

IMPORTANCE OF FOUR-WHEEL ALIGNMENT

Shop Manual
Chapter 9, page 360

A unitized body may be called a unibody.

Until the late 1970s, most vehicles in the United States were rear-wheel-drive. The majority of these vehicles had one-piece rear axle housings and frames. This type of vehicle design did not experience many rear wheel alignment problems and rear wheel alignment was usually not a concern in these years.

Beginning in the late 1970s, many domestic car makers began manufacturing front-wheel-drive cars with unitized bodies. The gasoline shortages in the 1970s and the introduction of federal corporate average fuel economy (CAFE) laws brought about a massive change to lighter weight, more fuel-efficient front-wheel-drive cars. A significant number of these cars had four-wheel independent suspension systems.

Cars with unitized bodies and independent or semi-independent rear suspension systems are more likely to experience rear wheel alignment problems compared with rear-wheel-drive vehicles with frames and one-piece rear axle housings. This is especially true after collision damage. Therefore, with the introduction of unitized bodies in massive numbers, four-wheel alignment became a necessity.

REAR WHEEL ALIGNMENT AND VEHICLE TRACKING PROBLEMS

Result of Proper Rear Wheel Alignment

The thrust line is an imaginary line at a 90-degree angle to the centerline of the rear wheels and projected forward.

The vehicle geometric centerline is an imaginary line through the exact center of the front and rear wheels.

A rear axle offset refers to a condition where the complete rear axle housing has rotated slightly, moving one rear wheel forward and the opposite rear wheel backward. Under this condition, the rear wheels are no longer parallel to the geometric centerline of the vehicle.

The driver uses the steering wheel to turn the front wheels and steer the vehicle in the desired direction. However, the rear wheels determine the direction of the vehicle to a large extent. When the **thrust line** is positioned at the **geometric centerline** and the front and rear wheels are parallel to the vehicle geometric centerline, the vehicle moves straight ahead with minimum guidance from the steering wheel (**Figure 9-41**).

Result of Improper Rear Wheel Alignment

Rear Axle Offset A rear axle offset may occur on a rear-wheel-drive vehicle with a one-piece rear axle housing or on a front-wheel-drive car with a trailing arm rear suspension. If the rear axle is offset, the thrust line is no longer at the vehicle centerline (**Figure 9-42**). The **thrust angle** is the angle between the geometric centerline and the thrust line.

With the rear axle offset problem shown in Figure 9-42, the thrust line is positioned to the left of the geometric centerline. Under this condition, the rear wheels steer the vehicle in a large clockwise circle. If the driver's hands are removed from the steering wheel, the steering pulls to the right. This steering action is similar to the rear wheels on a forklift. The rear wheels are used to steer the forklift because of the heavy load on the front wheels. If the rear wheels are pointed toward the left, the forklift turns to the right.

Figure 9-41 Front and rear wheels parallel to the vehicle centerline.

Figure 9-42 Rear axle offset and improperly positioned thrust line.

On a vehicle with this problem, the left front wheel toes out and the right front wheel toes in when the vehicle is driven straight ahead. This front wheel situation occurs because the front wheels try to compensate for the rear suspension defect. The front wheels are turned slightly to the left to compensate for the drift to the right caused by the rear suspension problem. Under this condition, both the front and rear tires may have feathered tire wear.

The ideal correction for this rear suspension problem is to reposition the rear suspension so the thrust line is at the vehicle centerline, then set the front wheel toe-in to the thrust line. The rear suspension problem that we have described also causes the steering wheel to be off-center when the vehicle is driven straight ahead.

Some of the causes of rear axle offset are:

1. A broken center bolt in rear leaf spring.
2. Worn shackles in rear leaf springs.
3. A bent frame.
4. A bent subframe or floor section, unitized body.
5. Worn trailing arm bushings.
6. Bent trailing arms.

Improper Rear Wheel Toe If the left rear wheel has excessive toe-out, the thrust line is moved to the left of the geometric centerline (**Figure 9-43**). This defect has basically the same effect on steering as rear axle offset. Improper toe on one rear wheel is most often encountered on vehicles with independent rear suspension.

Some of the causes of improper toe on one rear wheel are:

1. A bent one-piece rear axle.
2. A bent U section in trailing arm suspension.
3. A bent rear lower control arm.
4. A worn rear lower control arm bushing.
5. A bent rear spindle.
6. An improper rear toe adjustment.

Shop Manual
Chapter 9, page 375

Figure 9-43 Excessive toe-out on the left rear wheel moves the thrust line to the left of the geometric centerline.

AUTHOR'S NOTE It has been my experience that the effect of rear wheel alignment on steering pull is sometimes ignored when diagnosing a customer complaint of steering pull. If the rear axle is offset or the rear wheel toe is incorrect, the rear wheel position actually pushes the front wheels away from the straight-ahead position and causes the steering to pull to one side. This problem is more likely to occur on front-wheel-drive cars, especially those with independent rear suspension and a unitized body design. Therefore, it is very important that you perform a four-wheel alignment, and be sure the thrust angle is within specifications before aligning the front suspension.

 A BIT OF HISTORY

Early attempts at rear wheel alignment were slow and lacked precision. These attempts at rear wheel alignment included the use of a track bar and even backing the rear wheels of a car onto a front wheel aligner to align the rear wheels. To meet the need for fast, accurate front and rear wheel alignment, wheel alignment manufacturers designed computer wheel aligners. The technology in this equipment has greatly improved since the first models were introduced.

The necessary correction for this defect is to adjust the left rear wheel toe to the manufacturer's specifications. This adjustment moves the thrust line to the geometric centerline of the vehicle (**Figure 9-44**).

Rear Axle Sideset When **rear axle sideset** occurs, the rear wheels are parallel to each other, but the rear axle assembly has moved straight sideways so the geometric centerline is no longer at the proper vehicle centerline position (**Figure 9-45**). Because the rear wheels do not follow directly behind the front wheels when the vehicle is driven straight ahead, the vehicle dog tracks. Rear axle sideset does not affect steering pull as much as rear axle offset, but sideset will cause steering pull if it is severe.

Wheel Setback **Wheel setback** is a condition where one front or rear wheel is moved rearward in relation to the opposite wheel (**Figure 9-46**). This condition is usually caused by front-end collision damage, and thus it is most likely to be a front suspension problem. However, setback may occur on independent rear suspension systems. A severe setback condition causes steering pull.

Geometric centerline

Thrust line

Toe adjustment

Figure 9-44 Adjusting toe to specifications on the left rear wheel moves the thrust line to the geometric centerline.

Figure 9-45 Rear axle offset occurs when the rear axle assembly has moved straight sideways, and the geometric centerline is shifted away from the proper vehicle centerline position.

Figure 9-46 Front wheel setback occurs when one front wheel is moved rearward in relation to the opposite front wheel.

COMPUTER ALIGNMENT SYSTEMS

Computer Wheel Aligner Features

Some computer wheel aligners have four high-resolution digital cameras that measure wheel target position and orientation. The front and rear wheel alignment angles are sensed by the digital cameras and wheel targets and then displayed on the wheel alignment monitor. The vehicle is raised to a comfortable working height on the aligner lift, and two digital cameras are mounted in each end of a crossbar on a post in front of the vehicle (**Figure 9-47**). The post and crossbar height may be adjusted to match the vehicle height.

One digital camera is aimed at each wheel target (**Figure 9-48**). A target is mounted on each wheel with a self-centering adapter. The adapters are adjustable to fit rims up to 24 in. (60 cm) in diameter. The targets contain a polished aluminum faceplate that acts as a mirror (**Figure 9-49**). The targets do not contain any electronics, glass, or cables, and they do not require calibration. The targets are also lightweight and do not require any maintenance.

Figure 9-47 Computer wheel aligner with digital cameras and wheel targets.

Figure 9-48 One digital camera is aimed at each wheel target.

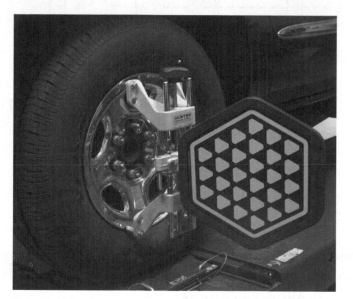

Figure 9-49 Each target contains a polished aluminum face plate with no electronics, glass, or cables.

This type of computer aligner is very simple to use. After the prealignment procedures and checks have been performed on the vehicle, and the vehicle is positioned at the desired height and position on the lift and turntables, the first step is to mount the wheel targets on the wheels. The second step is to roll the vehicle back a short distance until the onscreen indicators on the monitor turn green. The third step is to roll the vehicle forward and stop the vehicle so the front wheels are centered on the turntables. The fourth step is to observe all the front and rear camber and toe angles displayed on the monitor.

A remote control is wired into the monitor so the technician can perform the same functions when the monitor is not in the technician's view. A handheld remote control is used to measure the vehicle ride height (**Figure 9-50**). This remote is held against the

Figure 9-50 A handheld remote control transmits ride height measurements electronically to the console and monitor.

lower edge of the fender well to measure the ride height, and the remote transmits the ride height measurement electronically to the console and monitor.

On some computer wheel aligners, the **wheel sensors** contain a microprocessor and a **high-frequency transmitter** that acquire measurements and process data and then send these data to a **receiver** mounted on top of the wheel alignment monitor (**Figure 9-51** and **Figure 9-52**). This type of wheel sensor does not require any cables connected between the sensors and the computer wheel aligner. The data from this type of wheel sensor are virtually uninterruptible, even by solid objects. When these wheel sensors are stored on the computer wheel aligner, a "docking station" feature charges the batteries in the wheel sensors. The front wheel sensors have optical arms that project ahead of the tires to provide front toe readings. Some rear wheel sensors also have optical arms that project behind the rear tires (**Figure 9-53**). This type of rear wheel sensor measures rear wheel setback.

On some computer alignment systems, the reference signals between the wheel sensors are provided by light-emitting diodes (LEDs) or electronic signals. Cables are

Wheel sensors are mounted on each wheel, and these sensors transmit wheel alignment data to the computer wheel aligner.

A **high-frequency transmitter** is contained in many wheel sensors. This transmitter sends data to a receiver on top of the monitor.

A **receiver** on the computer wheel aligner receives signals from the wheel sensors.

Figure 9-51 Cordless wheel sensor containing a high-frequency transmitter.

Figure 9-52 Computer wheel aligner with data receiver on top of the monitor.

Figure 9-53 Front and rear wheel sensors with optical arms.

Figure 9-54 Front wheel sensor with optical arm and string connected to the rear wheel sensor.

connected from the wheel units to the computer aligner. Reference signals between the wheel units are provided by strings on some older model computer alignment systems (**Figure 9-54**). Most wheel sensors provide wheel runout compensation. Because the wheel unit is clamped to the rim, a bent rim will affect the alignment readings. On older computer wheel aligners, the technician had to perform a wheel runout check at each wheel. Newer wheel units use audible and visual prompts on the monitor screen to inform the technician if the wheel sensors require leveling or calibration. Wheel runout compensation may be completed by pressing a button on the wheel sensor. Wheel sensor leveling is done by adjusting the sensor level control.

Shop Manual
Chapter 9, page 363

Computer alignment systems have the capability to measure thrust angle and setback as well as other front and rear alignment angles. A typical computer wheel aligner has the following features:

1. Color monitor and computer (**Figure 9-55**).
2. Computer software program stored on a hard disc drive. This program contains vehicle four-wheel alignment procedures and diagnostic drawings.
3. Compact disc (CD) or digital video disk (DVD) drive. The vehicle specifications and alignment program may be contained on CDs or DVDs, and updates are possible by replacing the CDs or DVDs (**Figure 9-56**). Wheel alignment specifications include domestic and imported vehicles for the current year and a minimum of 10 previous years.

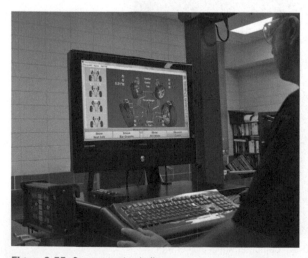

Figure 9-55 Computer wheel aligner.

Figure 9-56 CDs for wheel alignment program and specifications.

Figure 9-58 Tire inspection screen.

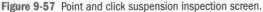

Figure 9-57 Point and click suspension inspection screen.

4. The software in some computer wheel aligners includes a CD or DVD image database containing 2100 digital photos of the adjustment and inspection points. In some computer aligners, a list of inspection points appears on the left of the screen. When the operator clicks on a listed inspection point, the matching photo appears, illustrating the selected inspection point (**Figure 9-57**). Special digital photos are available for inspection of the cradle and cradle-to-body alignment.

5. Live action inspection videos may be accessed instantly from the inspection screen. These inspection videos are part of the 142-video training library.

6. A tire inspection screen illustrates various types of tire tread wear and the related causes (**Figure 9-58**). This screen may be printed out and shown to the customer.

7. Measurement and adjustment screens illustrate suspension measurements. If a measurement is within specifications, the measurement is illustrated with a green bar and an arrow indicates the measurement. A red measurement bar indicates the adjustment is not within specifications (**Figure 9-59**). In the adjustment mode, the measurement remains on the screen, and as the necessary suspension adjustment is performed, the adjustment bar turns green when the arrow moves within specifications.

8. An alignment procedure bar is available in some computer aligners. This vertical bar contains an icon for each step in an alignment procedure (**Figure 9-60**). These

Figure 9-59 Wheel alignment adjustment screen.

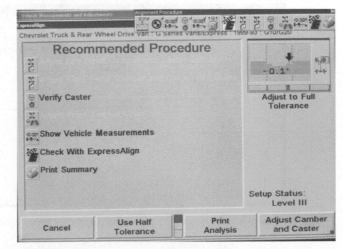

Figure 9-60 Wheel alignment procedure screen.

steps are arranged in the proper sequence. A check mark appears when a step has been completed, and the technician may select and click any step to move to that adjustment.

9. Correction kit videos may be accessed to illustrate the installation of aftermarket wheel alignment correction kits. Special tools videos are available in the computer wheel aligner software. These videos indicate the special wheel alignment adjustment tools required for the vehicle being aligned. Operation videos may be accessed to show the technician how to operate the computer wheel aligner.

10. Some computer wheel aligners have multilingual capabilities. These capabilities usually include three languages; however, up to 20 languages may be available from the equipment manufacturers.

11. Full-function keyboard allows the technician to enter such information as the customer's name and address and the make and year of the vehicle.

12. Set of four self-centering adjustable rim clamps and four-wheel sensors.

13. Full-function remote control that allows the technician to control the alignment program and see the results of suspension adjustments while working under the vehicle.

14. Color or laser printer provides detailed wheel alignment reports for the customer and specific alignment messages and diagrams for the technician.

15. In some computer wheel aligners, the **Motorist Assurance Program (MAP)** uniform inspection guidelines may be accessed for many steering, suspension, and brake components (**Figure 9-61**). The MAP program establishes uniform parts inspection guidelines to improve customer satisfaction with the automotive industry.

16. A variety of software programs are available with some computer wheel aligners. Much of the software may be extra-cost options. This software may include the following:

 ■ Wheel alignment software on CD or DVD: This software provides the technician with a very extensive vehicle information database as well as many patented adjustment and productivity features (**Figure 9-62**).

 ■ Specific computer software provides service bulletins and other information on domestic and import vehicles (**Figure 9-63** and **Figure 9-64**). Some computer software also provides labor estimates for automotive repairs.

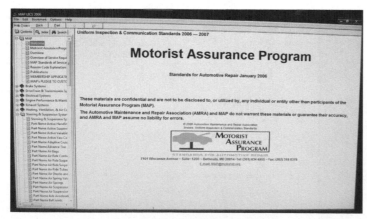

Figure 9-61 The Motorist Assurance Program (MAP) parts inspection guidelines may be accessed in computer wheel aligners.

Figure 9-62 Wheel Alignment software on CD or DVD for computer wheel aligners.

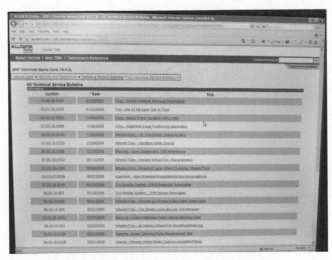

Figure 9-63 Some computer software provides service bulletin information for suspension and steering problems.

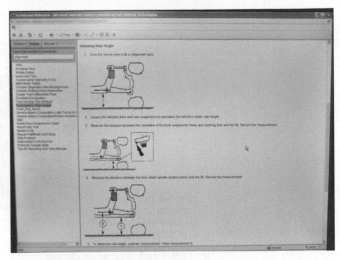

Figure 9-64 Specific computer software provides diagnostic and service information in some computer wheel aligners.

REAR WHEEL ALIGNMENT

Rear Wheel Alignment Diagnosis

Customer complaints that indicate rear wheel alignment problems are the following:

1. The front wheel toe is set correctly and the steering wheel is centered on the alignment rack, but the steering wheel is not centered when the vehicle is driven straight ahead.
2. The steering pulls to one side, but there are no worn suspension parts, defective tires, or improper front suspension alignment angles.
3. The vehicle is not overloaded and there are no worn suspension parts or improper front suspension alignment angles, but tire wear occurs.

Defects That Cause Incorrect Rear Wheel Alignment

These suspension or chassis defects cause incorrect rear wheel alignment:

1. Collision damage that results in a bent frame or distorted unitized body
2. A leaf-spring eye that is unwrapped or spread open
3. Leaf-spring shackles that are broken, bent, or worn
4. Broken leaf springs or leaf-spring center bolts
5. Worn rear upper or lower control arm bushings
6. Worn trailing arm bushings or dislocated trailing arm brackets
7. Bent components such as radius rods, control arms, struts, and rear axles

Rear Wheel Camber

Many front-wheel-drive vehicles have a slightly negative rear wheel camber that provides improved cornering stability. **Rear wheel camber** is basically the same as front wheel camber (**Figure 9-65**).

During rear wheel jounce on a multilink rear suspension system, the camber becomes more positive, and the camber moves toward a negative position during wheel rebound. Because camber changes during wheel jounce, and rebound causes tire tread wear, the rear suspension is designed to minimize camber change during wheel jounce and rebound.

> **Rear wheel camber** is the tilt of a line through the center of the rear tire and wheel in relation to the true vertical centerline of the tire and wheel.

Figure 9-65 Rear wheel camber.

Shop Manual
Chapter 9, page 375

Rear wheel toe is the distance between the front edges of the rear tires in relation to the distance between the rear edges of the rear tires.

Rear Wheel Toe

On a front-wheel-drive vehicle, driving forces tend to push back the rear wheel spindles. Therefore, these rear wheels are designed with zero toe-in or a slight toe-in depending on the vehicle (**Figure 9-66**). Correct **rear wheel toe** is important to obtain normal tire life.

On a multilink rear suspension system with the toe adjustment link mounted in a level position, the toe moves toward a toe-in position during wheel jounce or rebound. These toe changes cause rear tire tread wear.

Steering Angle Sensor Reset

The last step of an alignment procedure that must be performed on some vehicle platforms is a reset procedure for the steering angle sensor (SAS) and other related sensors to update the vehicle geometry information stored in the vehicle's software system after an alignment (**Figure 9-67**). This procedure can be performed with an enhanced computer scan tool,

Figure 9-66 Rear wheel toe.

Figure 9-67 On some vehicles the last step after an alignment procedure is to update vehicle geometry in the vehicle's software system.

depending on model, or your alignment equipment may incorporate this feature. Hunter Engineering has an interface called "CodeLink" that works with the alignment program to relearn the SAS and other related sensor information at the end of the alignment procedure (**Figure 9-68**). Reset requirements vary between vehicle manufacturer year, make, and model of the platform. Always refer to manufacturer relearn service procedures. Other sensors that may be part of this reset procedure are the yaw rate sensor, torque angle sensor, deceleration sensor, and lane departure among others, depending on vehicle manufacturer model. Refer to specific vehicle information for specific steps and requirements.

Figure 9-68 To update vehicle geometry in the vehicles software system a scan tool or interface module is connected to the vehicle's data link connector.

SUMMARY

- Front-wheel-drive cars with unitized bodies are more subject to rear wheel alignment problems compared with rear-wheel-drive cars with frames and one-piece rear axle housings.
- The vehicle geometric centerline is an imaginary line through the exact center of the front and rear wheels.
- The thrust line is an imaginary line at a 90-degree angle to the rear wheel centerline and projected forward.
- A rear axle offset occurs when the complete rear axle is rotated slightly, moving one rear wheel backward and the opposite rear wheel forward.
- Dog tracking is a term applied to a condition where the rear wheels are not directly following the front wheels.
- Rear axle offset or improper toe on one rear wheel causes steering pull to one side and tire tread wear.
- Rear axle sideset is a condition where the rear axle assembly has moved straight sideways and the geometric centerline is not positioned at the true vehicle centerline.
- Setback occurs when one front or rear wheel is moved backward in relation to the opposite wheel.
- In a geometric centerline front wheel alignment, the geometric centerline is used for a reference for front wheel toe. This type of alignment ignores thrust line position.
- In a thrust line front wheel alignment, the front wheel toe is checked using the thrust line as a reference, but the thrust line position may not be at the geometric centerline.
- In a total four-wheel alignment, the thrust line position is measured and the rear wheel toe is measured and adjusted as necessary so the thrust line is at the geometric centerline. The front wheel toe is measured using the common thrust line and geometric centerline as a reference.
- Computer alignment systems have a wheel unit mounted on a wheel clamp attached to each rim.
- Signals from the wheel sensors to the computer wheel aligner may be transmitted by a high-frequency transmitter in the wheel sensor.
- Computer alignment systems provide vehicle specifications plus diagrams of adjustment and inspection points.
- Directional stability is the tendency of a vehicle to travel straight ahead without being steered.

- Suspension and steering systems are designed to provide satisfactory vehicle control with acceptable driver effort and road feel and minimal tire tread wear.
- Proper wheel alignment and suspension component condition are extremely important to maintain vehicle driving safety.
- Many road variables such as bumps and holes, road crown, and heavy vehicle loads affect wheel alignment.
- Wheel alignment angles are designed to compensate for road variables. Camber is the tilt of a line through the center of the tire and wheel in relation to the vertical centerline of the tire and wheel.
- Positive camber is obtained when the centerline of the tire and wheel is tilted outward in relation to the vertical centerline of the tire and wheel.
- Negative camber is present when the centerline of the tire and wheel is tilted inward in relation to the vertical centerline of the tire and wheel.
- During wheel jounce, the top of the wheel moves outward and the camber becomes more positive.
- During wheel rebound, the top of the wheel moves inward and the camber becomes less positive or moves to a slightly negative position.
- While cornering at high speeds, centrifugal force attempts to move the vehicle to the outside of the turn. This force raises the front suspension on the inside of the turn while lowering the front suspension on the outside of the turn.
- While cornering at high speeds, the front wheel on the inside of the turn moves to a less positive camber angle, and the camber angle becomes more positive on the outside front wheel.
- Excessive positive or negative camber concentrates the vehicle weight on one side of the front tire. The tire edge on which the weight is concentrated has a smaller diameter compared with the other side of the tire. Because the side of the tire with the smaller diameter makes more revolutions to go the same distance, this side of the tire becomes worn and scuffed.
- A wheel turns in the direction it is tilted. Road crown causes the vehicle steering to drift to the right. The camber on the left front wheel may be adjusted so it is slightly more positive than the right front wheel camber to compensate for steering pull to the right caused by road crown.

- Caster is the tilt of a line that intersects the center of the lower ball joint and the center of the upper strut mount in relation to a vertical line through the center of the spindle and wheel as viewed from the side.
- Positive caster occurs when the centerline of the lower ball joint and upper strut mount is tilted rearward in relation to the centerline of the spindle and wheel, viewed from the side.
- Negative caster is obtained when the centerline of the lower ball joint and upper strut mount is tilted forward in relation to the centerline of the spindle and wheel.
- Positive caster increases directional stability, steering effort, and steering wheel returning force.
- Excessive positive caster results in harsh ride quality. Excessive positive caster may cause front wheel shimmy. Negative caster reduces directional stability and steering effort while improving ride quality.
- If the caster is different on the two front wheels, the steering pulls toward the side with the least positive caster.
- The caster adjustment on the left front wheel may be adjusted so it is less positive than the caster on the right front wheel to compensate for road crown.

- If the rear suspension height is lowered, the front wheel caster becomes more positive. When the front suspension height is lowered, the front wheel caster becomes more positive. Proper caster adjustment is very important to maintaining vehicle directional control and safety.
- Front wheel toe is the distance between the front edges of the tires compared with the distance between the rear edges of the tires.
- Toe-in is present when the distance between the front edges of the tires is less than the distance between the rear edges of the tires.
- Toe-out occurs when the distance between the front edges of the tires is greater than the distance between the rear edges of the tires.
- Most front-wheel-drive cars have a slight toe-out setting on the front wheels because drive axle forces tend to move the front wheels to a toe-in position.
- Most rear-wheel-drive cars have a slight toe-in setting because driving forces tend to move the front wheels to a toe-out position.
- The front wheel toe is adjusted with the vehicle at rest so the front wheels are straight ahead when the vehicle is driven.
- Improper toe adjustment results in feathered tire wear.

REVIEW QUESTIONS

Short Answer Essays

1. Explain why four-wheel alignment is essential on cars with semi-independent or independent rear suspension.

2. Explain why the rear wheel toe angle, thrust angle, and camber should be correct before adjusting the front suspension angles.

3. Describe the term total toe and what the difference is between a positive toe angle and a negative toe angle.

4. Explain what will happen if toe angles are outside the allowed tolerance.

5. Explain why excessive positive camber wears the outside edge of the tire tread.

6. Describe the type of tire tread wear caused by an improper toe setting.

7. Explain why positive caster provides increased directional stability.

8. Describe how positive caster provides increased steering wheel returning force.

9. Explain why positive caster causes harsh riding quality.

10. Define toe-in and toe-out.

Fill-in-the-Blanks

1. Excessive toe-out on the right rear wheel moves the thrust line to the _____ of the geometric centerline.

2. If the thrust line is positioned to the left of the geometric centerline so the thrust angle is more than specified, the steering pulls to the _____.

3. Directional stability refers to the tendency of a vehicle to travel straight ahead without being _____.

4. When the front wheel camber is negative, the centerline of the tire and wheel is tilted _____ in relation to the true vertical centerline of the tire and wheel.

5. Excessively _____ camber on a front wheel assembly will cause the tire to wear on the inside edge.

6. Negative front wheel caster decreases _____ _____ and steering wheel _____ _____.

7. Raising the rear suspension height above the manufacturer's specification may change the front wheel caster from _____ to _____.

8. Front wheel shimmy may be caused by excessive _____ caster.

9. Front wheel caster becomes more positive if the rear suspension height is _____.

10. If the front suspension height is lowered on a MacPherson strut suspension, the front wheel caster becomes _____ _____.

Multiple Choice

1. While diagnosing front wheel toe problems:
   A. The front wheels are set to a straight-ahead position on most front-wheel-drive cars.
   B. Driving forces tend to move the front wheels toward a toe-in position on a rear-wheel-drive car.
   C. Improper toe adjustment may cause feathered wear on the front tire treads.
   D. Sagged front springs may increase the front wheel toe-out on a short-and-long arm suspension system.

2. "Feathering" type wear of a rear tire is likely caused by:
   A. Improper rear wheel camber alignment.
   B. Improper rear tire inflation.
   C. Improper rear wheel balance.
   D. Improper rear wheel toe alignment.

3. A front-wheel-drive vehicle with an independent rear suspension pulls to the right when driving straight ahead. All the front suspension alignment angles are within specifications. The most likely cause of this problem is:
   A. Excessive toe-out on the right rear wheel.
   B. Excessive negative camber on the right rear wheel.
   C. Excessive toe-out on the left rear wheel.
   D. Excessive positive camber on the left rear wheel.

4. All of these statements about the vehicle thrust line and steering pull are true EXCEPT:
   A. Excessive toe-out on the left rear wheel moves the thrust line to the left of the vehicle centerline.
   B. Excessive toe-in on the right rear wheel moves the thrust line to the right of the vehicle centerline.
   C. If the thrust line is moved to the left of the vehicle centerline, the steering tends to pull to the right.
   D. Excessive toe-out on the right rear wheel causes the steering to pull to the left.

5. While diagnosing suspension and wheel alignment problems:
   A. Road crown has no effect on vehicle steering or wheel alignment.
   B. Vehicle loads have no effect on wheel alignment.
   C. Steering angles are designed to reduce tire wear and provide directional control.
   D. Wheel alignment angles do not affect riding quality.

6. While diagnosing the vehicle geometric centerline and thrust line:
   A. The front and rear wheels should be parallel to the geometric centerline.
   B. The thrust angle is the difference between the front wheel camber and steering axis inclination (SAI) angles.
   C. If the thrust angle is more than specified, front wheel shimmy may occur.
   D. A bent front cradle may cause the thrust angle to be more than specified.

7. While driving straight ahead, a front-wheel-drive car pulls to the right. The most likely cause of this problem is:

    A. More positive camber on the left front wheel compared to the right front wheel.

    B. Sagged front springs and improper front wheel toe setting.

    C. Less positive caster on the right front wheel compared to the left front wheel.

    D. The SAI on the right front wheel is 1.5 degree more than the SAI on the left front wheel.

8. While adjusting front wheel camber and diagnosing camber-related problems:

    A. During front wheel jounce travel, the positive camber increases.

    B. Excessive positive camber on a front wheel causes premature wear on the inside edge of the tire tread.

    C. If the right front wheel camber is 1.5 degrees and the left front camber is 0.5 degree, the steering pulls to the left.

    D. Excessive positive camber on both front wheels may cause front wheel shimmy.

9. A driver complains of harsh riding and the suspension height is normal. The most likely cause of this problem is:

    A. Excessive negative camber on both front wheels.

    B. Excessive positive caster on both front wheels.

    C. The left front wheel has more positive caster than the right front wheel.

    D. Both front wheels have negative camber and negative caster.

10. Excessive positive caster may cause all of these problems EXCEPT:

    A. Front wheel shimmy.

    B. Harsh ride quality.

    C. Excessive steering effort.

    D. Slow steering wheel return.

CHAPTER 10

FOUR-WHEEL ALIGNMENT, PART 2 DIAGNOSTIC ANGLES AND FRAME DAMAGE

Upon completion and review of this chapter, you should be able to understand and describe:

- The steering axis inclination (SAI) angle and the included angle.
- How SAI helps return the front wheels to the straight-ahead position.
- How SAI eliminates the need for excessive positive camber and caster.
- The effect that front suspension defects such as dislocated upper strut towers and bent struts or spindles have on SAI.
- Negative and positive scrub radius and the effect of each on steering quality.
- Setback on front suspension systems.
- How the front suspension system is designed to provide toe-out on turns.
- The purposes of a vehicle frame.
- Different types of frame construction.
- Different types of frame designs.
- Unitized body construction, and explain how this body design obtains its strength.

- Directional stability.
- Vehicle tracking, and explain how the four wheels on a vehicle must be positioned to obtain proper tracking.
- Wheelbase, and explain how the wheels on a vehicle must be positioned to provide correct wheelbase.
- Setback on a front wheel, and explain the effect that setback has on vehicle steering.
- Rear axle offset, and describe the effect of this problem on steering control.
- Rear axle sideset, and explain the effect of rear axle sideset on steering control.
- Side sway frame damage.
- Frame sag.
- Frame buckle.
- A diamond-frame condition.
- Frame twist.
- Aluminum space frame construction.

Terms To Know

Axle offset
Axle sideset
Axle thrust line
Channel frame
Complete box frame
Diamond-frame condition
Directional stability
Frame buckle
Frame sag
Frame twist

Geometric vehicle centerline
High-strength steels (HSS)
Hydro-forming
Included angle
King pin inclination (KPI)
Ladder frame design
Negative camber
Negative scrub radius
Perimeter-type frame
Positive scrub radius

Scrub radius
Side sway
Slip angle
Steering axis inclination (SAI)
Tracking
Tubular frame
Turning radius
Unitized body
Wheel setback
Wheelbase

INTRODUCTION

The frame and unitized body play an important role in vehicle geometry and alignment angles. It is important to understand the vehicle's structure elements and to be aware of how damage to the vehicle's structure elements can affect alignment angle. Many of the angles discussed in this chapter are not adjustable on their own but are used as diagnostic angles. These angles are often used in combination with one another as an aid in determining the root cause for the vehicle geometry being out of specification.

As an example, improper steering axis inclination (SAI) angles on either side of the front suspension may cause hazardous steering conditions while braking or accelerating. Therefore, technicians must be familiar with SAI and other related steering geometry if they are to diagnose routine drivability and handling issues.

Shop Manual
Chapter 10, page 431

STEERING AXIS INCLINATION DEFINITION

On front-wheel-drive vehicles with MacPherson strut front suspension systems, **steering axis inclination (SAI)** refers to the inward tilt of a line through the center of the top strut mount and the center of the lower ball joint in relation to the true vertical line through the center of the tire. These two lines are viewed from the front of the vehicle, and the SAI line always tilts inward in relation to the true vertical line (**Figure 10-1**).

Many rear-wheel-drive cars have a short-and-long arm front suspension system with unequal-length upper and lower control arms and a ball joint mounted in each control arm. The steering knuckle and spindle pivot on the ball joints as the wheels are turned. On this type of front suspension, the SAI line runs through the upper and lower ball joint centers. The **included angle** is the sum of the SAI angle and the camber angle (**Figure 10-2**).

Figure 10-1 Steering axis inclination angle.   Figure 10-2 Steering axis inclination and included angle.

If the camber angle is positive, this angle is added to the SAI angle to obtain the included angle. A negative camber angle must be subtracted from the SAI angle to calculate the included angle. On some truck front suspension systems, the steering knuckle pivots on a king pin that is mounted in an I-beam-type front axle. The SAI line on this type of suspension is referred to as a **king pin inclination (KPI)** line. This invisible line runs through the center of the king pin.

AUTHOR'S NOTE It has been my experience that after front-end collision damage, a common problem for cars with a unitized body is improperly positioned strut towers. Even though the cosmetic appearance of the car body remains good, the strut towers may be out of position, resulting in unequal SAI between the right and left front wheels. During severe braking, an unequal SAI causes steering pull and possible loss of steering control. Therefore, after a frontal collision on a unitized body car, it is very important that you measure the strut tower location and be sure this measurement is within specifications. It is also critical that you check the front and rear end wheel alignment and adjust the alignment angles as necessary to bring them within specifications.

SAI PURPOSE

When the SAI angle is tilted toward the center of the vehicle and the wheels are straight ahead, the height of the spindle is raised closer to the chassis. Thanks to gravity, this action lowers the height of the vehicle. When the front wheels are turned, each spindle moves through an arc that tries to force the tire into the ground. Because this reaction cannot take place, the chassis lifts when the wheels are turned. When the steering wheel is released after a turn, the vehicle weight has a tendency to settle to its lowest point. Therefore, SAI helps return the wheels to the straight-ahead position after a turn, and it also tends to maintain the wheels in the straight-ahead position. However, SAI does increase steering effort because the chassis has to lift slightly on turns.

The vehicle weight is projected through the SAI line to the road surface. Let us assume that a front suspension is designed with a vertical 0-degree SAI line. Under this condition, the vehicle weight is projected through the SAI line a considerable distance inside the true vertical tire centerline. With this type of front suspension, severe tire scuffing would occur because the tire and wheel pivot around the SAI line during a turn (**Figure 10-3**).

Pivot axis

℄

0°

Pivot point

Figure 10-3 Front suspension with a 0-degree SAI line.

Figure 10-4 If the SAI line meets the tire vertical centerline at the road surface, tire wear is reduced, steering effort is decreased, and stress on the steering components is diminished.

With a 0-degree SAI line, greater steering effort is required during a turn, and stress on the steering mechanism increases. This type of front suspension design causes excessive road shock and kick-back on the steering wheel during a turn, because the distance between the SAI line and the tire vertical line returns the wheels to the straight-ahead position.

When the SAI line meets the tire vertical centerline at the road surface, tire wear is reduced, steering effort is decreased, and stress on steering components is diminished (**Figure 10-4**).

The point at which the SAI line and the tire vertical centerline intersect could be positioned closer together by designing the front suspension with excessive positive camber. However, this design would result in rapid wear on the outside edge of the tire tread. A correct SAI line reduces the need for excessive positive camber.

MacPherson strut front suspension systems have a much greater SAI angle (12 to 18 degrees) compared with short-and-long arm front suspensions in rear-wheel-drive cars (6 to 8 degrees). Front-wheel-drive cars require the higher SAI angle for directional stability. Positive caster also provides directional stability.

SAI AND SAFETY FACTORS

On MacPherson strut suspension systems, an incorrect SAI angle may indicate that the upper strut mount is out of position, the lower control arm is bent, or the center cross member is shifted. Any of these defects may be caused by collision damage.

When the SAI angles are unequal on the left and right front suspension, serious handling problems may occur. These problems include torque steering during hard acceleration, steering pull during sudden stops, and bump steer. Torque steer is the tendency to pull to one side on hard acceleration because of unequal-length drive axles on a front-wheel-drive vehicle. Unequal SAI angles aggravate torque steer. Bump steer refers to unequal toe and/or camber changes that jerk the car to one side during front suspension jounce and rebound. Therefore, incorrect SAI angles may cause hazardous driving situations and contribute to serious accidents. Technicians must inspect the SAI angle during a wheel alignment.

If the SAI angle is correct and the included angle and camber are less than specified, bent components such as the strut or spindle are indicated. On the type of suspension that requires inward or outward upper strut movement to adjust camber, the SAI angle changes with this movement. When an eccentric bolt between the strut and steering knuckle is used for camber adjustment, the SAI angle will not change if the camber is adjusted.

It is very important to remember that a camber adjustment on many front suspension systems also changes the SAI angle. For example, if the upper control arm on a front suspension system with upper and lower ball joints is shimmed outward to increase positive camber, the SAI angle will change with the camber angle. Therefore, the included angle remains the same. The camber may be adjusted to specification, but the SAI angle and included angle could be out of specification. This is especially true on MacPherson strut suspension systems, where collision damage may bend front struts or shift the upper strut towers. A service technician who inspects and adjusts the camber angle while ignoring the SAI, and included angles may be overlooking serious and dangerous front suspension defects.

SCRUB RADIUS

Scrub radius is the distance between the SAI line and the true vertical centerline of the tire at the road surface.

Scrub radius affects steering quality related to stability and returnability. However, scrub radius is not an alignment angle and it cannot be measured on conventional alignment equipment. Scrub radius is the distance from the point where the tire vertical line contacts the road to the location where the line through the upper strut center and the ball joint center meets the road surface. **Positive scrub radius** occurs when the line through the strut and ball joint meets the road surface inside the tire vertical centerline. A **negative scrub radius** is when the line through the strut and ball joint centers meets the road surface outside the tire and hub centerline (**Figure 10-5**).

Conventional short-and-long arm front suspension systems usually have positive scrub radius. Many front-wheel-drive vehicles have negative scrub radius.

When negative scrub radius is used in front-wheel-drive vehicles, straight-line braking is ensured and directional stability is maintained. As the vehicle moves forward, negative scrub radius tends to turn the front wheels inward. This action causes unequal forces applied to the steering to act inboard of the steering axis and pull the vehicle from any induced swerve. In a front-wheel-drive vehicle, a swerve to one side may be caused by one front wheel being on an ice patch while the other front wheel is on dry pavement. A failure of one-half the diagonal brake system, a sudden blowout of one front tire, or a grabbing brake on one front wheel will also cause a vehicle to swerve.

Figure 10-5 Scrub radius.

If a vehicle has positive scrub radius and the right front brake grabs, both the positive scrub radius and the grabbing brake tend to turn the right front wheel outward, and the vehicle pivots around the right front wheel. This action induces a swerve to the right.

When a right front brake grab occurs with negative scrub radius, the brake grab causes the right front wheel to turn outward, and the vehicle tends to pivot around the right front wheel. However, the negative scrub radius tends to turn the right front wheel inward. The two forces on the right front wheel cancel each other to maintain directional stability.

Scrub Radius and Safety

If front tires that are larger in diameter than specified by the car manufacturer are installed, directional control may be affected. Large tires raise the chassis farther from the road surface, which changes the scrub radius. The installation of larger front tires could change a positive scrub radius to a negative scrub radius.

Reversing the front rims so they are inside-out creates a significant scrub radius change and adversely affects directional control. This practice is not recommended by car manufacturers.

⚠ **WARNING Installing larger tires, or different rims than the ones specified by the vehicle manufacturer, changes the scrub radius, which may result in reduced directional control, collision damage, and personal injury.**

WHEEL SETBACK

Wheel setback is a condition in which one wheel is moved rearward in relation to the other wheel (**Figure 10-6**). Setback will not affect handling unless it is extreme. Collision damage may drive one front strut rearward and cause extreme setback. Setback can also occur on rear wheels, but it is more likely to occur on front wheels because of collision damage. Some computer four-wheel aligners have setback measuring capabilities.

TURNING RADIUS

Front and Rear Wheel Turning Action

When a vehicle turns a corner, the front and rear wheels must turn around a common center with respect to the **turning radius** (**Figure 10-7**). On most front suspension systems, the front wheels pivot independently at different distances from the center of the

Turning radius may be referred to as cornering angle or toe-out on turns.

Setback

Figure 10-6 Wheel bearing and hub assembly.

Figure 10-7 Front and rear wheels turning around a common center.

turn, and therefore the front wheels must turn at different angles. The inside front wheel must turn at a sharper angle than the outside wheel. This is because the inside wheel is actually ahead of the outside wheel. When this turning action occurs, both front wheels remain perpendicular to their turning radius, which prevents tire scuffing.

The turning angles of the front wheels are determined by steering arm design, and these angles are not adjustable. If the steering angles are not correct, the steering arms may be bent.

Steering Arm Design

An understanding of a lever moving in a circle is necessary before an explanation of steering arm design and operation. If a lever moves from point A to B, it pivots around point O and moves through a horizontal distance A to B (**Figure 10-8**).

Lever movement through arc B to C is much greater than the movement through arc A to B. However, during arc B to C, the lever moves through horizontal distance B to C and this distance is the same as horizontal distance A to B.

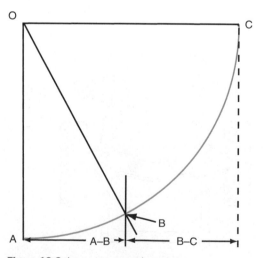

Figure 10-8 Lever movement in a circle.

The steering arms are connected from the tie rod ends to the steering knuckles. Steering arms and linkages maintain the front wheels parallel to each other when the vehicle is driven straight ahead. However, the steering arms are not parallel to each other. If the steering linkage is at the rear edge of the front wheels, the steering arms are closer together at the point where the tie rods connect than at their spindle pivot point (**Figure 10-9**). When the steering linkage is positioned at the front edge of the front wheels, the steering arms are closer together at their spindle pivot point than at the tie rod connecting point.

When the front wheels are turned on a vehicle with the steering linkage at the rear edge of the front wheels, the angle formed by the inside steering arm and linkage increases, whereas the angle of the outside steering arm and linkage decreases. The inside steering arm moves through the longer arc X, and the outside steering arm moves through shorter arc Z (**Figure 10-10**).

Therefore, the inside wheel turns at a sharper angle than the outside wheel. Because both steering arms are designed to have the same angle in the straight-ahead position, the inside front wheel always has a sharper angle regardless of the turning direction. The sharper inside wheel angle during a turn causes the inside wheel to toe out. Therefore, the term **toe-out on turns** is used for this steering action. If the front wheel turning angle is increased during a turn, the amount of toe-out on the inside wheel increases proportionally.

Toe-out on turns is the angle of the inside wheel in relation to the angle of the outside wheel during a turn.

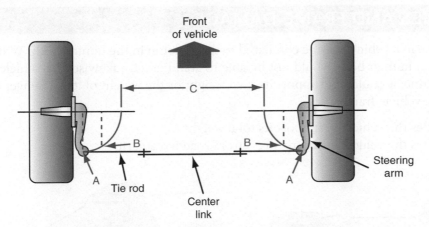

Figure 10-9 Steering arms are closer together at the point where the tie rods connect than at their spindle pivot points.

Figure 10-10 Steering arm operation while turning.

Slip Angle

During a turn, centrifugal force causes all the tires to slip a certain amount. The amount of tire slip increases with speed and the sharpness of the turn. This tire slip action causes the actual center of the turn to be considerably ahead of the theoretical turn center (**Figure 10-11**).

The **slip angle** on different vehicles varies depending on such factors as vehicle weight and type of suspension.

Slip angle is the actual angle of the front wheels during a turn compared to the turning angle of the front wheels with the vehicle at rest.

FRAMES AND FRAME DAMAGE

The frame in a vehicle may be compared to the skeleton in the human body. Without the skeleton, a human body would not be able to stand erect. Likewise, if a vehicle did not have a frame, it could not support its own weight or the weight of its passenger or cargo load. The vehicle frame:

1. Enables the vehicle to support its total weight.
2. Enables the vehicle to absorb stress from road irregularities.

Figure 10-11   Slip angle during a turn.

3. Enables the vehicle to absorb torque from the engine and drive wheels.
4. Provides a main member for attachment of body and other components.

The frame, together with the front and rear suspension systems, must position the wheels properly to minimize tire tread wear and provide accurate steering control.

TYPES OF FRAMES AND FRAME CONSTRUCTION

Frame Construction

Vehicle frame construction may contain three types of steel members: **channel** (partial box) frame, **complete box frame**, or **tubular frame** (**Figure 10-12**). On modern vehicles, most frames include all three types.

Compared to a channel frame or a tubular frame, the complete box frame has increased torsional rigidity and improved crash performance. Therefore, in many frames the front section is a complete box design and the other part of the frame is a channel design. Some frames have X-shaped bracing at the rear of the frame for increased strength. Many vehicles now have frames or partial frames that are manufactured by a **hydro-forming** process. Hydro-formed frames are considerably more rigid than frames manufactured by heat-treating. Therefore, hydro-formed frames reduce the flexing of the frame and body components, which reduces squeaks and rattles to decrease noise, vibration, and harshness (NVH). Hydro-formed frames also provide more stable wheel position, which improves steering quality and reduces tire tread wear.

Ladder-Type Frames

The **ladder frame design** is used on trucks and full-size vans. In this type of frame, the side rails have very little offset and are in a straight line between the front and rear wheels. The ladder-type frame has more crossmembers for increased load-carrying capacity and rigidity (**Figure 10-13**).

> In a **channel frame**, each side of the frame is made from a U-shaped steel channel.
>
> In a **complete box frame**, each side of the frame forms a metal box.
>
> A **tubular frame** member is formed in an oval shape or circle.
>
> **Hydro-forming** is the process of using extreme fluid pressure to shape metal.
>
> In a **ladder frame design**, crossmembers are mounted between the sides of the frame, and the crossmembers are similar to rungs on a ladder.

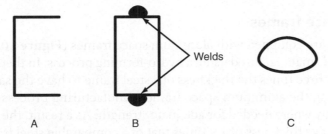

A   B   C

Figure 10-12 Types of frame construction: (A) channel or partial box, (B) complete box, (C) tubular crossmember.

Figure 10-13 Ladder-type frame.

Perimeter-Type Frames

The **perimeter-type frame** is used in some rear-wheel-drive cars. This type of frame forms a border around the passenger compartment. The frame rails are stepped inward at the cowl area to provide increased strength, which supports the engine mounts and front suspension. This inward step at the cowl area also provides room for movement of the front wheels. Lateral support is provided by crossmembers welded between the frame rails near the front and rear of the frame. A transmission support member is welded or bolted between the frame rails at the back of the transmission. The rear frame kickup side rails support the rear suspension and the rear portion of the body weight. A front torque box is positioned just ahead of the transmission support member, and a rear torque box is located in front of the rear frame kickup side rails (**Figure 10-14**). These torque boxes are designed to absorb most of the impact during a side collision and thus reduce damage to other body components. The torque boxes also provide some protection for the vehicle occupants during a side collision.

The body components are bolted to the frame, but rubber insulating bushings are positioned between the body and frame mounting locations. These bushings help prevent the transfer of road noise and vibration from the suspension and frame to the body and vehicle interior. Rubber insulating mounts are positioned between the engine and transmission and the frame mounting positions. These engine and transmission mounts reduce the transfer of engine vibration to the frame, body, and vehicle interior. Many suspension components are also connected to the frame through rubber insulating bushings to reduce the transfer of harshness and vibration from the suspension to the frame while driving over road irregularities.

During the manufacturing process, most manufacturers apply a special coating to the frame to help prevent rust and corrosion. For example, many light-duty truck frames are coated with epoxy or wax. An epoxy-coated frame is black and very smooth, whereas a wax-coated frame is gray and sticky. If these coatings are scratched or damaged, frame rusting and corrosion may occur.

Shop Manual
Chapter 10, page 440

Aluminum Space frames

Some sports cars are equipped with aluminum space frames (**Figure 10-15**). The aluminum space frame is manufactured using a hydro-forming process. In theory, an aluminum frame must have three times the thickness of a steel frame to have the same strength as a steel frame. During the aluminum space frame manufacturing process, a manufacturer adds thickness only where needed for adequate strength. As a result, the aluminum space frame has an average thickness of 1.9 times that of a comparable steel frame. This aluminum frame weighs 285 lb (129 kg), which is 136 lb (62 kg) lighter than a steel frame. Reducing vehicle weight improves fuel economy and vehicle performance.

Center frame rail
Front frame rail
Rear frame rail

Figure 10-14 Perimeter-type frame components.

Figure 10-15 Aluminum space frame.

Figure 10-16 Laser welding on aluminum space frame tunnel.

Laser welding is used extensively on the tunnel in the aluminum space frame rather than spot welding (**Figure 10-16**). Laser welding adds stiffness and provides an improved seal compared to spot welding. Laser welding uses less heat per length of weld, which results in reduced aluminum distortion.

 A BIT OF HISTORY

During the late 1940s, American Motors Corporation introduced the unitized body design in the United States. Some of these first-generation unitized bodies did not have partial frames. However, unitized bodies did not become popular until 1980 and 1981 with the introduction of General Motors X cars, Chrysler K cars, and other front-wheel-drive cars. These front-wheel-drive cars have partial frames and may be referred to as second-generation unitized bodies.

UNITIZED BODY DESIGN

Most front-wheel-drive cars have a **unitized body**. In this body design, the frame and body are combined as one unit, and the external frame assembly is eliminated. The strength and rigidity of the body is achieved by body design rather than by having a heavy steel frame to support the body. In the unitized body design, body sheet metal is fabricated into a box design. Body strength is obtained by shape and design in place of mass and weight of metal in vehicles with a separate frame.

All the members of a unitized body are load-carrying components. The floor pan, roof, inner aprons, quarter panels, pillars, and rocker panels are integrally joined to form a unitized body (**Figure 10-17**).

Most unitized bodies have bolt-on partial frames at the front and rear of the vehicle. Some of the components in a unitized body are made from **high-strength steels (HSS)** to provide additional protection in a collision. Ultra-HSS may be used in such unitized body components as door beams and bumper reinforcements. The unitized body has a complex design that spreads collision forces throughout the body to help protect the vehicle occupants.

VEHICLE DIRECTIONAL STABILITY

The front and rear suspension systems are attached to the frame, partial frame, or unitized body. Therefore, the frame or unitized body must support the suspension systems properly to provide directional stability. Vehicle **tracking** is the parallel relationship between the front and rear wheels during forward vehicle motion. To provide proper tracking, each front wheel must be at the same distance from the vehicle centerline, and each rear wheel must be at an equal distance from the same centerline. The distance between the front wheels and the distance between the rear wheels does not necessarily have to be the same, but all four wheels must have a parallel relationship to provide proper tracking (**Figure 10-18**).

The vehicle **wheelbase** is the distance between the centers of the front and rear wheels. To provide an accurate wheelbase measurement, the centers of the front spindles and the centers of the rear axles must be square with the centerline of the vehicle (**Figure 10-19**). For this condition to exist, the front and rear axle centers must be at a 90-degree angle in relation to the vehicle centerline, and the wheelbase measurements must be equal on each side of the vehicle. Equal wheelbase measurements on each side of the vehicle and proper tracking are absolutely essential to providing **directional stability**.

Shop Manual
Chapter 10, page 435

Directional stability refers to the tendency of a vehicle to remain in the straight-ahead position when driven straight ahead on a reasonably smooth, straight road.

The term dog tracking may be applied to axle offset or sideset, because in either of these conditions the rear wheels are not parallel to the front wheels. Similarly, many dogs run down the street with their rear ends out of line with their front ends.

Figure 10-17 Unitized body design.

Figure 10-18 Front and rear wheels must be at the same distance from the vehicle centerline to provide proper tracking.

Figure 10-19 To provide accurate wheelbase measurement, the front spindle centerlines and the rear axle centers must be at a 90-degree angle to the vehicle centerline.

VEHICLE TRACKING

Wheel Setback

Front wheel setback occurs when the spindle on one front wheel is rearward in relation to the other front wheel. The centerline of each front wheel is still at a 90-degree angle to the vehicle centerline, but the front wheels no longer share the same centerline (**Figure 10-20**).

When left front wheel setback occurs, the vehicle has a tendency to steer to the left as it is driven straight ahead. Under this condition, the driver has to continually turn the steering wheel to the right to keep the vehicle moving straight ahead.

Axle Offset

If the rear axle is rotated, the **axle thrust line** is no longer at a 90-degree angle to the **geometric vehicle centerline** (**Figure 10-21**). This condition is referred to as **axle offset**. When the left side of the rear axle is rotated rearward, the steering pulls continually to the right. Under this condition, the driver has to turn the steering wheel to the left to keep the vehicle moving straight ahead.

Axle Sideset

When **axle sideset** occurs, the rear axle moves inward or outward, but the axle and vehicle centerlines remain at a 90-degree angle in relation to each other (**Figure 10-22**).

> The **axle thrust line** is a line extending forward from the center of the rear axle at a 90-degree angle.
>
> The **geometric vehicle centerline** refers to the front-to-rear centerline of the vehicle body. The centerlines of the front and rear axles should be positioned on this geometric centerline.

Figure 10-20 Setback occurs when one front wheel is positioned rearward in relation to the other front wheel.

Figure 10-21 Rear axle offset occurs when the rear axle is rotated so the axle centerline and the vehicle centerline are no longer at a 90-degree angle.

Figure 10-22 Axle sideset occurs when the rear axle moves inward or outward and the axle centerline remains at a 90-degree angle in relation to the vehicle centerline.

Under this condition, the front-to-rear axle thrust line is no longer at the geometric vehicle centerline. This condition also causes steering pull. The vehicle frame or unitized body and the front and rear suspension systems must have proper tracking and equal wheelbase measurements on each side of the vehicle to provide directional stability and steering control.

TYPES OF FRAME DAMAGE

Side Sway

Side sway on the front or rear of a vehicle is usually the result of the vehicle being involved in a collision that pushes the front or rear frame sideways (**Figure 10-23**). Under this condition, the wheelbase on one side of the vehicle is longer than the opposite side. This side sway condition causes the steering to pull to the side with the shorter wheelbase.

> **AUTHOR'S NOTE** It has been my experience that frame damage is most commonly caused by abuse, and this problem is usually encountered on light-duty trucks or sport utility vehicles (SUVs). The frame damage may occur when the vehicle is overloaded and/or driven roughly on extremely uneven terrain. Another common cause of frame damage is from a vehicle collision. In this case, the frame damage was likely ignored or overlooked during the body repairs. Regardless of the cause, frame damage usually results in excessive tire tread wear and steering complaints.

Overall side sway occurs when the vehicle is hit directly on the side near the center in a collision. A vehicle frame is slightly V-shaped when it has overall side sway damage (**Figure 10-24**).

> **Side sway** is a lateral misalignment that affects the frame or body centerline.

> **Caution**
> When attaching any type of additional equipment to a vehicle frame, always try to use the holes that already exist in the frame. Do not drill holes in a vehicle frame except at locations specified by the vehicle manufacturer, and do not attach additional equipment to a vehicle frame unless this procedure is approved by the vehicle manufacturer.

> **Caution**
> Do not apply heat to a vehicle frame with an acetylene torch. When the metal is heated to a cherry red, the frame may be weakened.

Figure 10-23 Side sway frame condition caused by collision damage.

Figure 10-24 When a frame has overall side sway damage, it is slightly V-shaped.

⚡ **WARNING** Some types of frame damage cause steering pull when driving straight ahead, and this steering pull is increased during hard braking. Therefore, frame damage can create a safety hazard that leads to a collision involving personal injury and vehicle damage.

Sag

Frame sag usually occurs when the vehicle is involved in a direct front or rear collision. When this condition is present, the front and/or rear frame rails are moved upward in relation to the center of the frame (**Figure 10-25**). If one side of the vehicle sustained more collision force than the opposite side, the left and right wheel base measurements will also likely be different.

The front crossmember may receive sag damage in a collision. When this member is sagged, the upper control arms move closer together on a short-and-long arm suspension. If a MacPherson strut front suspension is sagged, the strut towers are moved closer together. In either type of front suspension, a sagged condition moves the top of the wheels inward to a **negative camber** position.

Frame Buckle

A buckle condition exists when the distance from the cowl to the front bumper is less than specified, or the measurement from the rear wheels to the rear bumper is less than specified (**Figure 10-26**). Frame buckle is caused by a direct front or rear collision. In many cases of **frame buckle**, the wheelbase is reduced on one or both sides of the vehicle. This type of collision damage may cause the sides of the vehicle to bulge outward, especially on unibody cars. Under this condition, the side rails and door openings are distorted.

Diamond-Frame Condition

A **diamond-frame condition** is present when collision damage causes a frame to be out of square. Under this condition, the frame is shaped like a parallelogram (**Figure 10-27**). If the right rear wheel is driven rearward in relation to the left rear wheel, the rear suspension steers the vehicle to the right, and this forces the front end of the vehicle to the left. Under this condition, the steering wheel must be held continually to the right to overcome

---

Negative camber occurs when the camber line through the center of the tire and wheel is tilted inward compared with the true vertical centerline of the tire and wheel.

Frame buckle is accordion-shaped damage on the front or rear of the frame, which causes the distance to be reduced between the cowl and front bumper or between the rear wheels and rear bumper. Frame buckle may be called frame crush or mash.

Frame rail
moved upward

Figure 10-25 Frame sag caused by direct front or rear collision damage.

Original dimension

Buckled dimension

Figure 10-26 Rear frame buckle.

Figure 10-27 Diamond-frame condition.

Frame corner
higher than all others

Figure 10-28 Frame twist.

the steering pull to the left. A diamond-frame condition usually occurs on vehicles with frames. Vehicles with unitized bodies seldom have this type of condition.

Frame Twist

A **frame twist** condition exists when one corner of the frame is higher than the other corners. When frame twist is present, the front or rear chassis does not sit level in relation to the road surface (**Figure 10-28**). Frame twist is usually caused by vehicle rollover.

SUMMARY

- Steering axis inclination (SAI) is the angle of a line through the center of the upper strut mount and lower ball joint in relation to the true vertical centerline of the tire viewed from the front of the vehicle.
- On short-and-long arm front suspension systems, the SAI line runs through the center of the upper and lower ball joints.
- The SAI line is always tilted toward the center of the vehicle. The included angle is the sum of the SAI and positive camber angle. If the camber is negative, the camber must be subtracted from the SAI angle to obtain the included angle.
- SAI causes the front spindles to move through an arc when the front wheels are steered to the right or left.
- Because the front spindles move through an arc, the chassis lifts as the front wheels are turned. This

- lifting action helps return the front wheels to the straight-ahead position after a turn.
- SAI also helps maintain the front wheels in the straight-ahead position. SAI reduces the need for excessive positive caster. SAI is not adjustable. If the SAI does not equal the manufacturer's specifications and the other steering angles are correct, some suspension component, such as a strut tower, is out of place.
- A suspension system with a 0-degree SAI line would have increased tire wear, greater steering effort, increased stress on suspension and steering components, and excessive road shock and kickback on the steering wheel.
- When the SAI line intersects the true vertical tire centerline at or near the road surface, tire life is

- improved, stress on steering and suspension components is reduced, steering effort is decreased, and road shock and kickback on the steering wheel are minimized.
- Because excessive positive camber could be used to bring the SAI lines and tire vertical centerlines closer together, SAI reduces the need for excessive positive camber.
- If SAI angles on both sides of the front suspension are unequal, excessive torque steer may occur on hard acceleration, and severe steering pull may be present during hard braking.
- Adjusting camber on some front suspensions creates a corresponding change in SAI angle, whereas on other front suspension systems adjusting the camber does not change the SAI angle.
- Scrub radius is the distance between the SAI line and the true vertical tire centerline at the road surface.
- A front suspension has positive scrub radius when the SAI line contacts the road surface inside the true vertical tire centerline.
- A front suspension has negative scrub radius when the SAI line contacts the road surface outside the true vertical tire centerline.
- If the front tires are larger than the original tires specified by the vehicle manufacturer, a change occurs in scrub radius that may affect steering control.
- The installation of larger-than-specified front tires may change positive scrub radius to a negative scrub radius.
- Reversing the front rims so they are inside out results in a significant scrub radius change that adversely affects directional control.
- Wheel setback is a condition where one wheel is moved rearward in relation to the opposite wheel.
- Toe-out on turns is the turning angle of the wheel on the inside of the turn compared with the turning angle of the wheel on the outside of the turn.
- When the front wheel on the inside of a turn has turned 20 degrees outward, the front wheel on the outside of the turn may have turned 18 degrees.
- Turning radius is the amount of toe-out on turns. Toe-out on turns prevents tire scuffing. This angle is determined by the steering arm design.

- During a turn, centrifugal force causes all the tires to slip a certain amount depending on vehicle speed and the sharpness of the turn.
- Because the tires slip during a turn, the actual vehicle turning center is moved ahead of the theoretical turning center.
- Slip angle is the actual angle of the front wheels during a turn compared with the turning angle of the front wheels with the vehicle at rest.
- The two rear wheel alignment angles are camber and toe. The frame enables a vehicle to support its weight and absorb stress and torque. It also provides a main member for attachment of other components.
- In a unitized body design, all body members are load-carrying components that are welded together.
- Proper tracking and wheelbase are essential to providing directional stability. Each front wheel and each rear wheel must be at an equal distance from the vehicle centerline to provide correct tracking.
- To provide proper wheelbase, the centers of the front and rear suspensions must be at a 90-degree angle to the vehicle centerline.
- Front wheel setback, rear axle offset, and rear axle sideset cause the steering to pull to one side.
- Regardless of the type of front or rear suspension system, the suspension system and the frame must position the wheels properly to provide correct tracking and wheelbase.
- Frame side sway occurs when the front suspension is forced sideways in a collision, and one front wheel is forced rearward in relation to the opposite front wheel. Side sway may also occur on the rear suspension.
- Frame sag occurs when the front or rear frame rails are bent upward in relation to the center of the frame.
- Frame buckle is accordion-shaped damage on the front or rear of the frame that causes the distance to be reduced between the cowl and front bumper or between the rear wheels and rear bumper.
- A diamond-frame condition is present when one side of the frame is driven rearward in relation to the opposite side of the frame, and the front and rear wheels on one side of vehicle are rearward in relation to the wheels on the other side.
- Frame twist occurs when one corner of the frame is bent up higher than the other frame corners.

REVIEW QUESTIONS

Short Answer Essays

1. Define steering axis inclination (SAI).

2. Explain the included angle.

3. Explain how SAI helps to return the steering wheel to the center position after a turn.

4. Define negative scrub radius and positive scrub radius, including the type of suspension system on which each condition is used.

5. Describe front wheel setback.

6. Explain the necessary wheel position to provide proper tracking.

7. Describe the necessary axle position to provide correct wheelbase.

8. Explain the turning angle of each front wheel during a turn.

9. Explain the effects of front wheel setback.

10. Describe the effects of rear axle offset.

Fill-in-the-Blanks

1. A 3-degree difference in the SAI angle on each side of the front suspension may cause _____ during hard braking.

2. A 3-degree difference in the SAI angle on each side of the front suspension may cause increased _____ during hard acceleration.

3. When a front suspension has a positive scrub radius, the SAI line meets the road surface _____ the true vertical centerline of the tire.

4. A negative scrub radius tends to turn the front wheels _____ when the car is driven.

5. Most rear-wheel-drive cars have a _____ scrub radius.

6. Wheelbase is the distance between the front and rear wheel _____.

7. Directional stability is the tendency of a vehicle to remain in the _____ position when driven straight ahead on a level road.

8. During a turn, the _____ front wheel turns at a sharper angle.

9. Tracking refers to the _____ relationship between the front and rear wheels.

10. During a turn, centrifugal force causes all the tires to slip a certain amount, and the actual center of the turn is shifted _____ of the theoretical center of the turn.

Multiple Choice

1. When measuring front wheel alignment angles, to calculate the included angle on the left front wheel when the camber on this wheel is positive:

   A. Add the camber to the toe setting.

   B. Add the camber to the SAI angle.

   C. Add the SAI to the caster angle.

   D. Subtract the SAI from the toe setting.

2. While diagnosing problems related to scrub radius:

   A. If the SAI line contacts the road surface inside the vertical tire and wheel centerline, the scrub radius is negative.

   B. Front-wheel-drive cars usually have a negative scrub radius and driving forces move the front wheels outward.

   C. If the SAI line contacts the road surface outside the tire and wheel vertical centerline, driving forces turn the front wheels outward.

   D. Larger-than-specified front tires may change the scrub radius from positive to negative.

3. While diagnosing a diamond-frame condition and frame twist:

   A. A diamond-frame condition causes the wheelbase to be unequal on the two sides of the vehicle.

   B. A diamond-frame condition does not affect steering pull and directional stability.

   C. Frame twist is usually caused when a vehicle is involved in a side collision.

   D. When frame twist occurs, the front or rear chassis does not sit level in relation to the road surface.

4. While diagnosing and adjusting turning radius:

   A. When a vehicle is making a left turn, the left front wheel turns at a sharper angle than the right front wheel.

B. Improper turning radius may be caused by a bent tie rod on a short-and-long arm suspension system.

C. Improper turning radius may be caused by an improperly positioned rack-and-pinion steering gear on a front-wheel-drive car.

D. Improper turning radius is adjusted by turning an eccentric strut-to-steering knuckle bolt.

5. All of these statements about unitized body design are true EXCEPT:

A. The frame and body are combined as one unit.

B. The external frame assembly is eliminated.

C. Some body members such as quarter panels do not contribute to body strength and rigidity.

D. Body strength is obtained by body shape and design.

6. A front-wheel-drive car has an improper toe-out on turns setting, and a visual check indicates all the steering linkage and suspension components are satisfactory. The most likely cause of this problem is:

A. A bent lower control arm.

B. A bent front strut.

C. A front strut tower that is out of position.

D. A bent steering arm.

7. All of these statements about SAI and front spindle movement are true EXCEPT:

A. When the steering wheel is turned, the front spindle movement is parallel to the road surface.

B. When the SAI angle is increased, the steering wheel returning force is increased.

C. The SAI angle tends to maintain the wheels in a straight-ahead position.

D. Greater SAI angle is necessary on front-wheel-drive cars to provide directional stability.

8. Front wheel setback occurs when one front wheel is:

A. Tilted inward from the true vertical position.

B. Moved rearward in relation to the opposite front wheel.

C. Tilted rearward from the true vertical position.

D. Inward from its original position.

9. A light-duty truck with a one-piece rear axle housing and a leaf-spring rear suspension has excessive toe-out on the left rear wheel and too much toe-in on the right rear wheel. The most likely cause of this problem is:

A. A broken center bolt in the left rear spring.

B. Both rear springs are sagged.

C. A bent rear axle housing.

D. Worn-out rubber bushings in the shock absorbers.

10. While discussing scrub radius:

A. Most front-wheel-drive cars have a positive scrub radius.

B. If the SAI line meets the road surface outside the tire vertical centerline, the scrub radius is positive.

C. Scrub radius is adjusted by shifting the upper strut tower on a MacPherson strut front suspension.

D. Incorrect scrub radius may be caused by larger-than-specified front tires.

CHAPTER 11

COMPUTER-CONTROLLED SUSPENSION SYSTEMS

Upon completion and review of this chapter, you should be able to understand and describe:

- The types of integrated computer networks used on vehicles.
- The operation of controller area network (CAN) system.
- The conditions that cause a programmed ride control (PRC) system to switch from the normal to the firm mode.
- How the firm ride condition is obtained in PRC struts.
- The major components in an electronic air suspension system.
- How air is forced into and exhausted from the air springs in an air suspension system.
- How an electronic air suspension system corrects low suspension trim height.
- The operation of an electronic air suspension system while driving the car with the doors closed and the brake pedal applied.
- The normal operation of the warning lamp in an electronic air suspension system.
- The three modes in the air suspension system on some modern four-wheel-drive sport utility vehicles (SUVs).
- How unnecessary rear suspension height corrections are prevented on irregular road surfaces with an air suspension system.

- The design of the struts and air springs in an automatic air suspension system.
- The design of an electronic rotary height sensor.
- Speed-leveling capabilities and the advantage of this function in a suspension control module.
- The operation of an automatic ride control (ARC) system in relation to transfer case modes.
- The inputs in an electronic suspension control (ESC) system.
- The advantages of an ESC system with magneto-rheological fluid in the shock absorbers compared to other computer-controlled suspension systems.
- The operation of the rear electronic level control system that is combined with the road sensing suspension system.
- The operation of the speed-sensitive steering system that is combined with the road sensing suspension system.
- The operation of a stability control system.
- The advantages of a traction control system.
- Various vehicle network systems.
- Active cruise control, lane departure warning, and collision-mitigation systems.
- The operation of active park assist system.

Terms To Know

Accelerometer	Air spring solenoid valve	Automatic level control (ALC)
Actuator	Air springs	
Adaptive cruise control system	Antilock brake system (ABS)	Brake pressure modulator valve (BPMV)

Brake pressure switch

City safety system

Controller Area Network (CAN)

Cross-axis ball joints

Damper solenoid valve

Data bus network

Electronic brake and traction control module (EBTCM)

Electronic rotary height sensors

Electronic suspension control (ESC) system

Firm relay

Hall element

Height sensors

Hold valve

Lane departure warning (LDW) system

Lateral accelerometer

Lift/dive input

Light-emitting diodes (LEDs)

Lower vehicle command

MagneRide system

Network Photo diodes

Programmed ride control (PRC) system

Raise vehicle command

Release valve

Soft relay

Speed-sensitive steering (SSS) system

Stabilitrak®

Steering sensor

Steering wheel position sensor (SWPS)

Throttle position sensor (TPS)

Traction control system (TCS)

Trim height

Vehicle dynamic suspension (VDS)

Vehicle speed sensor (VSS)

Vehicle stability control system

Vent valve

Wheel position sensor

Wheel speed sensors

Yaw rate sensor

INTRODUCTION

We are all aware of the ever-accelerating electronics revolution that began in the early 1980s as computers were integrated into vehicle powertrain management. Most industries have felt the impact of this revolution, and the automotive industry is no exception. Computers have greatly influenced the way vehicles are designed and built. Most systems on the automobile, including the suspension system, have been impacted by the computer. Many drivers like a soft, comfortable ride while driving normally on the highway. However, many of these same drivers prefer a firm ride during hard cornering, severe braking, or fast acceleration. A firm ride under these driving conditions reduces body sway and front end dive or lift. Prior to the age of electronics, cars were designed to provide either a soft, comfortable ride or a firm ride. Drivers who wanted a firm ride selected a sports car with a suspension designed to supply the type of ride and handling characteristics they desired. Car buyers who wanted a softer ride purchased a family sedan with a suspension designed to provide a softer, more comfortable ride.

 A BIT OF HISTORY

Suspension systems evolved slowly for many years. In the 1940s, many front suspensions were changed from I-beam to short-and-long arm. MacPherson strut suspensions replaced the short-and-long arm suspension systems on front-wheel-drive cars introduced in the late 1970s and 1980s. In the 2000s, we are experiencing an ever-expanding use of electronics technology in automotive suspension systems. Some computer-controlled suspension systems now have the capability to react to road or driving conditions in one millisecond to improve ride and handling quality. The use of electronics technology will continue to improve suspension systems so they provide better handling characteristics and safety with longer tire life.

Thanks to computer control, suspension system manufacturers can now provide a soft ride during normal highway driving, and then almost instantly switch to a firm ride during hard cornering, braking, fast acceleration, and high-speed driving. The computer-controlled suspension system allows the same car to meet the demands of both the driver who desires a soft ride, and the driver who wants a firm ride. Because computer-controlled suspension systems reduce body sway during hard cornering, these systems provide improved steering control.

Some computer-controlled suspension systems also supply a constant vehicle riding height regardless of the vehicle passenger or cargo load. This action maintains the vehicle's cosmetic appearance as the passenger and/or cargo load is changed. Maintaining a constant riding height also supplies more constant suspension alignment angles, which may provide improved steering control.

INTEGRATED ELECTRONIC SYSTEMS AND NETWORKS

Advantages of Integrated Electronic Systems and Networks

With the rapid advances in electronic technology, computer-controlled automotive systems have become integrated. Rather than having a separate computer for each electronic system, several of these systems may be controlled by one computer with control modules integrated. An example of this is the active park assist control module, which is integrated into the body control module (BCM). The active park assist system is an example of CAN data bus integration of various vehicle subsystems modules communicating with one another enabling a complex function. Vehicles without any integrated electronic systems may have many individual modules and computers. Because computers must have some protection from excessive temperature changes, extreme vibration, magnetic fields, voltage spikes, and oil contamination, it becomes difficult for engineers to find a suitable mounting place for this large number of computers. Integration of several electronic systems into one computer solves some of these computer mounting problems and reduces the length of wiring harness. The electronic suspension control (ESC) system explained in this chapter is another example of an integrated electronic system with suspension ride control, suspension level control, and speed-sensitive steering controlled by one computer.

Another method of reducing the number of wires on a vehicle is to interconnect many of the on-board computers with data links. A data link system may be referred to as a **network**. Some input sensor signals may be required by several computers. For example, on some vehicles the vehicle speed sensor (VSS) signal is required by the powertrain control module (PCM), suspension computer, transmission computer, cruise control module, and throttle control module. On many vehicles the VSS is hardwired to the PCM, and then the PCM relays the VSS signal to the other computers via the network.

Some vehicles use a front control module mounted near the front of the vehicle and a rear control module mounted near the rear of the vehicle. These vehicles also have a body computer module (BCM). The BCM, PCM, front and rear control modules, and other modules are interconnected by a network. The headlight switch may be hard-wired to the BCM. When the headlight switch is turned on, a voltage signal is sent to the BCM, and the BCM relays the appropriate LIGHTS ON message through the network to the front and rear control modules. These modules are hard-wired to the exterior lights. When the front and rear control modules receive a specific LIGHTS ON message, these modules turn on the appropriate front and rear lights. Connecting the BCM and the front and rear control modules via a network reduces the number of wires between the light switch and the front and rear lights. The headlight switch in no longer just a switch, but rather an input sensor to the BCM.

In a **network** several computers are interconnected by data links.

 A BIT OF HISTORY

In 1996, the typical vehicle had six electronic control units (ECUs). In 2012, some luxury vehicles had up to 120 modules, 5 main networks, and 12 to 14 subnetworks depending on the on-board electronic equipment.

Some vehicles equipped with power windows and power door locks have a module in each door. These modules are hard-wired to the window motor and door lock controls in each door. These door modules are connected by a network to the BCM that is usually mounted under the dash. The window and door lock control switches are hard-wired to the BCM. When a WINDOW DOWN signal is sent from a window switch to the BCM, the BCM relays this signal through the network to the proper door module. When the door module receives the WINDOW DOWN signal, it supplies voltage to the window motor in the proper direction to roll the window down. Connecting the door modules to the BCM by a network greatly reduces the number of wires connected from the door switches into each door, and this design reduces wiring harness size and weight. Networks also reduce some of the problems associated with wiring harnesses. The previous examples are the type of networking that we see in a Controller Area Network (CAN) described in more detail later in the chapter.

Types of Networks

One type of network system introduced in the early 1980s is the Chrysler Collision Detection (CCD) network. The CCD system has a twisted pair of wires connected between the PCM, BCM, transmission control module (TCM), air bag control module (ACM), electromechanical instrument cluster (MIC), and vehicle theft security system (VTSS) module. On some models the CCD system is also connected to the data link connector (DLC), allowing the computers in the system to communicate with a scan tool connected to the DLC. The network system may be called a data bus. The CCD system operates at 2.5 V.

Another type of network system introduced in the 1980s is the universal asynchronous receive and transmit (UART) system. The UART data links are connected between various on-board computers and the DLC. The UART system operates at 5 V and transmits data at 8.2 kilobits per second. When sending data, the UART system toggles the voltage from 5 V to ground at a fixed bit pulse width. At rest, the UART network system has 5 V.

With the implementation of on-board diagnostic II (OBD-II) systems in 1996, improved communication was required between the PCM, other computers, and the scan tool. Class 2 networks were installed to meet this demand. Class 2 networks transmit data at 10.4 kilobits per second, and transmit data by toggling the voltage from 0 to 7 V. At rest this network system has 0 V. The programmable communication interface (PCI) networks were also introduced on some vehicles to increase the data communication requirements. The PCI network system is a single wire system. The PCI system operates between 0 V and 6 to 8 V. Communication on a network system is accomplished by sending a group of 0 and 1 signals (**Figure 11-1**). A long voltage pulse at a high voltage and a short pulse at a low voltage represent a 0 signal. Conversely, a short pulse at a high voltage and a long pulse at a low voltage indicate a 1 signal (**Figure 11-2**).

Today's vehicles are now equipped with CAN systems. A local area network (LAN) system is similar to the CAN system, and the LAN system is used on a significant number of vehicles. Some vehicles have a low-speed LAN system and a separate high-speed LAN system (**Figure 11-3**). The low-speed LAN system interconnects modules for

Figure 11-1 A data link system transmits data by using a group of 0 and 1 signals.

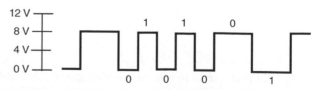

Figure 11-2 Voltage level and duration required for 0 and 1 signals.

Figure 11-3 Low-speed LAN and high-speed LAN data link systems.

applications such as door locks, window motors, HVAC, and radio. The low-speed LAN system is a single-wire system. The high-speed LAN system interconnects modules such as the PCM, transmission, antilock brake modules, and suspension modules. The high-speed LAN system is a two-wire system that operates at 500 kilobits per second. The greatly increased data transmission speed capabilities of the LAN system enhance the communication between various computers in the system and the scan tool. Other high-speed data link systems on modern vehicles include FlexRay and local interconnect network (LIN). The FlexRay data link system transmits data at 10 megabits per second (Mbps). The Byteflight network used on some luxury cars has much in common with the FlexRay network. The network systems on a current SUV are shown in **Figure 11-4**. Some networks, such as the one illustrated in Figure 11-4, have a gateway module. The gateway module changes and directs the signals to go to the appropriate network within the complete network system. The gateway module is often combined within one of the other network computers.

Some networks, such as CAN, contain terminators located at both ends of the network. The terminators are usually positioned inside some of the network computers. Terminators provide electrical resistance to absorb data and prevent this data from being transmitted back into the network.

Most luxury vehicles presently have a media-oriented systems transport (MOST) network system in which the computers are interconnected by fiber-optic data links. The MOST system transmits data at 150 megabits per second. The data transmission rate depends on the model year of the vehicle and data link system. For example, early model MOST systems transmitted data at 22.6 Mbps and current MOST systems transmit data at 150 Mbps. This high data transmission speed is required on vehicles with navigation systems, CD changers, video systems, and satellite radios. The MOST system greatly reduces the number of wires in the wiring harnesses, but is more expensive compared to a network interconnected by wires.

Networks have collision resolution (CR) capabilities to prevent data collisions. The CR system varies depending on the network. In a single-wire CAN system, the system voltage is high when not transmitting data. When any computer wants to transmit data, it initiates a low voltage condition to begin transmission. As explained previously in this chapter, the data is a series of low and high voltages. The low voltage signal is the dominant voltage bit and the high voltage signal is the recessive voltage bit. The CR system uses the dominant and recessive voltage bits to determine transmission priority. If two computers transmit data at the same time, the CR system will recognize the computer with the most dominant bits and give priority to that computer. The low-priority computer stops its communication, and the high-priority computer continues data transmission. In a typical network, air bag, antilock brake, and suspension computers have high priority compared with audio computers.

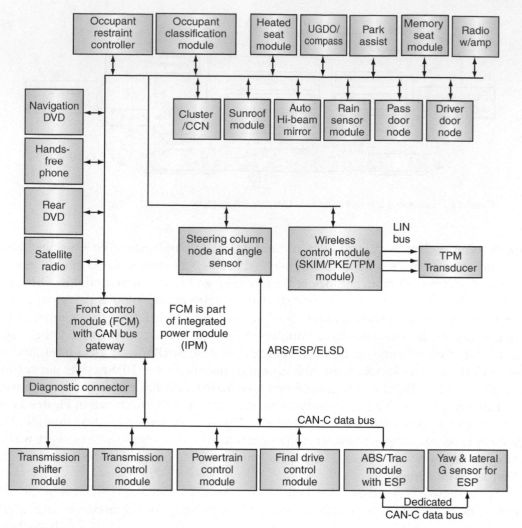

Figure 11-4 Data link systems on a current SUV.

DATA BUS NETWORK

Often information that is transmitted to and from a control module is in the form of a serial data stream. A serial data stream is a digitized code of ones and zeros that is known as a binary code. Each one or zero in a data stream is referred to as a binary digit or *bit* of information. A group of bits form a binary term or *byte*. The wire or wires over which serial data is transmitted is referred to as a **data bus network**. Often the data bus network is a twisted pair of wire (**Figure 11-5**). The twisted pair helps to eliminate the induction of electro-magnetic-interference (EMF), which could disrupt or cloud the data signal. Any microprocessor that communicates on the data bus network is referred to as a node. A node may only be capable of transmitting (sending) information, or may have bi-directional capabilities allowing it to both send and receive data on the network. The two wires on a CAN bus are the Can High and the CAN Low. On the CAN network, the entire CAN High and CAN Low pins of all the nodes are connected together. Both the CAN High and CAN Low wires transmit the same data across the network but in what amounts to a mirror image of the information on the two lines. In other words, they are a check and balance for each other. In the event of a network or connection problem the system will see a difference in the data packet over the two lines and set a Network diagnostic trouble code known as a U-code.

Control modules (nodes) that may be multiplexed on a data bus network allowing them to share information and sensor data between one another. Examples of control

Figure 11-5 Twisted wire pair used to share information between the controllers such as Node 2 the Powertrain Control Module (PCM), Node 1 the Body Control Module (BCM), and Node N the Traction Control Module.

modules that may share data on a data bus network include the PCM, BCM, TCM, instrument panel cluster (IPC), and electronic brake control module (EBCM) to name a few (**Figure 11-6**). The data bus network eliminates the need to run hard wire from each sensor to each control module, instead the information is shared on the bus network. The

Figure 11-6 Data bus network used to share information between the controllers, such as the powertrain control module (PCM) and the transmission control module (TCM), as well as the DLC.

Figure 11-9 Four different communication networks on one vehicle.

data flows on one network, Class B mid-speed data flows on a second network, while Class A low-speed data flows on a third network. As an example, the intake air temperature (IAT) sensor will place its data on the network data bus allowing any control module on the network direct access to the information without the need for one control module (i.e., BCM) requesting the information from another control module (i.e., PCM). The BCM has direct access to information without having to request it from the PCM.

In 1996 the EPA specified that all vehicles be able to transmit generic scan tool data. However, proprietary data, any data other than P0 codes, and data steams were free to use any other protocol the manufacturer chose. The CAN PCM still transmits data to the DLC in SAE's generic scan tool protocol as specified by the EPA, for generic scan tool data

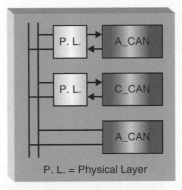

Figure 11-10 When multiple nodes are sending data simultaneously to the control module, the CAN system segments transmissions based on the priority identifier.

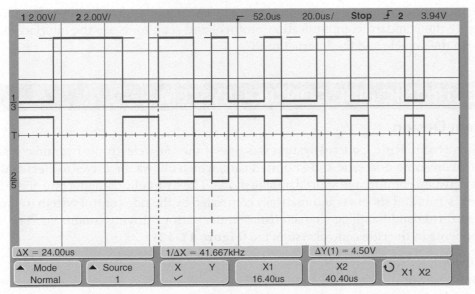

Figure 11-11 Waveform of the CAN+ and the CAN− twisted pair data bus network wires.

Figure 11-12 An example of a typical CAN bus network.

communication, such as generic DTC's. But in order to access all the functions available you will need to have a scan tool, which is compatible with CAN if that is the vehicle's network operation system. The EPA emission regulations for the 2008 model year have specified CAN as the new scan tool communications protocol for all vehicles sold in the United States, providing the repair technician more data for trouble shooting emission failures. With CAN the industry finally has a single standard for on-board diagnostic communication.

The CAN protocol still allows access to the typical DTC information and data streams, but with enhanced DTC detail. A scan tool is also capable of bi-directional communication directly with a smart sensor or actuator node as well as other control modules on the network. In addition, flash calibration for almost all nodes on the network will become common place. A smart sensor is capable of reporting the result of internal voltage drops, opens, grounds, and other self-test features. The network has the ability to take faulty sensors off line and can self-diagnose the difference between a faulty device or a circuit.

The EPA has required pass-thru flash programming using a standard PC connected to the Internet or using a data disk beginning in 2003 and required by 2008. The CAN is currently the standard communication protocol and scan tools are the interface device between the vehicle and the Internet or PC.

ELECTRONIC RIDE CONTROL SYSTEM

System Design

The **Electronic Ride Control system** was one of the early electronic hydraulic systems available on some domestic and imported cars, which allows the driver to electronically select between comfort and sport damping mode. The hydraulic damping rate of the front and rear struts and shocks is automatically controlled by the ride control system to provide improved ride and handling characteristics under various driving conditions. The main components in the ride control system are (**Figure 11-13**):

The **programmed ride control (PRC) system** adjusts shock absorber and strut damping.

1. Steering sensor
2. Brake sensor
3. Speed sensor

Figure 11-13 Programmed ride control (PRC) system components.

4. Struts and shocks with electric actuators
5. Control module
6. Powertrain control module (PCM)
7. Firm and plush shock relays
8. Mode select switch
9. Mode indicator light

Steering Sensor

The **steering sensor** is mounted on the steering column. This sensor contains a pair of **light-emitting diodes (LEDs)** and a matching pair of **photo diodes**. A slotted disc attached to the steering shaft rotates between the LEDs and photo diodes when the steering wheel is turned (**Figure 11-14**). This disc contains 20 slots spaced at 9-degree intervals. A signal is sent from the steering sensor to the control module in relation to the amount and speed of steering wheel rotation.

Brake Sensor

The brake sensor is a normally open (NO) switch mounted in the brake control valve assembly (**Figure 11-15**). When the brake fluid pressure reaches 400 pounds per square inch (psi) or 2758 kilopascals (kPa), the **brake pressure switch** closes and sends a signal to the control module.

Vehicle Speed Sensor

The vehicle speed sensor is usually mounted in the speedometer cable outlet of the transaxle or transmission (**Figure 11-16**). This sensor sends a vehicle speed signal to the control module. This signal is also used by the PCM.

Strut and Shock Actuators

An **actuator** is positioned in the top of each strut and shock (**Figure 11-17**). Each actuator contains a single pole armature, a pair of permanent magnets, and a position switch. When current is applied through the plush relay to the armature, the magnetic fields of the armature and the permanent magnets repel each other (**Figure 11-18**). This action causes clockwise armature rotation until the armature hits the internal stop. Under this condition, the leaf-spring switch is open in the position sensor circuit and no signal is returned to the ride control module.

An **actuator** in a strut or shock absorber varies damping action when activated and deactivated.

Slotted disc

Brake switch sensor

Brake control valve

Figure 11-14 Steering sensor.

Figure 11-15 Brake switch.

ELECTRONIC AIR SUSPENSION SYSTEM COMPONENTS

Air Springs

Air springs support the chassis weight in an air suspension system.

⚠ **Caution**

The ball joint studs in the upper ball joints do not have a press fit in the steering knuckle. When loosening these ball joint nuts, a hex holding feature on the ball joint stud prevents the stud from rotating when loosening the ball joint nuts. If the upper ball joint stud rotates in the aluminum knuckle, the knuckle opening may be damaged.

In an air suspension system, the **air springs** replace the coil springs in conventional suspension systems. These air springs have a composite rubber and plastic membrane that is clamped to a piston located in the lower end of the spring. An end cap is clamped to the top of the membrane and an air spring valve is positioned in the end cap. The air springs are inflated or deflated to provide a constant vehicle trim height. Front air springs are mounted between the control arms and the crossmember (**Figure 11-20**). The lower end of these air springs is retained in the control arm with a clip, and the upper end is positioned in a crossmember spring seat. The front shock absorbers are mounted separately from the air springs.

In some modern air suspension systems, the air springs are mounted and sealed on the shock absorbers (**Figure 11-21**). The lower end of the shock absorber is attached to the lower control arm through an insulating bushing, and the upper end of the shock absorber is attached to the chassis through an insulating mount. This type of air suspension system has aluminum front lower control arms and spindles, and forged steel upper control arms. Ball joints are mounted in the outer ends of the upper and lower control arms. The aluminum suspension components reduce the unsprung weight and improve ride quality. Reducing vehicle weight also improves fuel economy. The upper and lower ball joints are an integral part of the control arms, and these ball joints cannot be replaced separately.

Other vehicles have the air springs mounted over the front and rear struts, and these struts contain a solenoid actuator that varies the strut valve opening to control ride

Figure 11-20 Front air spring.

Figure 11-21 Air spring mounted on the shock absorber.

firmness (**Figure 11-22** and **Figure 11-23**). These strut actuators are similar to the ones explained previously in this chapter on the PRC system. Some of these air suspension systems that control ride firmness are not driver adjustable. The suspension module controls the strut firmness automatically in relation to the module inputs. This type of system may be called an automatic air suspension system. Some vehicles with solenoid actuators in the struts have up to four suspension modes that may be selected by the driver. One premium luxury car has these driver selectable suspension modes:

1. Comfort—provides a smooth luxurious ride.
2. Automatic—the suspension computer provides the best possible combination of comfort and handling based on speed, driver style, and road conditions.
3. Dynamic—stiffest, lowest, sportiest, most aerodynamic suspension mode.
4. Life—for rougher roads, steep approaches, and deep snow. This mode is used only for low-speed driving.

When the driver selects a suspension mode, the suspension module positions the strut actuators to provide the desired ride quality, and this module also adjusts the air spring pressure to provide the appropriate ride height.

This type of air suspension system dramatically reduces body roll and pitch that occurs during cornering and hard braking. Because this air suspension system lowers the ride height at higher speeds, it improves aerodynamic efficiency.

The rear air springs are similar to the front air springs and also have similar mountings. Some rear air springs are mounted between the rear suspension arms and the frame with the shock absorbers mounted separately from the air springs (**Figure 11-24**). Other rear air springs are mounted over the rear shock absorbers, and the shock absorbers are mounted between the lower control arms and the frame. Some modern air suspension systems on four-wheel-drive vehicles have rear knuckles with **cross-axis ball joints**. The lower cross-axis ball joint is a round insulating bushing mounted in the lower control arm, and a bolt attaches this bushing to the lower end of the knuckle. The upper cross-axis ball joint is a round insulating bushing mounted in the top of the knuckle, and a bolt attaches this bushing to the upper control arm (**Figure 11-25**). In this rear suspension system, the upper and lower control arms are made from aluminum. An adjustable toe link is

Cross-axis ball joints contain large insulating bushings in place of typical ball joints.

Figure 11-22 Front air spring and strut assembly.

Figure 11-23 Rear air spring and strut assembly.

connected from the knuckle to the frame to provide a rear toe adjustment. Rear wheel camber can be adjusted by installing a camber adjustment kit in place of the upper knuckle-to-control arm retaining bolt.

Air Spring Valves

An **air spring solenoid valve** allows air to flow into and out of an air spring.

An **air spring solenoid valve** is mounted in the top of each air spring (**Figure 11-26**). These valves are an electric solenoid-type valve that is normally closed. When the valve winding is energized, plunger movement opens the air passage to the air spring. Under this condition, air may enter or be exhausted from the air spring. Two O-ring seals are located on the end of the valves to seal them into the air spring cap. The valves are installed in the air spring cap with a two-stage rotating action similar to a radiator pressure cap.

Air Compressor

A single piston in the air compressor is moved up and down in the cylinder by a crankshaft and connecting rod (**Figure 11-27**). The armature is connected to the crankshaft, and therefore the rotating action of the armature moves the piston up and down. Armature rotation occurs when 12 V are supplied to the compressor input terminal. Intake and

Figure 11-24 Rear air spring mounting.

Figure 11-25 Rear air suspension system with cross-axis ball joints and air spring mounted over the shock absorber.

Figure 11-26 Air spring valve.

Figure 11-27 Air compressor.

discharge valves are located in the cylinder head. An air dryer that contains a silica gel is mounted on the compressor. This silica gel removes moisture from the air as it enters the system.

Nylon air lines are connected from the compressor outlets to the air spring valves. The compressor operates when it is necessary to force air into one or more air springs to restore the vehicle trim height.

An air **vent valve** is located in the compressor cylinder head (**Figure 11-28**). This normally closed electric solenoid valve allows air to be vented from the system. When it is necessary to exhaust air from an air spring, the air spring valve and vent valve must be energized at the same time with the compressor shut off. Air exhausting is necessary if the vehicle trim height is too high.

Figure 11-28 Air vent valve.

Figure 11-29 Compressor relay.

Compressor Relay

When the compressor relay is energized, it supplies 12 V through the relay contacts to the compressor input terminal (**Figure 11-29**). The relay contacts open the circuit to the compressor if the relay is de-energized. An electronic relay is used in some air suspension systems.

Control Module

The control module is a microprocessor that operates the compressor, vent valve, and air spring valves to control the amount of air in the air springs and maintain the trim height. The control module is located in the trunk (**Figure 11-30**) on some models. On other models, the module is mounted under the instrument panel above the parking brake.

Figure 11-30 Control module.

The control module turns on the suspension service indicator light in the roof panel or instrument panel to alert the driver when a suspension defect occurs. Diagnostic capabilities are designed into the suspension module. On some vehicles the suspension module is called a vehicle dynamics module (VDM). On many air suspension systems the suspension module is interconnected via data links to some of the other on-board computers and the DLC under the dash. A scan tool may be connected to the DLC to diagnose the air suspension and other electronic systems on the vehicle. The data links allow data transmission between the on-board computers and the DLC. For example, the vehicle speed sensor (VSS) signal may be sent through connecting wires to the PCM that controls engine functions. If the VSS signal is required by the suspension module, the PCM transmits the VSS signal through the data links. If an electronic defect occurs in a modern air suspension system, a diagnostic trouble code (DTC) is set in the suspension module memory. When a scan tool is connected to the DLC, the suspension module transmits the DTC through the data links to the DLC and scan tool.

On/Off Switch

The On/Off switch opens and closes the 12 V supply circuit to the suspension module. This switch is located in the trunk (**Figure 11-31**). Depending on the vehicle make and model year, one or two panels in the trunk may have to be removed to access the On/Off switch. The On/Off switch must be turned off before the vehicle is hoisted, jacked, or towed. Certain air suspension service procedures may require this switch to be placed in the Off position.

⚠ WARNING **If the vehicle is hoisted, jacked, or towed with the electronic air suspension switch in the On position, personal injury or vehicle damage may occur.**

Height Sensors

Height sensors send an electric signal to the control module in relation to curb riding height.

In the air suspension system, there are two front **height sensors** located between the lower control arms and the crossmember. A single rear height sensor is positioned between the suspension arm and the frame (**Figure 11-32**). Each height sensor contains a magnet slide that is attached to the upper end of the sensor. This magnet slide moves up and down in the lower sensor housing as changes in vehicle trim height occur (**Figure 11-33**). The lower sensor housing contains two electronic switches that are connected through a wiring harness to the control module.

NOTE: This wire must be grounded during diagnostics or air spring fill.

Figure 11-31 On/Off switch.

Figure 11-32 Height sensor.

Figure 11-33 Rear height sensor mounting.

Figure 11-34 Electronic rotary height sensor.

When the vehicle is at **trim height**, the switches remain closed and the control module receives a trim height signal. If the magnet slide moves upward, the above trim switch opens and a **lower vehicle command** is sent from the height sensor to the module. When this signal is received by the module, it opens the appropriate air spring valve and the vent valve. This action exhausts air from the air spring and corrects the above trim height condition. Downward magnet slide movement closes the above trim switch and opens the below trim switch. If this action occurs, the height sensor sends a **raise vehicle command** to the module. When the control module receives this signal, it energizes the compressor relay and starts the compressor. The control module opens the appropriate air spring valve, and this action forces air into the air spring to correct the below trim height condition. The height sensors are serviced as a unit.

Some air suspension systems have **electronic rotary height sensors**. Each rotary height sensor contains a permanent magnet rotor and a **Hall element** (**Figure 11-34**). An arm on the height sensor is attached to the rotor. The height sensor body is mounted on the chassis and a linkage is connected from the sensor arm to the suspension (**Figure 11-35**). Suspension movement rotates the permanent magnet rotor and changes the voltage signal in the Hall element. The voltage signal from a rotary height sensor is proportional to trim height, above trim height, and below trim height.

Warning Lamp

When the control module senses a system defect, the module turns on the air suspension warning lamp in the roof console or instrument panel to inform the driver that a problem exists (**Figure 11-36**). If the air suspension system is working normally, the warning lamp will be on for one second when the ignition switch is turned from the Off to the Run position. After this time, the warning lamp should remain off. This lamp does not operate with the ignition switch in the start position. The warning lamp is used during the self-diagnostic procedure and the spring fill sequence.

On some vehicles, the air suspension warning light is replaced with a CHECK SUSPENSION message in the instrument panel message center. The suspension module provides a CHECK SUSPENSION message if an electrical defect occurs in the air

Trim height refers to the distance between the chassis and the road surface measured at a specific location recommended by the vehicle manufacturer.

Electronic rotary height sensors have an internal rotating element and these sensors send voltage signals to the control module in relation to the curb riding height.

A **Hall element** is an electronic device that produces a voltage signal when the magnetic field approaches or moves away from the element.

⚠ **Caution**

Never attempt to probe the electronic switches in slide-type height sensors. This action may damage the sensor.

Figure 11-35 Height sensor mounting.

Figure 11-36 Air suspension warning light.

suspension system. An AIR SUSPENSION SWITCHED OFF message appears in the message center if the air suspension switch is in the Off position.

ELECTRONIC AIR SUSPENSION SYSTEM OPERATION

The operation of an air suspension system varies depending on the vehicle make and model year. The following section discusses typical air suspension operation.

Ignition Switch Off

The electronic air suspension system is fully operational for one hour after the ignition switch is turned from the Run to the Off position. During this time, lower vehicle commands are completed unless a height sensor was providing a high signal when the ignition

An electronically controlled suspension system may be called an active suspension system.

switch was turned off. After a one-hour period, raise vehicle commands are acted upon and lower vehicle commands are ignored. The air compressor run time is limited to 15 seconds for rear springs and 30 seconds for front springs.

Ignition Switch in Run Position

When the ignition system has been in the Run position for less than 45 seconds, raise vehicle commands are completed immediately, but lower vehicle commands are ignored. If the ignition switch has been in the Run position for more than 45 seconds, the operation is as follows:

1. If a door is opened with the brake pedal released, raise vehicle commands are completed immediately, but lower vehicle commands are serviced after the door is closed. This action prevents an open door from catching on curbs or other objects.
2. If the doors are closed and the brake pedal is released, all commands are serviced by a 45-second averaging method to prevent excessive suspension height corrections on irregular road surfaces.
3. If the brake is applied and a door is open, raise vehicle commands are completed immediately, but lower vehicle commands are ignored.
4. When the doors are closed and the brake pedal is applied, all commands are ignored by the control module. If a command to raise the rear suspension is in progress under these conditions, this command will be completed. This action prevents correction of front end jounce while braking.

General Operation

When a height sensor sends a raise vehicle command to the control module and the other input signals are acceptable, the module grounds the compressor relay winding and starts the compressor. The module also opens the appropriate air spring valve to allow air flow into the air spring (**Figure 11-37**).

On non-computer-controlled suspension systems, the trim height may be called the curb riding height.

The rear air valve solenoids always operate together, but the front solenoids may be energized independently. When the correct chassis trim height is obtained, the control module opens the circuit from the compressor relay winding to ground and de-energizes the air spring valve. This action shuts off the compressor and traps the air in the air spring.

If a lower vehicle request is sent from a height sensor to the control module and the other input signals are acceptable, the control module opens the air vent valve and appropriate air spring valve. When this action occurs, air is released from the air spring until the correct trim height is obtained. The trim height signal from the height sensor to the module causes the module to close the air vent valve and the air spring valve.

Shop Manual
Chapter 11, page 468

Specific Control Module Operation

Commands are completed by the module in this order: rear up, front up, rear down, front down. When the ignition switch is in the Run position and a command cannot be completed within three minutes, the module turns on the air suspension warning lamp. This lamp remains on until the ignition switch is turned off. On some older models, all the control module memory is erased when the ignition switch is turned off. Therefore, the warning lamp may not indicate a defect immediately if the ignition switch is turned from the Off to the Run position. When a system defect causes the module to illuminate the warning lamp with the ignition in the Run position, other commands may be completed by the module. Commands from the front and rear height sensors are never completed simultaneously.

If an electrical defect occurs in a modern air suspension system, a DTC is set and retained in the suspension module memory. In these systems, DTCs are transmitted via the data links to the DLC under the instrument panel, and the DTCs may be displayed on a scan tool connected to the DLC.

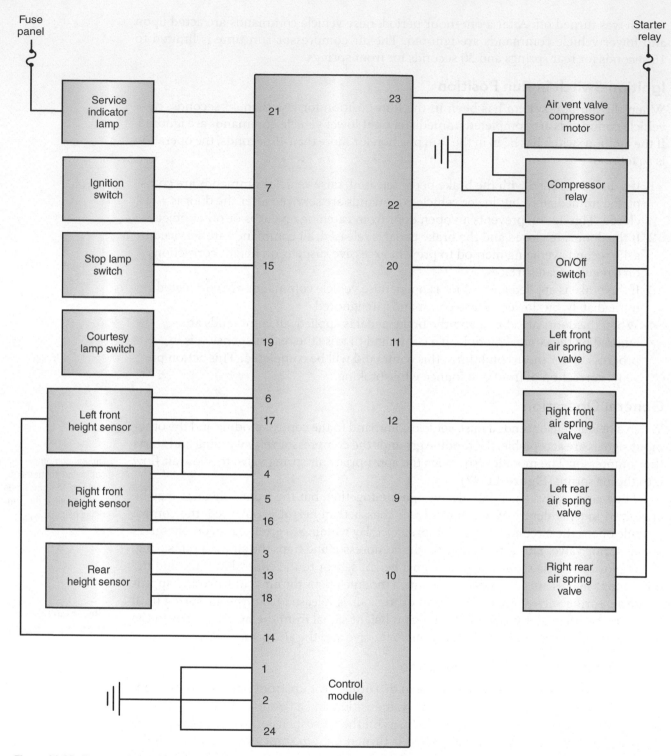

Figure 11-37 Air suspension wiring diagram.

AIR SUSPENSION SYSTEM DESIGN VARIATIONS

Rear Load-Leveling Air Suspension System

Some vehicles have air suspension on only the rear wheels, and these systems may be referred to as rear load-leveling air suspension systems. These air suspension systems have basically the same air springs, compressor and relay, On/Off switch, and rear height sensor as the air suspensions described previously. If additional weight is placed in the trunk, the rear height sensor signals the suspension control module to raise the rear suspension height to the specified trim height. When the extra weight is removed from the trunk, the suspension height is higher than specified, and the air suspension system lowers the rear trim height to the specified height.

Air Suspension System with Speed Leveling Capabilities

Some vehicles have an air suspension system on the front and rear wheels, which is similar to the air suspension systems described earlier in this chapter. However, these systems have an input signal from the vehicle speed sensor (VSS) to the suspension module. On some vehicles the signal from the VSS is sent to the powertrain control module (PCM) that controls engine functions. The PCM transmits the VSS signal through the data links from the PCM to the suspension module. When the VSS signal indicates the vehicle is traveling above 65 mph (105 km/h), the suspension module opens the vent valve and spring solenoid valves to lower the suspension height a specific amount. Under this condition, the vehicle is more dynamically efficient and this improves high-speed vehicle stability and fuel economy.

Special Air Suspension System Features on Four-Wheel-Drive Vehicles

Some four-wheel-drive vehicles with a conventional front torsion bar suspension and a leaf-spring rear suspension also have an air suspension system on all four wheels. These front and rear suspension systems maintain the specified ride height during two-wheel-drive operation. The air suspension system has the same components as described previously, except this air suspension system has front and rear air shock absorbers in place of air springs. An air solenoid is connected in the air line to each air shock absorber (**Figure 11-38**). The air shock absorbers also contain electrical solenoids that rotate the shock absorber valves and vary the ride firmness (**Figure 11-39**). This type of air suspension may be called an automatic ride control (ARC) system.

Figure 11-38 Electrical and air fill line connections to the shock absorber.

Figure 11-39 Strut actuator.

When driving on a smooth road surface in two-wheel drive, the suspension module positions the shock absorber valves to provide a soft ride. If the vehicle is driven on a rough road surface, the suspension's computer switches the shock absorber valves to the firm mode. When the driver selects four-wheel high mode, the suspension module starts the compressor and opens the air shock absorber valves, and increases the air shock absorber pressure until the suspension trim height is raised 1 in. (25.4 mm). If the vehicle speed exceeds 58 mph (93 km/h), the module returns the suspension height to the specified trim height. If the driver selects the four-wheel low mode, the suspension module operates the compressor and air shock absorber valves to increase the trim height 2 in. (50.8 mm). If the vehicle speed exceeds 30 mph (48 km/h in the four-wheel low mode), the module adjusts the suspension height to 1 in. (25.4 mm) above the specified trim height.

Shop Manual
Chapter 11, page 477

VEHICLE DYNAMIC SUSPENSION SYSTEM

Some SUVs are equipped with a **vehicle dynamic suspension (VDS)** system that is similar to the air suspension systems described previously in this chapter. The VDS system has these components:

1. Off/On service switch
2. Two front height sensors
3. One rear height sensor
4. Compressor with internal vent solenoid and air dryer
5. Control module
6. Air lines
7. Front and rear combined air springs and shock absorbers
8. Four air spring solenoids
9. Compressor relay

When increased air pressure is required in an air spring, the control module closes the compressor relay, starts the compressor, and opens the solenoid on the appropriate air spring. To vent air from an air spring, the control module must energize the air spring solenoid and the vent valve. The VDS system has three operating modes:

1. The kneel mode is provided by the suspension module when the ignition switch is in the Off or Lock position and all doors, liftgate, and liftgate glass are closed. In this mode the module opens the vent valve and air spring valves, and slowly reduces the suspension height to 1 in. (25.4 mm) below the specified trim height. This mode improves the ease of entering and exiting the vehicle.
2. When the ignition switch is On, and the transmission is initially shifted into Drive or Reverse, and all the doors, liftgate, and liftgate glass are closed, the VDS switches from the kneel mode to the trim mode, which provides the normal trim height. The VDS system also switches to the trim height mode if the module detects a vehicle speed above 15 mph (24 km/h). Transitions between modes require approximately 30 to 45 seconds.
3. On four-wheel-drive models, the off-road height mode is provided when the driver selects the four-wheel low mode and the vehicle speed is less than 25 mph (40 km/h). In this mode, the module starts the compressor and opens the air spring valve to increase the air spring pressure until the suspension height is 1 in. (25.4 mm) above the specified trim height.

The VDS system also maintains the specified trim height in relation to the weight placed in the vehicle. If a heavy load is placed in the rear of the vehicle, the rear height sensors transmit low trim height signals to the module, and the module opens the rear air spring solenoids and starts the compressor to restore the proper rear suspension trim height. The system stores front and rear trim height when a door or the rear liftgate is opened. The module maintains this suspension height even if weight is added to or removed from the vehicle. When all the doors, liftgate, and liftgate glass are closed, the system returns to normal operation. The VDS system makes limited height adjustments for 40 minutes after the ignition switch is turned Off.

ELECTRONIC SUSPENSION CONTROL (ESC) SYSTEM

General Description

The **electronic suspension control (ESC) system** design, which is favored by manufacturers, uses magneto-rheological fluid in the struts or shocks, which offers advanced damping technology. The magneto-rheological fluid offers rapid response to changing road conditions and driver requirements. Some manufactures have also integrated magneto-rheological fluid into engine mounts and other vehicle dampening mounts to improve driver comfort. The magneto-rheological system automatically controls damping forces in the front and rear struts or shock absorbers within milliseconds in response to various road and driving conditions. This system also offers improved performance of the stability control system on vehicles. The ESC system changes shock and strut damping forces in 1 to 12 milliseconds, whereas other suspension damping systems require a much longer time interval to change damping forces. It requires about 200 milliseconds to blink your eye. This gives us some idea how quickly the ESC system reacts. On some older models the ESC system may be called a continuously variable road sensing suspension (CVRSS).

The ESC module receives inputs regarding vertical acceleration, wheel-to-body position, speed of wheel movement, vehicle speed, and lift/dive (**Figure 11-40**). The CVRSS module evaluates these inputs and controls a solenoid in each shock or strut to provide suspension damping control. The solenoids in the shocks and struts can react much faster compared with the strut actuators explained previously in some systems.

The ESC module also controls the speed-dependent steering system called MagnaSteer® and the automatic level control (ALC). This MagnaSteer® system is similar to the electronic variable orifice (EVO) steering explained in Chapter 13 under

The powertrain control module (PCM) is a computer that controls such output functions as electronic fuel injection, spark advance, emission devices, and transaxle shifting.

The ALC system is similar to the rear load-leveling suspension system explained previously in this chapter.

Figure 11-40 Electronic suspension control (ESC) system.

"Conventional and Electronic Rack and-Pinion Steering Gears." The ALC system controls the air pressure in the rear air shock absorbers to maintain the proper rear suspension height.

Inputs

Position Sensors A **wheel position sensor** is mounted at each corner of the vehicle between a control arm and the chassis (**Figure 11-41** and **Figure 11-42**). These sensor inputs provide analog voltage signals to the ESC module regarding relative wheel-to-body movement and the velocity of wheel movement (**Figure 11-43**). The rear position sensor inputs also provide rear suspension height information to the ESC module, and this

Figure 11-41 Front wheel position sensor.

Figure 11-42 Rear wheel position sensor.

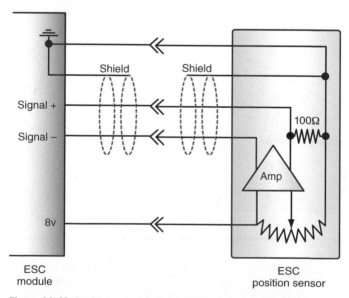

Figure 11-43 Position sensor internal design and wiring diagram.

information is used by the module to control the rear suspension trim height. All four position sensors have the same design. The wheel position sensors may be linear-type or rotary-type.

Accelerometer An **accelerometer** is mounted on each corner of the vehicle. These inputs send information to the ESC module in relation to vertical acceleration of the body. The front accelerometers are mounted on the strut towers (**Figure 11-44**), and the rear accelerometers are located on the rear chassis near the rear suspension support (**Figure 11-45**). All four accelerometers are similar in design, and they send analog voltage signals to the ESC module (**Figure 11-46**). On some later model vehicles, the four accelerometers are replaced by a single accelerometer under the driver's seat. On other late-model vehicles, the accelerometer(s) are eliminated.

Vehicle Speed Sensor The **vehicle speed sensor (VSS)** is mounted in the transaxle. This sensor sends a voltage signal to the PCM in relation to vehicle speed (**Figure 11-47**). The VSS signal is transmitted via data links from the PCM to the ESC module.

Lift/Dive Input The **lift/dive input** is sent from the PCM to the ESC module (**Figure 11-48**). Suspension lift information is obtained by the PCM from the throttle position, vehicle speed, and transaxle gear input signals. The PCM calculates suspension dive information from the rate of vehicle speed change when decelerating.

Figure 11-45 Rear accelerometer position.

Figure 11-44 Front accelerometer mounting location.

Figure 11-46 Accelerometer internal design and wiring diagram.

Figure 11-47 The vehicle speed sensor (VSS) signal is sent to the powertrain control module (PCM) and transmitted to the ESC module.

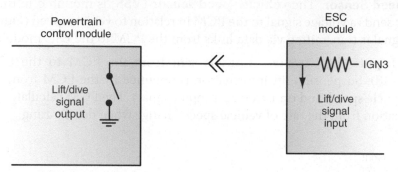

Figure 11-48 The lift/dive signal is sent from the powertrain control module (PCM) to the ESC module.

Electronic Suspension Control Module

Electronic suspension control module contains three microprocessors that control the ESC, MagnaSteer® system if equipped, and automatic level control (ELC). The ESC module is mounted on the right side of the electronics bay in the trunk. Extensive self-diagnostic capabilities are programmed into the ESC module.

Outputs

Shock Absorbers and Struts with Magneto-Rheological (MR) Fluid The shock absorbers or struts in the ESC system contain magneto-rheological (MR) fluid. This fluid is a synthetic fluid containing suspended iron particles. An electric winding is mounted in each shock absorber housing, and the ends of each shock absorber winding are connected to the ESC module. When the shock absorber windings are not energized by the ESC module, the iron particles in the MR fluid are dispersed randomly. Under this condition the MR fluid has a mineral oil–like consistency, and this fluid flows easily through the shock absorber orifices to provide soft ride quality.

When the ESC module energizes the shock absorber windings, the magnetic field around the winding aligns the iron particles in the MR fluid into thick fibrous structures. In this condition the MR fluid has a jelly-like consistency that does not flow easily through the shock absorber orifices (**Figure 11-49**). This fluid change provides firm ride quality. When a shock absorber coil is energized, the amount of attraction between the fibrous particles is proportional to the magnetic field strength surrounding the shock absorber winding, and this field strength is controlled by the current flow through the winding. The computer provides very precise control of the current flow through each shock absorber or strut winding to supply a very broad shock absorber and strut damping range. Based on the wheel position sensor and other inputs, the ESC module can energize each individual shock absorber winding many times per second with a **pulse width modulated (PWM)** signal. Therefore, the ESC system with magneto-rheological fluid in the shock

A **pulse width modulated** signal is a signal with a variable On time that may be used by a computer to control an output.

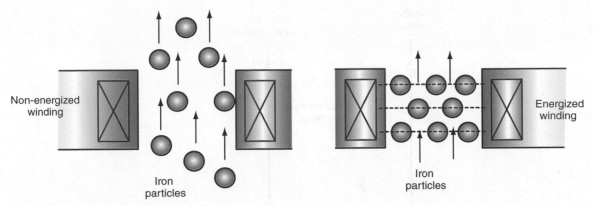

Non-energized
winding

Iron
particles

Energized
winding

Iron
particles

Figure 11-49 Magneto-rheological fluid action in a strut or shock absorber.

absorbers or struts provides an almost infinite variation in shock absorber damping. The ESC system can change the damping characteristics of the MR fluid in 1 millisecond. The ESC system may be called a **MagneRide system**.

The ESC system with MR fluid in the shock absorbers or struts provides greatly improved control of pitch and body roll motions, which supplies better road-holding capabilities, steering control, ride quality, and safety. Recent updates to the ESC system with MR fluid in the shock absorbers or struts include the following:

1. Improved computer software to provide a broader damping range and more precise control of the shock actuators.
2. Expanded sensor inputs including brake pressure, vehicle yaw rate, steering angle, and engine torque.
3. Improved computer networks to allow increased and faster data transmission.

These updates provide improved vehicle dynamics and ride quality without any increase in packaging or weight.

Damper Solenoid Valves On older designs and models that do not use magneto-rheological fluid, but instead use conventional hydraulic fluid, each strut or shock damper contains a solenoid that is controlled by the ESC module. Each **damper solenoid valve** provides a wide range of damping forces between soft and firm levels by increasing or decreasing fluid flow through or around the damper piston by decreasing or increasing the bypass passage size. While this system is responsive to changing road conditions and driving demands, it is not as responsive as the magneto-rheological fluid systems. Strut or shock absorber damping is controlled by the amount of current supplied to the damper solenoid in each strut or shock absorber. Battery voltage and ignition voltage are supplied through separate fuses to the ESC module (**Figure 11-50**). If the damper solenoids are not energized, the struts provide minimum damping force. When the damper solenoids are energized, the struts provide increased damping force for a firmer ride. The ESC module switches the voltage supplied to the damper solenoid in each strut on and off very quickly with a 2 kilohertz pulse width modulated action. If the ESC module keeps the damper solenoid in a strut energized longer on each cycle, current flow is increased through the strut damper solenoid. Under this condition, strut damping force is increased to provide a firmer ride. The ESC module provides precise, variable control of the current flow through each strut or shock damper solenoid to achieve a wide range of damping forces in the struts. The ESC system can change the shock absorber damping forces in 10 to 12 milliseconds.

Each damper solenoid is an integral part of the damper assembly and is not serviced separately. The ESC system operates automatically without any driver-controlled inputs.

Figure 11-50 Strut damper solenoid circuit.

The fast reaction time of the ESC system provides excellent control over ride quality and body lift or dive, which provides improved vehicle stability and handling. Since the position sensors actually sense the velocity of upward and downward wheel movements and the damper solenoid reaction time is 11 to 12 milliseconds, the ESC module can react to these position sensor inputs very quickly. For example, if a road irregularity drives a wheel upward, the ESC module switches the damper solenoid to the firm mode before that wheel strikes the road again during the downward movement.

Rear Automatic Level Control

The **automatic level control (ALC)** system maintains the rear suspension trim height regardless of the rear suspension load. If a heavy object is placed in the trunk, the rear wheel position sensors send below trim height signals to the ESC module. When this signal is received, the ESC module grounds the ALC relay winding and closes the relay contacts that supply voltage to the compressor motor (**Figure 11-51**).

Once the compressor starts running, it supplies air through the nylon lines to the rear air shocks and raises the rear suspension height (**Figure 11-52**). When trim height signals are received from the rear wheel position sensors, the ESC module opens the compressor relay winding circuit and stops the compressor.

If a heavy object is removed from the trunk, the rear wheel position sensors send above trim height signals to the ESC module. Under this condition, the ESC module energizes the exhaust solenoid in the compressor assembly, and this action releases air from the rear air shocks. When the rear wheel position sensors send rear suspension trim height signals to the ESC module, this module shuts off the exhaust solenoid.

An independent ALC system is used on cars without the ESC system. In these systems, the computer is not required and a single suspension height sensor is used. This height sensor contains electronic circuits that control the compressor relay and the exhaust solenoid. This electronic circuit limits the compressor run time and the exhaust solenoid On time to seven minutes.

Figure 11-51 Rear automatic level control (ALC).

Figure 11-52 Nylon air line and rear shock air line fitting.

Speed-Sensitive Steering System

On some models the ESC module operates a solenoid in the **speed-sensitive steering (SSS) system** to control the power steering pump pressure in relation to vehicle speed (**Figure 11-53**). This action varies the power steering assist levels.

The ESC module varies the on time of the steering solenoid. This action may be referred to as pulse width modulation. When the solenoid is in the Off mode, the power steering pump supplies normal power assist. Below 10 mph (16 km/h), the computer

Figure 11-53 Steering solenoid and ESC module wiring diagram.

Figure 11-54 Power steering assist in relation to vehicle speed.

operates the steering solenoid to provide full power steering assist (**Figure 11-54**). This action reduces steering effort during low-speed maneuvers and parking.

A **MagneRide** suspension system has magneto-rheological fluid in the shock absorbers.

As the vehicle speed increases, the ESC module operates the steering solenoid so the power steering assist is gradually reduced to provide increased road feel and improved handling.

On later model cars, the speed-sensitive steering is called speed-dependent steering or MagnaSteer® The module that controls the MagnaSteer® is contained in the electronic brake and traction control module (EBTCM).

VEHICLE STABILITY CONTROL

The National Highway Traffic Safety Administration (NHTSA) required that all new passenger vehicles sold in the United States to be equipped with electronic stability control (ESC) as of the 2012 model year. Some European models were equipped with ESC as early as 1995. A **vehicle stability control system** provides improved control if the vehicle begins to swerve sideways because of slippery road surfaces, excessive acceleration, or a combination of these two conditions. Therefore, a vehicle stability control system provides increased vehicle safety. Vehicle stability control systems have various brand names depending on the vehicle manufacturer. For example, on General Motors vehicles the vehicle stability control system is called **Stabilitrak®**. The Stabilitrak® system is available on many General Motors cars and some SUVs. The module that controls the Stabilitrak® system is combined with the **antilock brake system (ABS)** module and **traction control system (TCS)** module (**Figure 11-55**). This three-in-one module assembly is referred to as the **electronic brake and traction control module (EBTCM)**. The EBTCM is attached to the **brake pressure modulator valve (BPMV)**, and this assembly is mounted in the left front area in the engine compartment. A data link is connected between all the computers including the EBTCM and the ESC module (**Figure 11-56**). The combined EBTCM and ESC systems may be referred to as the integrated chassis control system 2 (ICCS2). Some sensors such as the **steering wheel position sensor (SWPS)** are hard-wired to both the EBTCM and the ESC module (**Figure 11-57**). The signals from other sensors may be sent to one of these modules and then transmitted to the other module on the data link. The data link also transmits data from these modules to the instrument panel cluster (IPC) during system diagnosis. This allows the IPC to display diagnostic information.

This book is concerned with suspension and steering systems. Because the stability control system operates in cooperation with the ABS and TCS systems, a brief description of these systems is necessary.

The **antilock brake system (ABS)** prevents wheel lockup during a brake application.

The **traction control system (TCS)** prevents drive wheel slippage.

The **electronic brake and traction control module (EBTCM)** controls ABS, TCS, and Stabilitrak® functions.

The **brake pressure modulator valve (BPMV)** controls brake fluid pressure to the wheel calipers or cylinders.

The **steering wheel position sensor** supplies a voltage signal in relation to the amount and speed of steering wheel rotation.

Figure 11-55 The electronic brake and traction control module (EBTCM) contains the antilock brake system (ABS), traction control system, and stability control modules.

Figure 11-56 Data link between the EBTCM and ESC modules.

Figure 11-57 The steering wheel position sensor (SWPS) is connected to both the ESC module and the EBTCM.

AUTHOR'S NOTE Statistics compiled by the National Highway Traffic Safety Administration (NHTSA) indicate that stability control systems can reduce the incidence of single-vehicle accidents in SUVs by 63 percent.

Antilock Brake System (ABS) Operation

Wheel speed sensors are mounted at each wheel. In this ABS system, the wheel speed sensors are integral with the front or rear wheel bearing hubs. These wheel bearing hubs with the integral wheel speed sensors are non-serviceable (**Figure 11-58**). Each wheel speed sensor contains a toothed ring that rotates past a stationary electromagnetic wheel speed sensor. This sensor contains a coil of wire surrounding a permanent magnet. As the toothed ring rotates past the sensor, an alternating current (AC) voltage is produced in the sensor. This voltage signal from each wheel speed sensor is sent to the EBTCM. As wheel speed increases, the frequency of AC voltage produced by the wheel speed sensor increases proportionally. During a brake application, the wheels slow down, and the frequency of AC voltage in the wheel speed sensors also decreases. If a wheel is nearing a lockup condition during a hard brake application, the frequency of the AC voltage from that wheel speed sensor becomes very slow. The EBTCM detects impending wheel lockup from the frequency of AC voltage signals sent from the wheel speed sensors.

The brake pressure modulator valve (BPMV) contains a number of electrohydraulic valves. These valves are operated electrically by the EBTCM. These valves in the BPMV are connected hydraulically in the brake system. A **hold valve** and a **release valve** are connected in the brake line to each wheel (**Figure 11-59**). If a wheel speed sensor signal indicates an impending wheel lockup condition, the EBTCM energizes the normally open hold solenoid connected to the wheel that is about to lock up. This action closes the solenoid and isolates the wheel caliper from the master cylinder to prevent any further

---

A **hold valve** opens and closes the fluid passage to each wheel caliper.

When energized, a **release valve** reduces pressure in a wheel caliper.

Figure 11-58 Wheel speed sensor.

increase in brake pressure. If the wheel speed sensor signal still indicates an impending wheel lockup, the EBTCM keeps the hold solenoid closed and opens the normally closed release solenoid momentarily. This action reduces wheel caliper pressure to reduce brake application force and prevent wheel lockup. The EBTCM pulses the hold and the release solenoids on and off to supply maximum braking force without wheel lockup.

When the hold and the release valves are pulsated during a prolonged antilock brake function, the brake pedal fades downward as brake fluid flows from the release valves into the accumulators. To maintain brake pedal height during an antilock brake function, the EBTCM starts the pump in the BPMV at the beginning of an antilock function. When the pump motor is started, the pump supplies brake fluid pressure to the hold valves and wheel calipers. Pump motor pressure is also supplied back to the master cylinder. Under this condition, the driver may feel a firmer brake pedal and pedal pulsations and may hear the clicking action of the hold and the release solenoids.

ANTILOCK and BRAKE warning lights are mounted in the instrument panel. Both of these lights are illuminated for a few seconds after the engine starts. If the amber ANTILOCK light is on with the engine running, the EBTCM has detected an electrical fault in the ABS system. Under this condition, the EBTCM no longer provides an ABS function, but normal power-assisted brake operation is still available. When the red BRAKE warning light is illuminated with the engine running, the parking brake may be on, the master cylinder may be low on brake fluid, or there may be a fault in the ABS system.

Traction Control System (TCS) Operation

The EBTCM detects drive wheel spin by comparing the two drive wheel speed sensor signals. Wheel spin on both drive wheels is detected by comparing the wheel speed sensor signals on the drive wheels and non-drive wheels. If a wheel speed sensor signal informs the EBTCM that one or both drive wheels are spinning, the EBTCM enters the traction control mode. First, the EBTCM requests the PCM to reduce the amount of engine torque supplied to the drive wheels. The PCM reduces engine torque by retarding the ignition spark advance and shutting off the fuel injectors for a very short time. The PCM then sends a signal back to the EBTCM regarding the amount of torque delivered to the drive

Figure 11-59 Brake pressure modulator valve (BPMV).

wheels. If one or both drive wheels continue to spin on the road surface, the EBTCM energizes the normally closed prime valve, closes the normally open isolation valve, and starts the pump in the BPMV. Under this condition, the prime valve opens and the pump begins to move brake fluid from the master cylinder through the prime valve to the pump inlet. The closed isolation valve prevents the pump pressure from being applied back to the master cylinder. Under this condition, the pump pressure is supplied through the normally open hold valve to the brake caliper on the spinning wheel. This action stops the wheel from spinning. If both drive wheels are spinning on the road surface, the EBTCM operates both prime valves and isolation valves to supply brake fluid pressure to both drive wheel brake calipers. During a TCS function, the EBTCM pulses the hold and the release solenoids on and off to control wheel caliper pressure. The EBTCM limits the traction control function to a short time period to prevent overheating brake components. During a TCS function, these messages may be displayed in the driver information center (DIC):

1. TRACTION ENGAGED is displayed after the TCS is in operation for three seconds.
2. TRACTION SUSPENDED is displayed if the EBTCM has discontinued the TCS function to prevent brake component overheating.
3. TRACTION OFF is displayed if the driver places the TCS switch on the instrument panel in the Off position.
4. TRACTION READY is displayed if the TCS switch is turned from Off to On.

During a TCS function, the EBTCM sends a signal through the data link to the PCM. When this signal is received, the PCM disables some of the fuel injectors to reduce engine torque. This action also helps to prevent drive wheel spin. The PCM disables the two injectors at the beginning of the firing order and in the center of the firing order. Depending on the speed of drive wheel spin, the PCM may disable every second injector in the firing order. The injectors are disabled for a very short time period. The TCS system improves drive wheel traction, and this system also prevents the tendency for the vehicle to swerve sideways when one drive wheel is spinning. Therefore, the TCS system increases vehicle safety.

Vehicle Stability Control

To provide stability control, the EBTCM uses two additional input signals from the **lateral accelerometer** and the **yaw rate sensor**. The lateral accelerometer is mounted under the front passenger's seat (**Figure 11-60**). The EBTCM sends a 5 V reference voltage to the lateral accelerometer. If the vehicle is driven straight ahead, the chassis has zero lateral acceleration. Under this condition, the lateral accelerometer provides a 2.5 V signal to the EBTCM. If the vehicle begins to swerve sideways because of slippery road conditions, high-speed cornering, or erratic driving, the lateral accelerometer signal to the EBTCM varies from 0.25 V to 4.75 V, depending on the direction and severity of the swerving action.

The yaw rate sensor is mounted under the rear package shelf (**Figure 11-61**). Some yaw rate sensors contain a precision metal cylinder whose rim vibrates in elliptical shapes. The vibration and rotation of this metal cylinder is proportional to the rotational speed of the vehicle around the center of the cylinder. On some models, the lateral accelerometer and yaw rate sensors are combined in a single sensor. The sensor mounting location varies depending on the vehicle make and model year.

The EBTCM sends a 5 V reference voltage to the yaw rate sensor. If the vehicle chassis experiences zero yaw rate, the yaw rate sensor sends a 2.5 V signal to the EBTCM module.

A **lateral accelerometer** supplies a voltage signal in relation to sideways movement of the chassis.

Yaw is erratic, side-to-side deviation from a course. The **yaw rate sensor** supplies a voltage signal in relation to rotational chassis speed during a sideways swerve.

Figure 11-60 Lateral accelerometer.

Figure 11-61 Yaw rate sensor.

If the vehicle begins to swerve sideways, the yaw rate sensor provides a 0.25V to 4.75V signal to the EBTCM, depending on the direction and severity of the swerve. The EBTCM also uses the wheel speed sensor signals for stability control. If the vehicle begins to swerve sideways, the EBTCM energizes the normally closed prime valve and closes the normally open isolation valve connected to the appropriate front wheel; then it starts the pump in the BPMV. Under this condition, the prime valve opens and the pump begins to move brake fluid from the master cylinder through the prime valve to the pump inlet. The closed isolation valve prevents the pump pressure from being applied back to the master cylinder. Under this condition, the pump pressure is supplied through the normally open hold valve to the brake caliper on the appropriate front wheel. Applying the brake on the front wheel pulls the vehicle out of the swerve and prevents the complete loss of steering control. If the EBTCM detects an electrical fault in the stability control system, STABILITY REDUCED is displayed in the DIC. If the EBTCM enters the stability control mode, STABILITY ENGAGED is indicated in the DIC.

In **Figure 11-62**, two vehicles driving side-by-side are negotiating a lane change to the left. The vehicle on the right has vehicle stability control, and the vehicle on the left

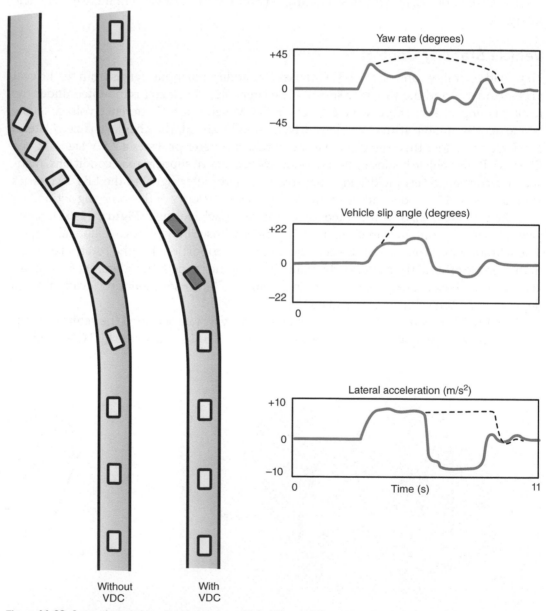

Figure 11-62 Comparison during a turn between a vehicle with a stability control system and a vehicle with no stability control system.

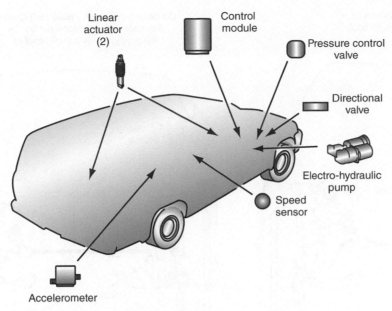

Figure 11-63 Active roll control system components.

does not have this system. As the vehicle on the left begins to turn, the rear of the vehicle begins to swing around. When the vehicle on the right starts to turn, the rear of the vehicle swerves slightly and the right front brake is applied by the vehicle stability control system to prevent this swerve. Further into the turn, the driver attempts to steer the car on the left, but this car enters into an uncontrolled swerve with loss of steering control. As the car on the right continues into the turn, the rear of the vehicle swerves slightly, but the vehicle stability control system again applies the right front brake momentarily to prevent this swerve. The car with the stability control system completes the turn while maintaining directional stability, but the vehicle without stability control goes into an uncontrolled swerve with complete loss of directional control. The vehicle stability control system improves vehicle safety! The other charts in **Figure 11-63** indicate that yaw rate, vehicle slip angle, and lateral acceleration are greatly reduced on a vehicle with a stability control system.

ACTIVE ROLL CONTROL SYSTEMS

Two independent automotive component manufacturers have developed active roll control systems in response to concerns about SUVs, which roll over more easily compared with cars because of their higher center of gravity. The active roll control system contains a control module, accelerometer, speed sensor, fluid reservoir, electrohydraulic pump, pressure control valve, directional control valve, and a hydraulic actuator in both the front and rear stabilizer bars (**Figure 11-63**). The accelerometer and speed sensor may be common to systems other than the active roll control. The electrohydraulic pump may also be used as the power steering pump. The active roll control system may be called a roll stability control (RSC) system or a dynamic handling system (DHS). Some active roll control systems have a gyro sensor that provides a voltage signal to the control module in relation to the vehicle roll speed and angle.

When the vehicle is driven straight ahead, the active roll control system does not supply any hydraulic pressure to the linear actuators in the stabilizer bars. Under this condition, both stabilizer bars move freely until the linear actuators are fully compressed. This action provides improved individual wheel bump performance and better ride quality. If the chassis begins to lean while cornering, the module operates the directional valve

Cornering roll—no system.
Stabilizer bar deflects due to body
roll motion.

Cornering roll—with Active Roll Control.
Actuator deflects stabilizer bar by
extending. Body roll eliminated.

Cornering
force

Stabilizer bar

Cornering
force

Stabilizer bar

Actuator

Figure 11-64 Active roll control system operation while cornering.

so it supplies fluid pressure to the linear actuators in the stabilizer bars. This action stiffens the stabilizer bar and reduces body lean (**Figure 11-64**). The active roll control system increases safety by reducing body lean, which decreases the possibility of a vehicle rollover.

One vehicle manufacturer has chosen to reduce the yaw force on a new model by reducing the weight at the front and rear of the vehicle and shifting the weight toward the midpoint on the car. This was accomplished by manufacturing the hood, trunk lid, and bumper beams from aluminum. Engine weight was reduced by installing hollow camshafts, and using a lighter weight block and cylinder heads. These changes reduced the weight of the vehicle by 180 lb (80 kg), which reduced the yaw force and also improved fuel economy and performance.

ADAPTIVE CRUISE CONTROL (ACC) SYSTEMS

Adaptive cruise control (ACC) is available on some current vehicles. The ACC system has a long-range radar sensor mounted behind the front bumper. Radar stands for Radio Detection and Ranging. Radar is an electronic system developed for establishing the position of an object, and is based on the principle that electromagnetic waves are reflected by surfaces of an object and the waves returned are detected as an "echo" of the object. The ACC system measures the distance to the vehicle in front and the relative speed of that vehicle (**Figure 11-65**) using the radar. The radar emits a frequency modulated continuous wave at a 200 megahertz modulation and a carrier signal with a frequency of 76.5 gigahertz (**Figure 11-66**). The ACC sensor monitors the time it takes for the signal to be returned. The further the distance, the longer it takes for the signal to be returned (**Figure 11-67**). This radar sensor sends signals to the ACC computer. To determine the speed of the vehicle in front, a phenomenon known as the Doppler effect is used. An example of the Doppler effect can be demonstrated, as an emergency vehicle approaches, the siren sounds to be a constant high-pitch (high-frequency) noise. But as the emergency vehicle passes and moves farther away the tone now sounds lower (lower frequency). If the distance between the transmitter and the object increases, the frequency of the

Figure 11-65 The radar sensor measures the distance to the vehicle in front and the relative speed of this vehicle.

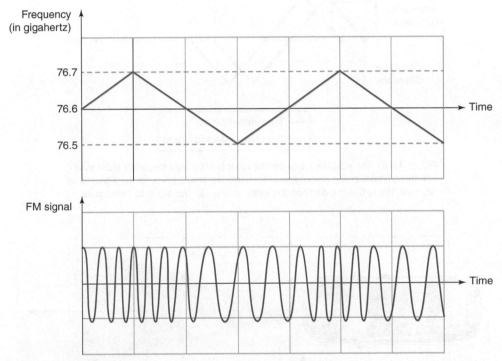

Figure 11-66 The adaptive cruise control radar emits a high-frequency radio wave modulated continuous at 200 megahertz and a carrier signal with a frequency of 76.5 gigahertz.

reflected wave decreases, and if the distance between the transmitter and the object decreases, the frequency of the reflected wave increases. The adaptive cruise control determines the vehicle in front has increased its speed by a decreasing frequency signal (ΔfD) received. The result is a differential frequency signal difference between the transmitted leading signal ($\Delta f1$) and the received trailing edge signal ($\Delta f2$). This signal difference is analyzed by the control module to determine vehicle in front's speed .The radar signal produced is a lobe-shaped signal pattern whereby the signal strength decreases as the distance from the transmitter is increased (**Figure 11-68**). The ACC system uses a three-beam radar to determine the angle the vehicle in front is traveling. The three radar receivers compare the ratio of the signal strengths (amplitude) returned to calculate angular information (**Figure 11-69**). The ACC computer uses the CAN network to remain in constant contact with the PCM and ABS computers. The ACC system can detect objects more than 330 ft. (100 m) ahead of the vehicle.

Figure 11-67 The adaptive cruise control radar emits a high-frequency radio wave and the adaptive cruise control sensor monitors the time it takes for the signal to be returned. The further the distance, the longer it takes for the signal to be returned.

Figure 11-68 The adaptive cruise control radar signal strength decreases with increased distance from transmitter both in height and length.

A dash control allows the driver to set the distance between the vehicle and the vehicle ahead. A 1-second time delay behind the vehicle in front is used for sport driving mode and is best suited for stimulating driving and for slow-moving traffic. A 1.3- to 1.8-second delay behind the vehicle in front is used for standard driving mode, and is best suited for relaxed free-flowing traffic. A 2.3-second delay behind the vehicle in front is used for touring driving mode and is best suited for country roads or when hauling a trailer.

Figure 11-69 The adaptive cruise control uses a three-beam radar system in order to calculate angular information about the travel direction of the vehicle in front.

If the ACC system detects a slower-moving vehicle ahead and the distance to that vehicle becomes less than the driver-adjusted setting, the ACC system uses the throttle control and limited brake application, with a maximum possible braking action of 25 percent of full braking potential, to slow down the vehicle. When this action is taken, a small green car icon is illuminated in the head up display (HUD). If the distance to the vehicle in front is still decreasing after this action is taken, the icon in the HUD turns from green to yellow. When the ACC system senses that a collision is imminent, a large red car icon with yellow flashes is illuminated in the HUD and the ACC system activates a beeper.

The ACC system uses input information from the wheel speed sensors, steering angle and turning sensor to evaluate bends in the road. This information along with an average lane width programmed into the system is used to determine the fictitious lane the vehicle is traveling in and determine if there are vehicles around or in the travel lane (**Figure 11-70**).

B = Average lane width
R = Cornering radius

Figure 11-70 The adaptive cruise control system uses information from steering input sensors to determine bends in the road based on an average lane width, and monitors the road ahead.

When the lane ahead of the vehicle is clear, the ACC system maintains the cruising speed set by the driver. The ACC system can detect another vehicle crossing from an adjacent lane into the lane in front of the vehicle and take appropriate action if there is not sufficient distance between the two vehicles. The radar signals in the ACC system provide excellent performance even in adverse weather conditions.

Some of the limitations of the ACC system are as follows:

- Functional speed regulation limitations 18-124 mph (30-200 km/h)
- Radar operation may be impaired by snow, rain, freezing rain, and spray from other vehicles.
- Tight corning conditions will affect the operation of the system because of limited radar detection range.
- The adaptive cruise control system does not react to stationary objects. The driver must pay attention to and react to road hazards.
- The adaptive cruise control system is not a safety system and does not provide full autonomous driving control. The vehicle operator is fully responsible for the safe operation of the vehicle at all times. Regardless of whether advanced driver assistance systems are engaged.

LANE DEPARTURE WARNING (LDW) SYSTEMS

Some vehicles are equipped with **lane departure warning (LDW) systems**. The LDW system is just one example of driver assistance systems that have been integrated into vehicles and discussed in this chapter. A dash switch allows the driver to turn the LDW system on or off.

The introduction of electromechanical power steering systems that use an electric motor for power steering assist instead of hydraulic pressure and electronic throttle control has allowed for advancements in active and direct intervention with driving behavior. Advanced software systems have been developed to make use of existing vehicle sensors and actuators used already on other systems along with CAN networks to further increase driver safely and assist the driver with difficult driving maneuvers. In fact, because the LDW system is only available when bundled with other advanced driver control systems, the only additional hardware required are the LDW control unit with camera mounted in rearview mirror, and a warning lamp for the lane departure system; the rest of the system relies on existing components and the addition of software (**Figure 11-71**).

The LDW system integrates functions that were once limited only to humans or robots in science fiction. But they are here and include:

- Seeing (optical perception of a situation)
- Thinking (evaluation of a situation)
- Acting (reacting to a situation)

The LDW system camera and control module (**Figure 11-72**) are mounted in the rearview mirror assembly. The camera captures a gray scale image with a resolution of 640 x 480 pixels and color depth of 4096 gray scale, at a capture rate of 25 images per second allowing for virtual lane calculations even at higher speeds. The digitized road image is recorded within a range of 18–197 ft. (5.5–60 m) in front of the vehicle. For the system to function correctly, the recordable lane width must be between 8–15 ft. (2.45–4.6 m) and the distance between two successive lane marking stripes may not exceed twice the length of the marking itself (**Figure 11-73**). The lane departure system uses two trapezoidal image areas on the left and right side of the image to analyze lane markings (**Figure 11-74**). The lane departure system analyzes the camera images to create a virtual travel lane that the vehicle is traveling on (**Figure 11-75**). By looking for one or more gray scale values increasing on both the left and right lane line markings the system can calculate

Legend

Lane departure warning system
J759 Lane departure warning control unit

K240 Lane departure warning lamp

Z67   Windscreen heater for lane departure warning

a   Gray scale camera in the lane departure
      warning control unit

Electromechanical power steering system
G269 Steering moment sender

J500 Power steering control unit

V187 Electromechanical power steering motor

Dash panel and steering column electronics
E2   Turn signal switch
E617 Button for driver assistance systems

H3   Buzzer and gong

J119 Multifunction display
J285 Control unit in dash panel insert
J527 Steering column electronics control unit

Brake system
F   Brake light switch

J104 ABS control unit

b   Speed sensors on the wheels
c   Available traction control systems,
      for example ESP

Engine management
G28   Engine speed sender
G79   Accelerator position sender

J623 Engine control unit

Cruise control
G550 Sensor for automatic distance control

J428 Adaptive cruise control unit

Further components
G17   Ambient temperature sensor

J393 Convenience system central control unit
J519 On-board supply control unit
J533 Data bus diagnostic interface

d   Windscreen wiper functional readiness

Figure 11-71 The lane departure warning system is another example of a driver assistance system that relies on existing sensors and actuators from other vehicle systems and advanced CAN network communication.

Interior rearview mirror base designed cover

Camera lens in the lane departure warning control unit

Figure 11-72 The lane departure warning system camera and control unit are mounted in the rearview mirror assembly.

vehicle position with precision. The system now calculates the true course of the road integrating in safety limits for vehicle position and the lateral orientation of the vehicle position within this virtual lane is determined (**Figure 11-76**). If the vehicle comes close to or crosses the virtual lane lines, steering correction is implemented. The system also has the ability to detect if the driver's hand is on the steering wheel. All roads have imperfections and when a driver grips a steering wheel these road imperfections cause the tires to move, which in turn applies torque to the steering linkage and steering gear. If the driver

Evaluation of individual image lines

Detection of gray scale value leaps

Digitization

Figure 11-73 The lane departure warning system camera digitizes the image to analyze vehicle lane position.

Selection of defined lines in the trapezoidal detection range

Figure 11-74 The lane departure system camera captures 25 frames per second of two trapezoidal images to analyze lane markings and vehicle position.

Definition of marking points for the true course of the road

Figure 11-75 The lane departure system analyzes the camera images to create a virtual travel lane that the vehicle is traveling on.

Figure 11-76 The lane departure system next calculates the virtual position of the vehicle with the travel lane.

Legend for examples 1 and 2

G269 - Steering moment sender
J759 - Lane departure warning control unit
J500 - Power steering control unit
V187 - Electromechanical power steering motor
a - Electronic warning sound
b - Warning vibration
c - Warning message in the dash panel insert display
d - Sensor signal

Figure 11-77 The lane departure warning system can sense that the driver is holding onto the steering wheel from torque information received by the steering column torque angle sensor.

is holding onto the steering wheel, the steering column torque angle sensor will detect this torque input (**Figure 11-77**). If the driver lets go of the steering wheel, the system senses the lack of torque resistance data received from the steering column torque angle sensor (**Figure 11-78**).

As advancements continue in the miniaturization of electronics and as computer artificial intelligence continues to improve, we will see more advanced driver assistance systems. One day the driver may be eliminated altogether.

Legend for examples 1 and 2

G269 - Steering moment sender
J759 - Lane departure warning control unit
J500 - Power steering control unit
V187 - Electromechanical power steering motor
a - Electronic warning sound
b - Warning vibration
c - Warning message in the dash panel insert display
d - Sensor signal

Figure 11-78 The lane departure warning system can sense that the driver is no longer holding onto the steering wheel from the lack of torque information received by the steering column torque angle sensor.

The LDW system has a video camera mounted behind the rearview mirror (**Figure 11-79**). This camera uses software to monitor highway lane markings (**Figure 11-80**). The camera measures the distance from the vehicle to the lane markings and the lateral velocity of the vehicle in relation to the lane markings to determine if the vehicle is moving out of the lane. Signals from the camera are sent to the LDW module. The vehicle speed sensor information is also sent to the LDW module. The LDW system does not operate below 45 mph (72 km/h). If the vehicle speed is above 45 mph (72 km/h) and the camera signal indicates the vehicle is moving out of the lane, the LDW module operates a buzzer and an indicator warning in the instrument panel. Activation of the turn signals temporarily disables the LDW system to prevent incorrect LDW warnings.

Figure 11-79 Components in the lane departure warning (LDW) system.

Figure 11-80 The LDW camera measures the distance from the vehicle to the highway lane markings.

Figure 11-81 The lane departure warning system will stay engaged even while cornering if the cornering maneuver is completed in less than 100 seconds.

Some LDW systems have lane-departure prevention capabilities. If the vehicle is moving toward the lane markings on the left, the ABS and stability control computer lightly pulse the brakes on the right side of the vehicle. When the wheels on the right side slow down, the wheels on the left side turn faster than the wheels on the right side, and this action steers the car away from the highway markers on the left. Other LDW systems activate the electronic (electromechanical) steering on the vehicle to steer the vehicle away from the lane markers if a lane departure is starting to occur. The LDW system will engage the electromechanical power steering system to first cause a slight vibration of the steering wheel to warn driver to take corrective action to maintain lane position and then take corrective action to maintain lane position. If the driver exerts more than 2.21 ft.-lb (3 Nm) of pressure on the steering wheel, the LDW system will disengage to allow a lane change without using turn signal. The LDW system will continue to function even while corning as long as the corning maneuver is completed in less than 100 seconds (**Figure 11-81**). The lane departure system can be adversely affected by weather and road surfaces. While the system is effective at detecting lane lines on asphalt, it can have issues detecting lane lines on concrete roads on bright sunny days. The system can also be affected by some wet surfaces due to reflection or snow-covered surfaces that cover lane lines. A dirty windshield that affects camera image will also cause lane departure system issues and will set a diagnostic trouble code "Lane Assist – no sensor visibility present."

AUTHOR'S NOTE Statistics compiled by one of the major car manufacturers indicate that 30 percent of all vehicle accidents in the United States involve rear-end collisions. In half of these collisions, the driver did not apply the brakes before collision occurred. The statistics also indicate that 75 percent of these rear-end collisions occur below 19 mph (30 km/h).

COLLISION MITIGATION SYSTEMS

Some vehicles are equipped with a collision mitigation system that uses radar and camera information to detect a vehicle in front. The collision mitigation computer is in constant communication with the PCM and ABS computers via the vehicle network. If the radar and camera signals indicate a vehicle in front is too close and a collision is imminent, the collision mitigation computer warns the driver with a visual and audible warning. If the driver does not apply the brakes, the collision mitigation computer applies the brakes aggressively with a 0.5 g force to slow down the vehicle. The vehicle manufacturer's information indicates that the collision mitigation system will avoid rear-end collisions up to 9 mph (19 km/h) and mitigate the effects of a collision up to 19 mph (30 km/h). On some vehicles, the collision mitigation system is called a **city safety system**.

ACTIVE PARK ASSIST

> ⚡ **Caution**
>
> The active park assist system sensors rely on ultrasonic waves. Sensors may not detect objects in heavy rain, objects with surfaces that absorb ultrasonic waves, or cause disruptive reflections.

Some vehicle manufacturers have integrated an active park assist system as a convenience to their drivers to help aid in parallel parking. In order for a vehicle to be equipped with active parking assist, the vehicle must be equipped with an electronic power steering assist motor as well as other systems such as ABS, traction control, and backup sensors. When the active park assist system is engaged it will search for a suitably sized space to park and display a graphic and a message on the display screen indicating that the system is searching for a space (**Figure 11-82**). The left or the right turn signal may be engaged to tell the system which side of the vehicle to search for a parking space. If the turn signal is not engaged, the system default is to search on the passenger side (**Figure 11-83**). Once the system locates a suitable space the display screen will notify the driver with an audible signal and message and direct the driver to pull forward slowly and stop when the next audible signal is heard and the screen displays a message to stop vehicle (**Figure 11-84**). When the vehicle is shifted into reverse, the system will re-verify that the parking spot is large enough to accept the vehicle. The drive controls the accelerator, gear shift, and the brakes. The active park assist system controls vehicle steering hands-free, steering the vehicle into the space. The system uses both audible and visual directions to direct driver's response (**Figure 11-85**). The active park assist system is designed to assist with parallel parking, but does not replace driver's attention and judgment. The active park assist system also has a park out feature that will automatically steer the vehicle out of the parking space.

Figure 11-82 When engaged the active park assist system will direct the driver with both audible and visual signals to detect a suitable space that will fit vehicle.

Figure 11-83 The active park assist system will by default search for a space on the passenger side.

Figure 11-84 The active park assist system will detect a suitable space that will fit the vehicle and direct driver to pull forward.

Figure 11-85 The active park assist system will direct the driver to place the vehicle in reverse and let go of the steering wheel.

The following conditions will deactivate the active park assist system:

- Holding on to the steering wheel
- Driving above 6 mph (10 km/h) during parking maneuver
- Driving above 0 mph (80 km/h) for 10 seconds during active parking spot search
- Turning off traction control
- Antilock brake activation or failure
- Traction control activation on a gravel surface or slippery road

The active park assist control module uses inputs from parking aid sensors, active park assist sensors, steering angle and torque sensor, as well as traction control module data and an advanced geometrical equation to analyze if vehicle will fit in a parking space and assist in parking vehicle. The active park assist system is an example of CAN data bus integration of various vehicle subsystems modules communicating with one another enabling a complex function (**Figure 11-86**). The subsystems that work with active park assist include the electromechanical power steering, powertrain control module, ABS, traction control, BCM, and steering column electronics. The active park assist sensors are narrow beam ultrasonic sonar sensors that continuously send out an inaudible ultrasonic sound waves at a consistent speed to detect the distance of other vehicles and obstacles around the vehicle and identify open parking spaces (**Figure 11-87**). The active park assist system ultrasonic sensors contain an emitter that sends out an ultrasonic sound wave beam and a piezoceramic receiver element to detect the return signal that has bounced off a detected object (**Figure 11-88**). The active park assist module calculates the echo time, the time it takes for the sound wave to return, to determine the distance of the object from the sensor (**Figure 11-89**). At sea level, atmospheric pressure of 14.7 psi (1 bar), sound with travel through 68°F (20°C) air at 1125 ft./sec (343 m/s). The speed at which sound travels depends on the density of the air, therefore the system uses information from the ambient air temperature sensor as a correction value in the distance calculations.

Fortunately, as with many advanced computer-controlled features on vehicles, a sophisticated self-diagnostic software system will set problem-specific diagnostic trouble codes to aid in pinpointing a system fault. These codes will be set as body (B) codes.

Figure 11-86 The active park assist system uses many inputs from various vehicle sensors.

☐ **A** Actuation of both side sensors for parking space search
■ **B** Actuation of the ten sensors for PARKTRONIC functionality

Figure 11-87 The active park assist system uses ultrasonic sensors that are always sending out a narrow ultrasonic beam to detect objects around the vehicle.

Figure 11-88 The active park assist system's ultrasonic sensors contain an emitter that sends out an ultrasonic beam and a piezoceramic receiver element; the active park assist module calculates the echo time to judge distance of objects.

Test setup for distance measurement

Transmission of an ultrasonic signal

Test measurement of reflected sound waves

Figure 11-89 The active park assist system's ultrasonic sensors measure the amount of time it takes for the sound wave the sensor generated to be reflected off an object and return to the sensor to calculate the distance an object is from the vehicle.

TELEMATICS

The use of telematics is expected to increase significantly in the next few years. Vehicle manufacturers and electronic communications companies have not decided the exact information that will be transmitted to vehicles through telematics, but the following are some possible options.

Vehicle manufacturers could use telematics to inform vehicle owners when their vehicles require specific service. This is being done presently to some extent by General Motors through their OnStar system. Telematics could be used for emission testing. At present, 33 states in the United States have emissions test centers, and vehicles must be taken to these centers for emissions testing at specific intervals. Using telematics, the emissions test center could monitor the vehicle emissions levels, and if the emissions standards are exceeded, the test center could inform the vehicle owner or driver that the vehicle must be taken to a certified repair center for emissions service.

Telematics may also be used to improve vehicle safety. At present, a Vehicle Infrastructure and Integration (VII) system has installed two test projects, one in California and another one in the Detroit area. In these test projects, roadside beacons communicate with vehicle networks. In each project, approximately 25 vehicles supplied by car manufacturers are driven on a 75-mi. (120-km) stretch of highway where the communication beacons are installed. At hidden intersections, monitoring stations inform the test vehicles that vehicles are approaching the intersection. Other inputs inform the test vehicles regarding heavy traffic congestion or black ice.

Telematics may be defined as wireless communication between the telephone system and vehicle networks.

⚡ Interesting Fact

Each year in the United States, approximately 43,000 deaths are caused by traffic accidents. This is approximately the equivalent of a 747 aircraft crashing each week.

AUTHOR'S NOTE The following are some interesting facts about the OnStar communications system:

1. OnStar is a factory-installed option on General Motors vehicles. One-year OnStar service is included in the new GM vehicle purchase. There are approximately 7 million OnStar subscribers.
2. A sophisticated module in each vehicle allows communication between the vehicle and an OnStar center at the push of a button in the vehicle. This communication is enabled by the vehicle battery and a special external antenna on each vehicle that combine cell reception, global positioning system (GPS), and XM satellite radio to provide range and performance that far exceed handheld cell phones and GPS devices.
3. A customer contacts OnStar every two seconds, which adds up to about 5 million calls a month. At the beginning of an OnStar call, the driver must select the English, Spanish, or French language option, and a special routing system directs the call to an advisor speaking the selected language.
4. Ninety-five percent of emergency calls are answered in five seconds. The OnStar module on the vehicle notifies a center if the air bags deploy. An advisor calls the vehicle in an attempt to contact the driver or vehicle occupants. If the advisor cannot contact the driver or occupants, the advisor notifies emergency response personnel regarding the vehicle location. The OnStar system handles approximately 2200 vehicle crash responses per month.
5. The OnStar system processes about 600,000 requests for driving directions each month. OnStar subscribers purchase approximately 31 million minutes of OnStar time per month.
6. Current OnStar modules are connected to the CAN data network on the vehicle. This connection allows the OnStar module to monitor the powertrain, ABS,

traction control, and air bag systems. On current vehicles, the OnStar module can read about 1600 DTCs and monitor the engine oil life, emissions levels, and tire pressure. About 3 million OnStar subscribers have signed up for a monthly vehicle condition report that informs the subscriber regarding vehicle defects in any of the monitored systems. The monthly vehicle condition report also informs the subscriber regarding any recall repairs that have not been completed on their vehicle. Approximately, 61,000 calls are processed monthly for instant remote diagnosis.

7. If an OnStar subscriber is locked out of one's vehicle, he or she can phone an OnStar advisor and provide his or her OnStar number. The advisor can then unlock the doors on the subscriber's vehicle. If an OnStar advisor is informed that a vehicle is stolen, the advisor can use GPS to provide the vehicle location. Select 2009 General Motors vehicles have OnStar's Stolen Vehicle Slowdown service. If police are following a stolen vehicle, they can inform an OnStar advisor and request that the vehicle be stopped. The advisor will flash the vehicles external lights without the driver's awareness. When police confirm that the vehicle lights have flashed, the OnStar advisor will send a signal to the stolen vehicle that causes the electronic throttle control to remain continually in the idle position. All other vehicle functions such as steering and brakes operate normally. The OnStar subscriber cannot request that the vehicle be stopped. At present, OnStar receives about 700 stolen vehicle calls per month. It is estimated that there are 30,000 high-speed chases in the United States each year, and 25 percent of these end in injury; 300 deaths result from these chases.

8. OnStar has expanded beyond the United States and Canada and is now offered in China and Mexico.

SUMMARY

- In a programmed ride control (PRC) system, the steering sensor informs the control module regarding the amount and speed of steering wheel rotation.
- The PRC system switches from the normal to the firm mode during high-speed operation, braking, hard cornering, and fast acceleration.
- The struts and shock absorbers in some PRC systems provide three times as much damping action in the firm mode as in the normal mode.
- The accelerometer in a CCR system contains a mercury switch and this accelerometer sends a vehicle acceleration signal to the control module.
- In a CCR system, the accelerometer signal or the vehicle speed signal may inform the control module to switch from the normal to the firm mode.
- An electronic air suspension system maintains a constant vehicle trim height regardless of passenger or cargo load.
- To exhaust air from an air spring, the air spring solenoid valve and the vent valve in the compressor head must be energized.
- To force air into an air spring, the compressor must be running and the air spring solenoid valve must be energized.

- The air spring valves are retained in the air spring caps with a two-stage rotating action much like a radiator cap.
- An air spring valve must never be loosened until the air is exhausted from the spring. Voltage is supplied through the compressor relay points to the compressor motor. This relay winding is grounded by the control module to close the relay points.
- The On/Off switch in an electronic air suspension system supplies 12 V to the control module. This switch must be off before the car is hoisted, jacked, towed, or raised off the ground.
- If a car door is open, the control module does not respond to lower vehicle commands in an electronic air suspension system.
- When the brake pedal is applied and the doors are closed in an electronic air suspension system, the control module ignores all requests from the height sensors.
- In an electronic air suspension system, if the doors are closed and the brake pedal is released, all requests to the control module are serviced by a 45-second averaging method.
- If the control module in an electronic air suspension system cannot complete a request from a

height sensor in 3 minutes, the control module illuminates the suspension warning lamp.

- In an automatic air suspension system, the control module controls suspension height and strut damping automatically without any driver-controlled inputs.
- Rotary height sensors in automatic air suspension systems contain Hall elements. These sensors send voltage signals to the control module in relation to the amount and speed of wheel jounce and rebound.
- Some air suspension systems reduce trim height at speeds above 65 mph (105 km/h) to improve handling and fuel economy.
- The air suspension system on some four-wheel-drive vehicles increase suspension ride height when the driver selects four-wheel-drive high or four-wheel-drive low.
- The air suspension system on some four-wheel-drive vehicles have the capability to provide firmer shock absorber valving in relation to transfer case mode selection, vehicle speed, and operating conditions.

- The electronic suspension control system changes shock and strut damping forces in 10 to 12 milliseconds.
- In the electronic suspension control system, the module controls suspension damping, rear electronic level control, and speed-sensitive steering automatically without any driver-operated inputs.
- A vehicle stability control system applies one of the front brakes if the rear of the car begins to swerve out of control. This action maintains vehicle direction control.
- An adaptive cruise control (ACC) system applies the vehicle brakes lightly and warns the driver if the system senses inadequate distance to the vehicle in front.
- A lane departure warning (LDW) system warns the driver if the vehicle begins to leave the current lane. Some LDW systems apply the vehicle brakes on one side or steers the vehicle back into the current lane when lane departure begins to occur.
- A collision mitigation system warns the driver and applies the vehicle brakes aggressively if a rear-end collision is about to occur with the vehicle in front.

REVIEW QUESTIONS

Short Answer Essays

1. Describe the operation of the steering sensor in a programmed ride control (PRC) system.

2. Describe the purpose of the vehicle speed sensor signal in a programmed ride control (PRC) system.

3. Explain how air is forced into an air spring in a rear load-leveling air suspension system.

4. Describe the action taken by the control module if the control module in an electronic air suspension system receives a lower vehicle command from a rear suspension sensor with the doors closed, the brake pedal released, and the vehicle traveling at 60 mph (100 km/h).

5. Describe the action taken by the control module if the engine is running with a door open, and the control module receives a lower vehicle command from the height sensor in a rear load-leveling air suspension system.

6. List the conditions when the On/Off switch in an electronic air suspension system must be turned off.

7. Describe the conditions required for the control module to turn on the suspension warning lamp continually with the engine running in an electronic air suspension system.

8. Explain why the control module in an electronic air suspension system services all commands by a 45-second averaging method when the doors are closed and the brake pedal is released.

9. Explain why the suspension warning lamp in an electronic air suspension system may not indicate a defect immediately when the engine is started.

10. Explain how a vent solenoid is damaged by reversed battery polarity in a rear load-leveling air suspension system.

Fill-in-the-Blanks

1. The armature response time is _____ milliseconds in a programmed ride control (PRC) system strut.

2. In a programmed ride control (PRC) system, if the car is accelerating with the throttle wide open, the PRC system is in the _____ mode.

3. When the programmed ride control (PRC) mode switch is in the Auto position, the control module changes to the firm mode if lateral acceleration exceeds_____.

4. In a programmed ride control (PRC) system, lateral vehicle acceleration is sensed from the _____ sensor.

5. In a rear load-leveling air suspension system, the control module energizes the compressor relay when a _____ command is received.

6. Two height sensors are mounted on the _____ suspension in an electronic air suspension system.

7. In an electronic air suspension system two hours after the ignition switch is turned off, _____ _____ commands are completed, but _____ commands are ignored.

8. In a rear load-leveling air suspension system, if the On/Off switch in the trunk is off, the system is_____.

9. An electronic rotary height sensor contains a_____ _____.

10. In a continuously variable road sensing suspension system, the module senses vehicle lift and dive from some of the_____ _____ inputs.

Multiple Choice

1. While discussing a programmed ride control (PRC) system:
   A. The brake system pressure must be 300 psi (2068 kPa) before this mode change occurs.
   B. A PRC system switches from the auto mode to firm mode if the vehicle accelerates with 90 percent throttle opening.
   C. The PRC system switches to the firm mode if lateral acceleration exceeds 0.25 g.
   D. The mode indicator light in the tachometer is illuminated in the plush ride mode.

2. To increase the rear trim height in an electronic air suspension system, the control module:
   A. Starts the compressor and opens the vent valve.
   B. Starts the compressor and closes the rear air spring valves.
   C. Stops the compressor and opens the rear air spring valves.
   D. Starts the compressor and opens the rear air spring valves.

3. All these statements about air springs, shock absorbers, and struts are true EXCEPT:
   A. Air springs can be mounted separately from the shock absorbers.
   B. Air springs can be mounted over the front and rear struts.
   C. Some struts contain a solenoid actuator that controls strut firmness.
   D. Some air suspension systems have seven driver selectable operating modes.

4. The magneto-rheological fluid in the shock absorbers or struts in an ESC system contains:
   A. Automatic transmission fluid.
   B. Suspended iron particles.
   C. Power steering fluid.
   D. 5W-30 engine oil.

5. To sense the distance to the vehicle in front, an adaptive cruise control system uses:
   A. A video camera.
   B. A short-range radar signal.
   C. A digital camera.
   D. A long-range radar signal.

6. In an ESC system with magneto-rheological fluid in the shock absorbers or struts, the control module can vary the shock absorber firmness in:
   A. 1 millisecond.
   B. 5 milliseconds.
   C. 10 milliseconds.
   D. 152 milliseconds.

7. All of these statements about network systems are true EXCEPT:

   A. A network system can be a single-wire or dual-wire system.

   B. Some network systems operate between 0 V and 12 V.

   C. Some vehicles have two network systems.

   D. A fiber-optic network system has a very high data transmission rate.

8. In some traction control systems if the control module senses drive wheel spin, the first action taken by the control module is to:

   A. Retard the spark advance and shut off the fuel injectors.

   B. Apply the brakes on both non-drive wheels.

   C. Apply the brake on the spinning drive wheel.

   D. Apply the brake on the non-spinning drive wheel.

9. On a vehicle with a stability control system, if icy road conditions cause the vehicle to begin swerving sideways, the stability control module:

   A. Applies the brakes on both rear wheels.

   B. Applies the brakes on both front wheels.

   C. Applies the brake on one front wheel.

   D. Applies the brake on one front wheel and the opposite rear wheel.

10. All of these statements about a rear load-leveling suspension system are true EXCEPT:

   A. The On/Off switch is mounted in the vehicle trunk.

   B. The control module operates the compressor relay.

   C. If a door is open, the control module completes the lower suspension height commands.

   D. The rear suspension has one, non-serviceable suspension height sensor.

CHAPTER 12
POWER STEERING PUMPS

Upon completion and review of this chapter, you should be able to understand and describe:

- The difference between a conventional V-belt and a serpentine belt.
- The advantages of a serpentine belt compared with a conventional V-belt.
- The main components in a power-assisted rack and pinion steering system, and explain the steering gear mounting position.
- Two different types of power steering pump reservoirs.
- The difference between a hydro-boost power steering system and an integral power steering system.
- Three different types of power steering pump rotor designs.

- The power steering pump operation while driving with the front wheels straight ahead.
- The power steering pump operation while the vehicle is turning a corner.
- The power steering pump pressure relief operation, and explain when this operation occurs.
- The design and purpose of an electro-hydraulic power steering module.
- Electrohydraulic power steering (EHPS) systems.
- Hybrid electric vehicle (HEV) operation.

Terms To Know

Cam ring	Outlet fitting venturi	Remote fluid reservoir
Flow control valve	Parallel HEVs	Series HEVs
Hydro-boost power-assisted steering system	Pressure relief ball	Vane-type power steering pumps
Integral fluid reservoir	Rack and pinion power steering system	Venturi
Integral power-assisted steering system	Regenerative braking	

INTRODUCTION

Power steering systems have contributed to reduced driver fatigue and made driving a more pleasant experience. Nearly all power steering systems at present use fluid pressure to assist the driver in turning the front wheels. Since driver effort required to turn the front wheels is reduced, driver fatigue is decreased. The advantages of power steering have been made available on many vehicles, and safety has been maintained in these systems.

There are several different types of power steering systems, including integral, rack and pinion, hydro-boost, and linkage type. In any of these systems, the power steering pump is the heart of the system because it supplies the necessary pressure to assist steering.

The power steering pump drive belt is a simple, but very important, component in the power steering system. A power steering pump in perfect condition will not produce the required pressure for steering assist if the drive belt is slipping. Various types of steering systems, drive belts, and pump designs are described in this chapter.

POWER STEERING PUMP DRIVE BELTS

Many power steering pumps are driven by a multi-ribbed belt that surrounds the crankshaft pulley and the power steering pump pulley. If a multi-ribbed belt is used to drive other components in addition to the power steering pump, such as the water pump, alternator, or air-conditioning compressor, it is referred to as a serpentine belt (**Figure 12-1**). The serpentine belt is much wider than a conventional V-belt, and the underside of the belt has a number of small ribbed grooves. Many serpentine belts have spring-loaded automatic belt tensioners that eliminate periodic belt tension adjustments. Since the serpentine belt may be used to drive all the belt-driven components, these components are placed on the same vertical plane, which saves a considerable amount of underhood space. The smooth back of the serpentine belt may also be used to drive one of the components. Regardless of the type of belt, the belt tension is critical. A power steering pump will never develop full pressure if the belt is slipping.

Serpentine belts may be constructed of either Neoprene or ethylene propylene diene M-class rubber (commonly called EPDM). The majority of belts are constructed of EPDM, but vehicles produced prior to 2001 or inexpensive aftermarket replacement belts may still be made of neoprene. Visually it is difficult to tell the two materials apart. Older neoprene belts were designed to last between 50,000–60,000 miles and often showed signs of wear by that time. EPDM belts are designed to last 80,000–100,000 miles and seldom show signs of outward visual wear unless there is a problem. The EPDM belt is more elastic than a standard neoprene belt and resists cracking even at higher mileage. A better indicator of when to replace EPDM belts is rib wear. Belts are designed to have clearance between the rib peaks and the pulley grooves. All belts are exposed to dirt, grit, rocks, road salt, and water. Over time, these contaminants cause the EPDM belt to gradually lose material on the ribbing similar to the way a tire wears out causing the belt to ride deeper in the grooves. A 5 to 10 percent loss of material is enough to cause belt slippage, overheating, and hydroplaning. Hydroplaning occurs when the belt ribs sit deeper in the pulley grooves

Figure 12-1 Serpentine belt.

due to wear and there is not enough room for water to escape. Instead, the water is trapped between the belt ribs and the pulley grooves lifting the belt away from the pulley and causing slippage. A slipping serpentine belt can cause check engine lights associated with misfire codes, air-conditioning compressor codes, reduced alternator output, reduced engine cooling, and poor air-conditioning performance to name a few. Belts should be checked for wear beginning at 50,000 miles. See Chapter 3 in the Shop Manual for further detail on diagnosing both Neoprene and EPDM belts, and how to measure for belt grove wear. It is wise to replace the drive belt when any pulley driven component is replaced.

Another change that has occurred in belts is the use of stretch fit belts that self tension on the drive pulleys. They do not require any mechanical adjustment and are used in limited applications, such as a single drive belt for air-conditioning compressor. As the name implies, they contain an elastomeric material and with the use of a special tool they are stretched on to the pulley system, there is no mechanical means to loosen or tighten the belt and as with the standard EPDM belt are designed to last 80,000–100,000 miles.

Some current vehicles are equipped with a stretchy belt that does not require a belt tensioner or belt adjustment. The stretchy belt has a similar appearance compared to a conventional serpentine belt. However, stretchy belts have tensile cords made from a polyamide material that is three times more elastic compared with the cords in a conventional serpentine belt. In a stretchy belt, the rubber layers around the cord layer have superior durability and improved adhesion to the cord layer to accommodate the stretching action. Eliminating the mechanical belt tensioner saves weight, space, and financial cost. The stretchy belt is usually installed only around two pulleys such as the air-conditioning compressor and crankshaft pulleys, and the remaining belt-driven components have a serpentine-belt drive.

If a stretchy belt requires replacement, the vehicle manufacturer recommends cutting the old belt to remove it. A special tool is used to lift and guide the new stretchy belt onto the pulley.

An earlier drive belt design was the V-belt. The sides of a V-belt are the friction surfaces that drive the power steering pump (**Figure 12-2**). If the sides of the belt are worn and the lower edge of the belt is contacting the bottom of the pulley, the belt will slip. The power steering pump pulley, crankshaft pulley, and any other pulleys driven by the V-belt *must be properly aligned*. If these pulleys are misaligned, excessive belt wear occurs.

⚠ WARNING **Always keep hands, tools, and equipment away from rotating belts and pulleys. If any of these items become entangled in rotating belts, personal injury and equipment damage may result.**

⚠ WARNING **Always keep long hair tied back while working in the automotive shop! If long hair is entangled in rotating belts and pulleys, personal injury will result.**

Shop Manual
Chapter 12, page 512

AUTHOR'S NOTE It has been my experience that the most common cause of power steering pump complaints is the pump drive belt. A loose, worn, or dry belt may cause a chirping noise at idle or squealing during acceleration. A loose or worn belt may cause a humming-type vibration noise. Intermittent or continual hard steering may be caused by a loose pump drive belt. A loose power steering pump belt may even cause a complaint of hard steering when driving on wet days, because the belt has more tendency to slip when it is wet. A loose belt may cause low power steering pump pressure when testing the pump. Therefore, when diagnosing power steering pump problems, you should always inspect the pump drive belt and measure the belt tension before any other diagnostic tests are performed.

Figure 12-2 Conventional V-belt.

TYPES OF POWER-ASSISTED STEERING SYSTEMS

Rack and Pinion Steering System

The **rack and pinion power steering system** is used on most front-wheel-drive cars. In this steering system, the power steering pump is bolted to a bracket on the engine, and the pump is driven by a belt from the crankshaft. In most front-wheel-drive cars, the engine is mounted transversely, and the steering gear is mounted on the cowl behind the engine or on the crossmember below the engine (**Figure 12-3**).

In many power steering systems, an **integral fluid reservoir** is attached to the power steering pump. A dipstick is mounted in the reservoir for fluid level checking (**Figure 12-4**). A high-pressure hose and a low-pressure return hose are connected from the pump to the steering gear (**Figure 12-5**). The high-pressure hose is usually a steel-braided hose

Figure 12-3 Power steering gear mounting.

Figure 12-5 Power steering pump to gear hoses.

Figure 12-4 Power steering reservoir and dipstick.

with appropriate fittings in each end. The return hose is a rubber-braided hose that is clamped to the fittings on the pump and gear.

Some power steering pumps have a **remote fluid reservoir**. This type of system is commonly used on small cars with limited underhood space, where it may be difficult to access an integral reservoir on the power steering pump. The remote reservoir is placed in a convenient position, and the return hose is connected from the steering gear to the reservoir. A second return hose is routed from the reservoir to the pump (**Figure 12-6**).

A **remote reservoir** is mounted externally from the power steering pump.

> ## A BIT OF HISTORY
>
> The number of independent automotive repair facilities and the number of individuals employed by these shops have increased steadily in the last decade. In 1995, there were 63,844 independent automotive mechanical repair facilities in the United States, and they employed 244,430 workers. In 2010, 723,400 technicians and mechanics were employed in the United States, with an expected growth of 17 percent (or 124,800 jobs) over the next 10 years.

Integral Power-Assisted Steering System

In the **integral power-assisted steering system**, the pump is bolted to a bracket on the engine, and the recirculating ball steering gear is mounted on the frame beside the engine. This type of steering system is used on many rear-wheel-drive cars and light-duty trucks. The pump is driven by a belt from the crankshaft and an integral reservoir is mounted on the pump. A high-pressure hose and a return hose are connected from the pump to the steering gear (**Figure 12-7**).

In an **integral power-assisted steering system**, the control valve and power cylinder are contained in the steering gear.

Hydro-Boost Power-Assisted Steering System

The **hydro-boost power-assisted steering system** is used on many light-duty trucks with diesel engines and on some cars. Since the hydro-boost system does not use manifold vacuum for brake assist, this system is suitable for diesel engines, which have reduced intake manifold vacuum compared with gasoline engines. The power steering on these systems is similar to an integral unit, but the power steering pump pressure is also applied to the hydro-boost unit in the brake master cylinder. Hydraulic lines are routed from the power steering pump to the steering gear, and another set of lines is connected from the pump to the master cylinder (**Figure 12-8**). In the hydro-boost system, the power steering

Figure 12-6 Remote reservoir on power-assisted rack and pinion steering system.

Figure 12-7 Integral power-assisted steering system.

Figure 12-8 Hydro-boost power-assisted steering and brake system.

pump pressure applied to the master cylinder pistons acts as a brake booster to assist the driver in applying the brakes. The hydro-boost system does not have a conventional vacuum booster on the master cylinder.

POWER STEERING PUMP DESIGN

Various types of power steering pumps have been used by car manufacturers. Many **vane-type power steering pumps** have flat vanes that seal the pump rotor to the elliptical pump cam ring (**Figure 12-9**). Other vane-type power steering pumps have rollers to seal the rotor to the cam ring. In some pumps, inverted, U-shaped slippers are used for this purpose. The major differences in these pumps are in the rotor design and the method used to seal the pump rotor in the elliptical pump ring. The operating principles of all three types of pumps are similar.

A balanced pulley is pressed on the steering pump drive shaft. This pulley and shaft are belt-driven by the engine. A spring-loaded lip seal at the front of the pump housing prevents fluid leaks between the pump shaft and the housing. The oblong pump reservoir is made from steel or plastic. A large O-ring seals the front of the reservoir to the pump housing (**Figure 12-10**).

Smaller O-rings seal the bolt fittings on the back of the reservoir. The combination cap and dipstick keep the fluid reserve in the pump and vent the reservoir to the atmosphere. Some power steering pumps have a variable assist steering actuator in the back of

Shop Manual
Chapter 12, page 526

Figure 12-9 Power steering pump rotor and vanes.

Figure 12-10 Power steering pump housing and reservoir.

the pump housing. The PCM operates this actuator to provide increased steering assist at low vehicle speeds.

The rotating components inside the pump housing include the shaft and rotor with the vanes mounted in the rotor slots. A seal between the output shaft and the housing prevents oil leaks around the shaft. As the pulley drives the pump shaft, the vanes rotate inside an elliptical opening in the **cam ring**. The cam ring remains in a fixed position inside the pump housing. A pressure plate is installed in the housing behind the cam ring (**Figure 12-11**).

A spring is positioned between the pressure plate and the end cover, and a retaining ring holds the end cover in the pump housing. The flow control valve is mounted in the pump housing, and a magnet is positioned on the pump housing to pick up metal filings rather than allowing them to circulate through the power steering system (**Figure 12-12**).

The flow control valve is a precision-fit valve controlled by spring pressure and fluid pressure. Any dirt or roughness on the valve results in erratic pump pressure. The flow control valve contains a pressure relief ball (**Figure 12-13**). High-pressure fluid is forced past the control valve to the outlet fitting. A high-pressure hose connects the outlet fitting to the inlet fitting on the steering gear. A low-pressure hose returns the fluid from the steering gear to the inlet fitting in the pump reservoir.

⚠ Caution

Avoid prying on the power steering pump reservoir, because this action can damage or puncture the reservoir.

Figure 12-11 Power steering pump housing with shaft and rotor, vanes, cam ring, and pressure plate.

Figure 12-12 Power steering pump housing assembly with end cover, flow control valve, and magnet.

Figure 12-13 Flow control valve.

POWER STEERING PUMP OPERATION

Vane and Cam Action

As the belt rotates the rotor and vanes inside the cam ring, centrifugal force causes the vanes to slide out of the rotor slots. The vanes follow the elliptical surface of the cam. When the area between the vanes expands, a low-pressure area occurs between the vanes and fluid flows from the reservoir into the space between the vanes. As the vanes approach the higher portion of the cam, the area between the vanes shrinks and the fluid is pressurized.

High-pressure fluid is forced into two passages on the thrust plate. These passages reverse the fluid direction. Fluid is discharged through the cam crossover passages and pressure plate openings to the high-pressure cavity and **flow control valve**. The fluid is discharged from the flow control valve through the outlet fitting (**Figure 12-14**). Though most of the fluid is discharged from the pump, some fluid returns to the bottom of the vanes past the flow control valve and through the bypass passage.

A machined **venturi** orifice is located in the outlet fitting. Fluid passage applies pressure from the venturi area to the spring side of the flow control valve. When fluid flow in the venturi increases, the pressure in this area decreases.

> The **flow control valve** controls the fluid from the pump to the steering gear.

> A **venturi** is a narrow area in a pipe through which a liquid or gas is flowing. When liquid or gas flow increases in the venturi, the pressure drops proportionally. If liquid or gas flow decreases, the pressure in the venturi increases.

Power Steering Pump Operation at Idle Speed

At idle speed with the wheels straight ahead, the flow of fluid from the pump to the steering gear is relatively low. Under this condition, the steering gear valve is positioned so the fluid discharged from the pump is directed through the valve and the low-pressure hose to the pump vane inlet. With low fluid flow, the pressure is higher in the venturi and control orifice. This higher pressure is applied to the spring side of the flow control valve and helps the valve spring keep this flow control valve closed (**Figure 12-15**). Under this condition, pump pressure remains low, and all the fluid is discharged from the pump through the steering gear valve and back to the pump inlet.

Power Steering Pump Operation at Higher Engine Speeds

When engine speed is increased, the power steering pump delivers more fluid than the system requires. This increase in fluid flow creates a pressure decrease in the **outlet fitting venturi**. This pressure reduction is sensed at the spring side of the flow control valve, which allows the pump discharge pressure to force the flow control valve partially

> The **outlet fitting venturi** controls the pressure supplied to the flow control valve.

Figure 12-14 Power steering pump fluid flow to control valve and outlet fitting.

Figure 12-15 Flow control valve operation under various conditions.

open. Under this condition, the excess fluid from the pump is routed past the flow control valve to the pump inlet. If the steering wheel is turned, the fluid discharged from the pump rushes into the pressure chamber in the steering gear.

Power Steering Pump Operation During Pressure Relief

If the steering wheel is turned fully in either direction until the steering linkage contacts the steering stops, the rack piston is stopped. Under this condition, pump pressure could become extremely high and damage hoses or other components. When the rack piston stops, the flow from the pump decreases, but the high pump pressure is still directed through the steering gear valve to the rack piston. Since the flow of fluid through the venturi and control orifice is reduced, a higher pressure is present in this area. This extremely high pressure is also supplied to the spring end of the flow control valve. At a predetermined pressure, the **pressure relief ball** in the center of the flow control valve is unseated and the flow control valve moves to the wide-open position. This action allows some of the high pressure in the steering system to return to the pump inlet, which limits pump pressure to a maximum safe value.

The **pressure relief ball** limits the maximum power steering pump pressure.

Shop Manual
Chapter 12, page 523

Shop Manual
Chapter 12, page 531

HYBRID VEHICLES AND POWER STEERING SYSTEMS

Advantages and Types of Hybrid Electric Vehicles

Hybrid electric vehicles (HEVs) have two power sources: typically, a small displacement gasoline engine and a high-voltage battery pack that supplies voltage and current to an electric drive motor(s).

The two most common types of hybrid vehicles are **series HEVs** and **parallel HEVs**. In a series HEV powertrain, the mechanical power from engine is combined with electric power from the generator and/or battery pack to drive the vehicle (**Figure 12-16**). In a parallel HEV powertrain, the mechanical power from the engine or electric power from the generator and/or battery pack may be delivered separately to the drive wheels or the power from both of these sources may be combined and delivered to the drive wheels (**Figure 12-17**). In a parallel HEV, electric power alone may be supplied to the

In a **series HEV**, electric power from the battery pack and/or generator and mechanical power from the engine are supplied together to the drive wheels.

Figure 12-16 Series HEV.

Figure 12-17 Parallel HEV.

drive wheels to drive the vehicle, whereas in a series HEV, electric power cannot drive the vehicle.

The advantages of an HEV are increased fuel economy and reduced emissions. It is extremely important that all other systems and vehicle design are engineered to help achieve these objectives. Many HEVs have a smaller displacement engine compared with an equivalent-size non-hybrid vehicle. In a typical HEV, the smaller displacement engine coupled with the electric propulsion motor provides equivalent, or nearly equivalent, power compared with that provided in a non-hybrid vehicle. The smaller displacement engine in the HEV supplies improved fuel economy compared with that provided by the larger engine in the non-hybrid vehicle. Improved fuel economy is also achieved because the electric propulsion motor supplies some of the power to the drive wheels. In many HEVs, under specific operating conditions, the engine is shut off and power is supplied to the drive wheels only by the electric motor to provide additional fuel savings.

On many HEVs, the engine is stopped to conserve fuel and reduce emissions when the vehicle is standing still and the engine is warmed up and idling. Under this condition, the power steering must be active and ready to provide power steering assist if the driver turns the steering wheel. To meet this requirement, electrohydraulic power steering pumps are presently installed on some HEVs.

In a **parallel HEV**, electric power from the battery pack and/or generator and mechanical power from the engine may be supplied separately to the drive wheels or these two power sources may be combined.

HYBRID POWERTRAIN COMPONENTS

Engine to Transaxle Coupling

In some HEVs, the engine and propulsion motors are coupled through a torque splitting planetary gear set to an electronically controlled continuously variable transaxle (CVT). The engine is coupled to the carrier in the planetary gear set (**Figure 12-18**). The propulsion motor is connected to the planetary ring gear, and the generator is attached to the planetary sun gear. The drive gear that transfers torque to the transaxle is attached to the electric motor drive shaft. This arrangement allows torque from the engine, the propulsion motor, or both the engine and the propulsion motor to be transferred to the transaxle and drive wheels. The sun gear is the smallest gear in the planetary gear set.

Propulsion Motor/Generator

The propulsion motor is an alternate current (AC) synchronous permanent magnet, liquid-cooled type. Some propulsion motors are rated at 40 hp (30 kW) at 940 to 2,000 rpm

Figure 12-18 Planetary gear coupling between engine, electric motor, and transaxle.

Figure 12-19 Electric drive motor (MG-2) and generator (MG-1).

and 225 ft.-lb (305 Nm) peak torque at 0 to 940 rpm. During deceleration and braking, the propulsion motor acts as a generator to recharge the batteries. This action is called **regenerative braking**. In **Figure 12-19**, the propulsion motor is identified as MG-2 and the generator is identified as MG-1. Some propulsion motors have higher torque and horsepower output depending on the motor design. Later model propulsion motors are capable of turning at speeds in excess of 12,000 rpm.

Regenerative braking occurs during vehicle deceleration when the drive motor becomes a generator and supplies current to recharge the batteries.

Generator/Starter

The generator is an AC synchronous type that produces voltage and current to recharge the batteries and power the electrical system. When starting the engine, current is supplied through the generator stator windings, and the generator cranks the engine. Under normal driving conditions, current from the generator is routed to the electric drive motor to increase the torque supplied to the drive wheels. In later model HEVs, generator design has been improved to allow higher generator rpm and improved generator output. In some current HEVs, the generator is capable of turning at 10,000 rpm, whereas some early model generators turned at 6,500 rpm.

Inverter

The inverter contains the power control unit that controls the voltage and current supplied by the batteries to the propulsion motor. The battery pack supplies high direct current (DC) voltage to the inverter and the inverter converts the DC voltage to three-phase AC voltage that is supplied to the propulsion motor. The inverter also controls the voltage and current supplied from the propulsion motor to the batteries during regenerative braking. Because the voltage and current supplied from the batteries to the propulsion motor may be very high, the inverter contains insulated gate bipolar transistors (IGBT) to control this circuit (**Figure 12-20**). This high voltage and current controlled by the inverter produces a considerable amount of heat, and the inverter is cooled by a dedicated cooling system to dissipate this heat. Heavy cables are connected between the batteries, inverter, and the propulsion motor. For easy identification, these cables and connectors have orange insulation.

Figure 12-20 High-voltage power circuit in the inverter.

Figure 12-21 Inverter.

⚡ **WARNING** The high voltage in circuit between the batteries, inverter, and propulsion motor could electrocute a technician. It is very important to follow the vehicle manufacturer's recommended procedures when diagnosing and servicing HEVs.

⚡ **WARNING** Never touch, cut, pierce, or open any orange high-voltage cable, or high-voltage component. This action may cause severe electrical shock or electrocution.

The AC voltage and current supplied from the generator to the 12-V battery and electrical accessories is converted to DC voltage by the inverter and controlled in 14 V–15 V range. The inverter is illustrated with the cover removed in **Figure 12-21**.

Battery Pack and Related Cables

Some battery packs contain 240 cylindrical nickel-metal-hydride cells connected in series. Each cell has 1.2 V for a total of 288 V. Some HEVs have a higher voltage depending on the number of cells in the battery. The battery cells are connected in groups, with six cells in each group. The battery pack is installed behind the rear seat in some HEVs. The electrolyte is a gel in the nickel-metal-hydride battery pack; therefore, leakage is not a great concern. A cooling fan helps cool the battery pack. In the unlikely case of battery overcharging, a battery pack vent hose allows vapors from the battery pack to be vented outside the vehicle trunk.

Many battery packs have a long warranty period such as 10-year, 90,000 mi. (150,000 km). Positive and negative high-voltage cables are connected between the battery pack and the inverter. These cables are routed under the vehicle floor pan and they are completely insulated from the chassis to avoid any possibility of electrical shock when touching the chassis.

HEV components increase vehicle weight, which reduces vehicle performance and increases fuel consumption. Therefore, it is very important to reduce the weight of HEV components and also design other vehicle components with less weight.

Some HEVs are currently equipped with lithium-ion batteries, which are 40 percent lighter and take up 24 percent less space compared with a nickel-metal-hydride battery with the same power output.

AUTHOR'S NOTE One of the major Japanese vehicle manufacturers has just introduced a lithium-ion battery with a new internal design that has only one-half the volume but one and one-half times the power output compared with previous lithium-ion batteries.

HEV Indicators

Some HEVs have normally open battery-pack relays that open and close the circuit between the battery pack and the inverter. The vehicle computer controls these relays.

A READY indicator in the dash informs the driver when the relays are closed and the batteries are connected to the inverter. When the ignition switch is off, the relays in the high-voltage circuit are open and the READY indicator is off. If the ignition switch is on and the relays are closed, voltage is supplied from the battery pack to the inverter, and the READY indicator is illuminated (**Figure 12-22**). The engine may stop and start any

High-Voltage System—Vehicle Shut Off (READY -off)

High-Voltage System—Vehicle On and Operational (READY -on)

Figure 12-22 High-voltage electrical system.

Figure 12-23 Master warning light.

Figure 12-24 Hybrid warning light.

time while the READY indicator is illuminated. Never assume the electrical system is off just because the engine is stopped. The electrical system is off when the READY indicator is off. If the vehicle is involved in a collision, and the air bags are deployed, the battery pack relays automatically open and disconnect the voltage supplied from the battery pack to the inverter. When the engine is running, a ground fault monitor continuously monitors the high-voltage system for high voltage leakage to the metal chassis. If leakage occurs, the vehicle computer illuminates the master warning light in the instrument cluster (**Figure 12-23**) and the hybrid warning light in the liquid crystal display (LCD) (**Figure 12-24**). The conventional 12-V lead acid battery is located in the truck near the battery pack. The positive cable from the 12-V battery is routed to the engine compartment with the high-voltage cables from the battery pack. HEV systems and indicators vary depending on the vehicle make and model year.

HEV Operation

When the engine is at normal operating temperature, and the vehicle is starting off from a stop or operating at very low speed and light load, only electrical power from the battery pack is supplied to the propulsion motor to drive the vehicle (**Figure 12-25**).

During normal driving, mechanical power from the engine and electric power supplied from the generator to the propulsion motor are used to drive the vehicle (**Figure 12-26**).

If the engine is operating at wide throttle opening, mechanical power from the engine and electric power from the generator and battery pack are supplied to drive the vehicle (**Figure 12-27**).

During vehicle deceleration, the propulsion motor acts as a generator and supplies current to charge the batteries (**Figure 12-28**).

If the battery state of charge becomes low, more of the electric power from the generator is transmitted to the battery pack for recharging (**Figure 12-29**).

Power Steering and Hybrid Vehicles

Many HEVs have a stop/start function that allows the engine to be shut off when the vehicle has come to a complete stop, with the engine idling. When the ignition switch is on and the driver steps on the accelerator pedal, the vehicle starts off, and the engine may

(1) Starting out or moving under very low load

Figure 12-25 HEV starting off or operating at low speed and light load.

(2) Normal driving

Figure 12-26 HEV operation during normal driving.

(3) Full-throttle acceleration

Figure 12-27 HEV operation at wide-open throttle.

(4) Deceleration of braking

Figure 12-28 HEV operation during deceleration.

(5) Charging the batteries

Figure 12-29 HEV operation with low battery state of charge.

Figure 12-30 Electronic power steering (EPS).

restart at any time depending on throttle opening and other factors. The power steering must be operational even during the engine stop part of the stop/start function. To maintain power steering operation during the stop function, some vehicle manufacturers use an electronic power steering (EPS) gear. Some of these EPSs have an electric motor driving the steering gear pinion (**Figure 12-30**). A power steering control module supplies current to the EPS motor to supply the proper amount of steering assist. The EPS is fully operational when the ignition switch is turned on.

Other HEVs have an electrohydraulic power steering pump to maintain power steering pump operation during the stop function. This type of pump has an electric motor driving the pump rather than a belt drive (**Figure 12-31**). The electrohydraulic power steering pump contains the pump and electric drive motor, reservoir with filler cap, and the control module. Voltage signals are sent from the steering wheel position sensor in the steering column and the brake pedal position sensor actuated by the brake pedal to the control module. In response to these signals, the control module determines the amount of power steering assist required. In some electrohydraulic power steering pumps, 42 V are supplied from the inverter in the HEV to the power steering pump.

When a higher voltage is supplied to the power steering pump motor, less current is required to drive the motor. For example, electrical energy is measured in watts (W) and watts are calculated by multiplying the amperes by the volts. When 12 V are supplied to a headlight, 3 amperes may be required to illuminate the light. The electrical energy

Shop Manual
Chapter 12, page
531

Electrohydraulic
Power Steering
(EHPS)

Figure 12-31 Electrohydraulic power steering (EHPS) module.

required to illuminate the light is 12 × 3 = 36 W. If 18 V are supplied to the headlight, only 2 amperes are required to illuminate the light. The advantage of increasing the voltage is that the same watts can be obtained with fewer amperes.

AUTHOR'S NOTE In 2008, one leading manufacturer's worldwide sales of HEVs surpassed 1.5 million units. The average silicon content on a non-hybrid vehicle consumes approximately 30 percent of an 8 in. (20.32 cm) silicon wafer. The silicon content of a hybrid vehicle requires the full 8 in. (20.32 cm) wafer. Approximately 50 percent of the cost of a hybrid vehicle is in the hardware and software. Toyota has taken 70 percent of the cost of the hybrid technology out of the hybrid Prius, improved its fuel economy, and increased the size of the chassis/body.

VARIOUS TYPES OF HEVs

Belt-Driven HEVs

Some HEVs have a belt-driven motor/generator with an improved drive belt and tensioner compared with a conventional generator (**Figure 12-32**). The latest belt-driven HEV system features a stop/start function, regenerative braking, intelligent battery charging, and 4 kW of energy are supplied from the battery pack to the motor/generator for up to 5 seconds during wide throttle, heavy load driving. Some belt-driven HEVs supply a modest electric-only propulsion below 8 mph (12.9 km/h). Lithium-ion batteries will soon be available in some belt-driven HEV systems.

Plug-In HEVs

Several major vehicle manufacturers are scheduled to introduce plug-in HEVs (PHEVs) in their showrooms during 2010. A PHEV may be driven only by an electric motor(s), but a small gasoline or diesel engine is installed in the vehicle for extra power or to drive a generator and recharge the batteries. A typical PHEV may be driven for 40 mi. (64 km) using only electric power. When the batteries reach a specific state of discharge, the engine starts and drives the generator to recharge the batteries.

Figure 12-32 Belt-driven hybrid system components.

Extended Range Electric Vehicle

An extended range electric vehicle (EREV) is similar to a PHEV. The EREV has a small gasoline or diesel engine that drives a generator to recharge the batteries. The engine only drives the generator; it cannot supply power to the drive wheels. The Chevrolet Volt is an EREV.

FUEL CELL VEHICLES

Fuel cell vehicles (FCVs) are not presently available in the automotive showrooms. However, many car manufacturers around the world are working on this technology. Several vehicle manufacturers have designed and produced FCVs, and a significant number of these vehicles have been released to various fleet owners and officials to gain field experience and determine how these vehicles perform in the real world.

An FCV has an electric-drive motor(s), and the fuel cell supplies electricity to this drive motor(s). The fuel cell must be supplied with hydrogen as a fuel, and the fuel cell produces electricity as the hydrogen and air flow through the cell. There are various types of fuel cells, but the most common type is the proton exchange membrane (PEM) fuel cell. The fuel cell actually contains a large group of cells. The only by-product from fuel cell operation is water vapor. A PEM fuel cell is compared to an internal combustion engine in **Figure 12-33**. The FCV requires a source of high voltage and current such as a battery pack and/or ultra-capacitor to supply extra current to the drive motor during operation at wide throttle openings. The disadvantages of FCVs are as follows:

1. The size, weight, and cost of the FCV components. Fuel cells are very expensive at present.
2. The lack and cost of hydrogen refueling facilities.
3. The cost and availability of FCVs.
4. Difficulty in storing hydrogen on-board vehicles; liquid hydrogen must be stored under pressure in special insulated tanks. It is difficult to store a sufficient amount of gaseous hydrogen on-board a vehicle.

The advantages of FCVs are as follows:

1. Reduced dependency on high-priced hydrocarbon fuels.
2. Hydrogen may be obtained from several different sources.
3. Greatly reduced tailpipe and evaporative emissions.

Single-cylinder IC engine   **Single-cell Ballard Fuel-cell engine**

Figure 12-33 Comparison between an internal combustion engine and a fuel cell.

SUMMARY

- A multi-ribbed belt or a V-belt on older vehicles may be used to drive the power steering pump.

- A serpentine belt is a multi-ribbed belt that is used to drive all the belt-driven components. This allows these components to be on the same vertical plane.

- Proper belt tension is extremely important for adequate power steering pump operation. Many belts today are made of EPDM. A rack and pinion steering gear is usually mounted on the cowl or the front crossmember. Rack and pinion steering gears are used in many front-wheel-drive cars. Integral power steering gears are usually mounted on the vehicle frame. Integral power steering systems are found on many rear-wheel-drive cars and light-duty trucks. A power steering pump may have an integral or a remote fluid reservoir. In a hydro-boost power-assisted steering system, fluid pressure from the power steering pump is applied to the steering gear and the brake master cylinder. The power steering pump pressure acts as a booster to assist the driver in applying the brakes.

- A power steering pump may have a vane, roller, or slipper-type rotor assembly, but all three types of pumps operate on the same basic principle.

- The flow control valve in a power steering pump is moved toward the closed position by spring pressure and fluid pressure from the venturi in the pump outlet fitting.

- When a vehicle is driven at low speeds with the front wheels straight ahead, the power steering pump fluid flow is lower. Under this condition, the pressure in the outlet fitting venturi is higher. This higher pressure together with the spring tension keeps the flow control valve closed. In this valve position, a small amount of fluid is routed past the flow control valve to the pump inlet.

- If engine speed is increased, the pump delivers more fluid than the system requires. This increase in pump flow reduces pressure in the outlet fitting venturi. This pressure decrease is applied to the spring side of the flow control valve. This action allows the flow control valve to move toward the partially open position, and excessive pump flow is returned past the flow control valve to the pump inlet.

- If the driver turns the steering wheel, fluid rushes from the pump outlet into the steering gear pressure chamber. Under this condition, the flow control valve moves toward the closed position to maintain pump pressure and flow to the steering gear.

- If the front wheels are turned all the way against the stops, the power steering pump pressure could become high enough to damage hoses or other components. Under this condition, the high pump pressure unseats a pressure relief ball in the center of the flow control valve. This action allows some pump flow to move past the pressure relief ball to the pump inlet, which limits pump pressure.
- Most HEVs are powered by a small gasoline engine and a battery pack, generator, and electric motor.
- In a series HEV, mechanical power from the engine is combined with electric power from the battery pack and generator and supplied to the drive wheels.
- In a parallel HEV, electric power only from the battery pack and generator may be supplied to the drive wheels or a combination of electric power and engine power may be sent to the drive wheels.
- The stop/start function in an HEV allows the engine to stop when the vehicle is not moving and the engine is at normal temperature and idling. When the driver steps on the accelerator pedal, vehicle operation is immediately restored.
- A belt-driven HEV system has a belt-driven motor/generator and provides a stop/start function, regenerative braking, and electric power boost to the drive wheels.
- In a PHEV, only electric power is supplied to the drive wheels, and a small gasoline or diesel engine drives a generator to recharge the battery pack.
- An FCV has only electric power supplied to the drive wheels. A fuel cell and battery pack supply the electricity to drive the vehicle, and hydrogen and air are supplied to the fuel cell to create the electricity.

REVIEW QUESTIONS

Short Answer Essays

1. Describe the types of materials a multi-ribbed or serpentine belt may be constructed of.
2. Explain two advantages of a serpentine belt.
3. List two possible locations for a power steering pump reservoir.
4. How is the power steering pump output effected as engine speed is increased but no additional power steering assist is required, such as in straight-ahead driving?
5. Describe how the brake boost pressure is obtained in a hydro-boost power steering system.
6. Explain how the fluid pressure is produced in a vane-type power steering pump.
7. List the forces that move the flow control valve toward the closed position in a power steering pump.
8. Describe the operation of a venturi.
9. Explain the operation of the flow control valve when the steering wheel is turned.
10. Describe the operation of the power steering pump when the steering wheel is rotated all the way to the right or left until the front wheels contact the stops.

Fill-in-the-Blanks

1. On a multi-ribbed belt made of EPDM a _____ percent loss of material is enough to cause belt_____, _____, and _____.
2. Excessive belt wear occurs if pulleys are _____.
3. Many serpentine belts have a spring-loaded _____.
4. The master cylinder in a hydro-boost system does not have a conventional _____.
5. The purpose of the vanes in the power steering pump rotor is to _____ the rotor in the elliptical cam ring.
6. As the power steering pump shaft and rotor turn, _____ _____ causes the vanes to move outward against the cam ring.
7. A machined venturi is located in the power steering pump _____ _____.
8. When fluid movement through the venturi increases, the pressure in the venturi _____.
9. With the engine running and the front wheels straight ahead, the power steering pump pressure is _____.

10. The pressure relief ball in a power steering pump is forced open when the front wheels are turned against the _____.

Multiple Choice

1. The most likely result of a power steering drive-belt that is worn is:
   A. Belt slipping.
   B. Belt breaking.
   C. Belt contamination.
   D. Wear on the power steering pump pulley.

2. All of these statements about a serpentine belt are true EXCEPT:
   A. A serpentine belt can be used to drive all the belt-driven components.
   B. The back of a serpentine belt can drive one component.
   C. A serpentine belt made of EPDM is susceptible to cracking.
   D. A serpentine belt can have an automatic belt tensioner.

3. A hydro-boost power steering system:
   A. Is used on many small front-wheel-drive vehicles.
   B. Uses pressure from the power steering pump to provide brake boost.
   C. Has a conventional vacuum brake booster.
   D. Has a very high-capacity power steering pump.

4. All of the following statements about multi-ribbed belts made of EPDM are true EXCEPT:
   A. EPDM belt is more elastic than a standard neoprene belt and resists cracking even at higher mileage.
   B. EPDM belts are designed to last 50,000–60,000 miles.
   C. Belts are designed to have clearance between the rib peaks and the pulley grooves.
   D. EPDM belt to gradually lose material on the ribbing similar to the way a tire wears.

5. All of these statements about integral-type power steering pump sealing are true EXCEPT:
   A. A large O-ring seal is positioned between the reservoir and the pump housing.
   B. A lip seal on the pump drive shaft prevents oil leaks around the shaft.
   C. Lip seals are mounted on bolt fittings on the back of the reservoir.
   D. The cap and dipstick provide a seal on the top of the reservoir neck.

6. A power steering system experiences repeated belt failures with normal power steering system operation. The most likely cause of this problem is:
   A. A misaligned power steering pump pulley.
   B. A loose power steering pump belt.
   C. Low fluid level in the power steering reservoir.
   D. A worn cam ring in the power steering pump.

7. During normal power steering pump operation:
   A. The flow control valve spring moves this valve toward the open position.
   B. The fluid pressure from the outlet fitting venturi moves the flow control valve toward the open position.
   C. Pump flow increases as the engine speed increases.
   D. The pressure relief valve opens when the engine speed is increased with the front wheels straight ahead.

8. All of these statements about power steering pump design and operation are true EXCEPT:
   A. An increase in fluid movement through the outlet fitting venturi causes a pressure decrease in the venturi.
   B. An increase in the flow control valve opening allows more fluid to return to the pump inlet.
   C. The rotor is attached to the pump shaft and must rotate with this shaft.
   D. Fluid pressure from the pump vanes moves the flow control valve toward the closed position.

9. All of the following statements about hybrid electric vehicles that have an electrohydraulic power steering pump are true EXCEPT:

   A. The power steering pump has an electric-driven motor.

   B. A control module determines the amount of power assist required.

   C. An electric motor drives the steering gear.

   D. Some are sullied 42 V from the HEV inverter.

10. A vehicle experiences repeated ruptures of the high-pressure power steering hose. The most likely cause of this problem is:

    A. A pressure relief ball sticking closed.

    B. A flow control valve sticking open.

    C. Worn pump vanes and cam ring.

    D. A partially restricted power steering return hose.

CHAPTER 13

RACK AND PINION STEERING GEARS AND FOUR-WHEEL STEERING

Upon completion and review of this chapter, you should be able to understand and describe:

- The advantages of a rack and pinion steering system compared to a recirculating ball steering gear and parallelogram steering linkage.
- How the tie rods are connected to the rack.
- The purpose of the rack bearing and adjuster plug.
- Two possible mounting positions for the rack and pinion steering gear.
- The fluid movement in a rack and pinion steering gear during a right turn.
- The fluid movement in a rack and pinion steering gear during a left turn.
- The operation of the spool valve and rotary valve.
- The operation of the power rack and pinion steering gear when hydraulic pressure is not available.
- The purpose of the breather tube.
- The main differences between a Saginaw and a TRW power rack and pinion steering gear.
- The advantages of an electronic variable orifice (EVO) steering system.
- The input sensors in an EVO steering system.

- The driving conditions when an EVO steering system provides increased power steering assistance.
- The operation of a rack-drive electronic power steering system during right and left turns.
- The operation of the steering shaft torque sensor in a column-drive electronic power steering system.
- The operation and advantages of active front steering (AFS).
- The advantages of four-wheel steering (4WS).
- How the rear steering rack is driven.
- The two inputs used by the control unit to operate the rear steering system.
- Negative phase and positive phase steering.
- The input sensors in an electronically controlled four-wheel steering system and give the location of each sensor.
- The type of signal produced by a wheel speed sensor.
- The steering action of the rear wheels in relation to vehicle speed in an electronically controlled 4WS system.
- The operation and advantages of rear active steering (RAS) and four-wheel active steering (4WAS).

Terms To Know

Analog voltage signal

Breather tube

Car access system (CAS)

Center take-off (CTO)

Column-drive EPS

Digital motor electronics (DME)

Electric locking unit (ELU)

Electronic control unit (ECU)

Electronic variable orifice (EVO) steering

Electronically controlled orifice (ECO) valve

Electronically controlled 4WS system

Encoded disc

End take-off (ETO)

Feel of the road

Four-wheel steering (4WS)

Four-wheel active steering (4WAS)

Helical gear teeth

Lock-to-lock

Magnasteer system

Neutral phase steering

Pinion

Pinion angle sensor

Pinion-drive EPS

Positive phase steering

Pulse width modulated (PWM) voltage signal

Rack

Rack-drive EPS

Rack piston

Rack seals

Rear active steering (RAS)

Rear steering actuator

Rotary valve

Safety and gateway module (SGM)

Servotronic valve

Sideslip

Spool valve

Spur gear teeth

Steering angle

Steering angle sensor

Steering kickback

Steering ratio

Steering wheel rotation sensor

Stop-to-stop

Summation sensor

Torsion bar

Transistor

Turning circle

Variable effort steering (VES)

Vehicle speed sensor (VSS)

Yaw forces

Yaw motion

INTRODUCTION

During the late 1970s and 1980s, the domestic automotive industry converted much of its production from larger rear-wheel-drive cars to smaller, lightweight, and more fuel-efficient front-wheel-drive cars. These front-wheel-drive cars required smaller, lighter components wherever possible. Manual and power rack and pinion steering gears are lighter and more compact than the recirculating ball steering gears and parallelogram steering linkages used on most rear-wheel-drive cars. Therefore, rack and pinion steering gears are ideally suited to these compact front-wheel-drive cars.

Steering systems have not escaped the electronics revolution. Many cars are presently equipped with electronic variable orifice (EVO) steering, which provides greater power assistance during low-speed cornering and parking for increased driver convenience. Some cars are now equipped with electronic power steering. In these systems, an electric motor in the steering gear provides steering assist.

MANUAL RACK AND PINION STEERING GEAR MAIN COMPONENTS

Shop Manual
Chapter 13, page 556

Spur gear teeth are cut so they are parallel to the gear rotational axis.

Helical gear teeth are curved to increase the amount of tooth contact between a pair of meshed gears. Helical gears tend to operate more quietly than spur gear teeth and provide increased strength compared with spur gear teeth.

Rack

The **rack** is a toothed bar that slides back and forth in a metal housing. The steering gear housing is mounted in a fixed position on the front crossmember or on the firewall. The rack takes the place of the idler and pitman arms in a parallelogram steering system and maintains the proper height of the tie rods so they are parallel to the lower control arms. The rack may be compared to the center link in a parallelogram steering linkage. Bushings support the rack in the steering gear housing. Sideways movement of the rack pulls or pushes the tie rods and steers the front wheels (**Figure 13-1**).

Pinion

The **pinion** is a toothed shaft mounted in the steering gear housing so the pinion teeth are meshed with the rack teeth. The pinion may contain **spur gear teeth** or **helical gear teeth**. The upper end of the pinion shaft is connected to the steering shaft from the steering column. Therefore, steering wheel rotation moves the rack sideways to steer the front wheels. The pinion is supported on a ball bearing in the steering gear housing.

Figure 13-1 Manual rack and pinion steering gear components.

Tie Rods and Tie Rod Ends

The tie rods are similar to those used on parallelogram steering linkages. A spring-loaded ball socket on the inner end of the tie rod is threaded onto the rack. When these ball sockets are torqued to the vehicle manufacturer's specification, a preload is placed on the ball socket. A bellows boot is clamped to the housing and tie rod on each side of the steering gear, and these boots keep contaminants out of the ball socket and rack.

A tie rod end is threaded onto the outer end of each tie rod. These tie rod ends are similar to those used on parallelogram steering linkages. A jam nut locks the outer tie rod end to the tie rod.

Rack Adjustment

A rack bearing is positioned against the smooth side of the rack. A spring is located between the rack bearing and the rack adjuster plug that is threaded into the housing. This adjuster plug is retained with a locknut. The rack bearing adjustment sets the preload between the rack and pinion teeth, which affects **steering kickback**, harshness, and noise.

Steering kickback is the movement of the steering wheel caused by a front wheel striking a road irregularity.

STEERING GEAR RATIO

When the steering wheel is rotated from **lock-to-lock** or **stop-to-stop**, the front wheels turn about 30° each in each direction from the straight-ahead position. Therefore, the total front wheel movement from left to right is approximately 60°. With a steering ratio of 1:1, 1° of steering wheel rotation would turn the front wheels 1°, and 30° of steering wheel rotation in either direction would result in lock-to-lock front wheel movement. This steering ratio is much too extreme because the slightest steering wheel movement would cause the vehicle to swerve. The steering gear must have a ratio that allows more steering wheel rotation in relation to front wheel movement.

A **steering ratio** of 15:1 is acceptable, and this ratio provides 1° of front wheel movement for every 15° of steering wheel rotation. To calculate the steering ratio, divide the lock-to-lock steering wheel rotation in degrees by the total front wheel movement in degrees. For example, if the lock-to-lock steering wheel rotation is 3.5 turns, or 1,260°, and the total front wheel movement is 60°, the steering ratio is $1,260/60 = 21:1$. As a general rule, large, heavy cars have higher numerical steering ratios than small, lightweight cars.

MANUAL RACK AND PINION STEERING GEAR MOUNTING

Large rubber insulating grommets are positioned between the steering gear and the mounting brackets. These bushings help prevent the transfer of road noise and vibration from the steering gear to the chassis and passenger compartment. The rack and pinion steering gear may be attached to the front crossmember or to the cowl. Proper steering gear mounting is important to maintain the parallel relationship between the tie rods and the lower control arms. The firewall is reinforced at the steering gear mounting locations to maintain the proper steering gear position.

ADVANTAGES AND DISADVANTAGES OF RACK AND PINION STEERING

As mentioned earlier, the rack and pinion steering gear is lighter and more compact than a recirculating ball steering gear and parallelogram steering linkage. Therefore, the rack and pinion steering gear is most suitable for front-wheel-drive unibody vehicles.

Since there are fewer friction points in the rack and pinion steering than in the recirculating ball steering gear with a parallelogram steering linkage, the driver has a greater feeling of the road with rack and pinion steering gear. However, fewer friction points reduce the steering system's ability to isolate road noise and vibration. Therefore, drivers of a vehicle with a rack and pinion steering system may have more complaints of road noise and vibration transfer to the steering wheel and passenger compartment.

POWER RACK AND PINION STEERING GEARS

Design and Operation

A power-assisted rack and pinion steering gear uses the same basic operating principles as a manual rack and pinion steering gear, but in the power-assisted steering gear, hydraulic fluid pressure from the power steering pump is used to reduce steering effort. A **rack piston** is integral with the rack, and this piston is located in a sealed chamber in the steering gear housing. Hydraulic fluid lines are connected to each end of this chamber, and **rack seals** are positioned in the housing at ends of the chamber. A seal is also located on the rack piston (**Figure 13-2**).

Figure 13-2 Hydraulic chamber in a power rack and pinion steering gear.

The following description assumes that the power steering rack is mounted behind the front wheel spindle, such as on the vehicle firewall or cradle assembly. When a driver is completing a left turn, fluid is pumped into the left side of the fluid chamber and exhausted from the right chamber area. This hydraulic pressure on the left side of the rack piston helps the pinion move the rack to the right (**Figure 13-3**, view A).

When a right turn is made, fluid is pumped into the right side of the fluid chamber, and fluid flows out of the left end of the chamber. Thus, hydraulic pressure is exerted on the right side of the rack piston, which assists the pinion gear in moving the rack to the left (**Figure 13-3**, view B). Since the steering gear is mounted behind the front wheels, rack movement to the left is necessary for a right turn, whereas rack movement to the right causes a left turn.

Rotary Valve and Spool Valve Operation

Fluid direction in the steering gear is controlled by a rotary valve attached to the pinion assembly (**Figure 13-4**). A stub shaft on the pinion assembly is connected to the steering shaft and wheel. The pinion is connected to the stub shaft through a **torsion bar** that twists when the steering wheel is rotated and springs back to the center position when the wheel is released. A rotary valve body contains an inner **spool valve** that is mounted over the torsion bar on the pinion assembly.

When the front wheels are in the straight-ahead position, fluid flows from the pump through the high-pressure hose to the center **rotary valve** body passage. Fluid is then routed through the valve body to the low-pressure return hose and the pump reservoir (**Figure 13-5**).

Many power steering systems contain a fluid cooler connected in the high-pressure hose between the pump and the steering gear. The fluid cooler is like a small radiator. Air

Torsion bar twisting during steering wheel rotation moves the spool valve in relation to the rotary valve.

The **rotary valve** is mounted over the spool valve on the steering gear pinion.

Pressurized fluid
pumped into right chamber

Pressurized fluid
pumped into left chamber

Figure 13-3 Rack movement during left and right turns.

Stub shaft

Torsion bar
(inside stub shaft)

Pinion

Outer valve body
(spool valve contained inside)

Figure 13-4 Pinion assembly for a power rack and pinion steering
gear.

flows through the fins on the cooler and cools the fluid flowing through the internal cooler passages. Some power steering systems have a remote fluid reservoir connected in the low-pressure hose between the steering gear and the pump (**Figure 13-5**). Many power steering systems have a fluid filter, and this filter is often mounted in the remote reservoir.

Teflon rings or O-rings seal the rotary valve ring lands to the steering gear housing. A lot of force is required to turn the pinion and move the rack because of the vehicle weight on the front wheels. When the driver turns the wheel, the stub shaft is forced to turn. However, the pinion resists turning because it is in mesh with the rack, which is connected to the front wheels. This resistance of the pinion to rotation results in torsion bar twisting. During this twisting action, a pin on the torsion bar moves the spool valve with a circular motion inside the rotary valve. If the driver makes a left turn, the spool valve movement aligns the inlet center rotary valve passage with the outlet passage to the left side of the rack piston. Therefore, hydraulic fluid pressure applied to the left side of the rack piston assists the driver in moving the rack to the right.

Item	Part Number	Description
1	3A697	Fluid reservoir (early build vehicles)
2	3A697	Fluid reservoir (late build vehicles)
3	3691	Supply hose (reservoir-to-pump) (early build vehicles)
4	3A713	Return hose (cooler-to-reservoir) (early build vehicles)
5	3691	Supply hose (reservoir-to-pump) (late build vehicles)
6	3A713	Return hose (cooler-to-reservoir) (late build vehicles)
7	3489	Fluid reservoir bracket (early build vehicles)
8	3A733	Pulley
9	3A764	Power steering pump
10	3504	Steering gear
11	3A131	Tie rod end
12	3280	Inner tie rod
13	3332	Inner tie rod boot
14	3A713	Return line (gear-to-cooler)
15	3D746	Fluid cooler
16	3A719	Pressure line (pump-to-gear)
17	9F274	Return hose retainer clip
18	3F886-AA/ 3F886-BA	O-ring (one each required)

Figure 13-5 Power rack and pinion steering gear with connecting hoses and lines.

When a right turn is made, twisting of the torsion bar moves the spool valve and aligns the center rotary valve passage with the outlet passage to the right side of the rack piston (**Figure 13-6**). Under this condition, hydraulic fluid pressure applied to the rack piston helps the driver move the rack to the left. The torsion bar provides a **feel of the road** to the driver.

When the steering wheel is released after a turn, the torsion bar centers the spool valve and power assistance stops. If hydraulic fluid pressure is not available from the pump, the power steering system operates like a manual system, but steering effort is higher. When the torsion bar is twisted to a designed limit, tangs on the stub shaft engage with drive tabs on the pinion. This action mechanically transfers motion from the steering wheel to the rack and front wheels. Since hydraulic pressure is not available on the rack piston, greater steering effort is required. If a front wheel raises going over a bump or drops into a hole, the tie rod pivots along with the wheel. However, the rack and tie rod still maintain the left-to-right wheel direction.

When the driver turns the steering wheel, the amount of feeling that he or she senses regarding front wheel turning is called **feel of the road**.

In power rack and pinion steering gears, a condition that causes excessive steering wheel turning effort when the vehicle is first started may be called "morning sickness."

Figure 13-6 Spool valve movement inside the rotary valve.

Figure 13-7 Breather tube and boot.

The rack boots are clamped to the housing and the rack. Since the boots are sealed and air cannot be moved through the housing, a **breather tube** is necessary to move air from one boot to the other when the steering wheel is turned (**Figure 13-7**). This air movement through the vent tube prevents pressure changes in the bellows boots during a turn.

AUTHOR'S NOTE It has been my experience that one of the most common problems with power rack and pinion steering gears is a condition that causes excessive steering wheel turning effort when the vehicle is first started in the morning. After the steering wheel is turned several times, the condition disappears. This problem is caused by grooves worn in the aluminum pinion housing by the seals on the control valve (**Figure 13-8**). When this condition is present, steering gear replacement is usually required. As an automotive technician, you will be required to diagnose and correct this problem.

Figure 13-8 Score marks in a rack and pinion steering gear housing.

TYPES OF POWER RACK AND PINION STEERING GEARS

Power Rack and Pinion Steering Gear

Many vehicles have a rack and pinion steering gear manufactured by Saginaw (**Figure 13-9**). Some vehicles are equipped with a TRW rack and pinion steering gear. This type of steering gear is similar to the Saginaw gear except for the following differences:

1. Method of tie rod attachment
2. Bulkhead oil seal and retainer
3. Pinion upper and lower bearing hardware

On both the Saginaw and TRW power rack and pinion steering gears, the tie rods are connected to the ends of the rack. This type of steering gear may be referred to as **end take-off (ETO)**. On other power rack and pinion steering gears, the rack piston and cylinder are positioned on the right end of the rack and the tie rods are attached to a movable sleeve in the center of the gear (**Figure 13-10**). This type of steering gear may be called **center take-off (CTO)**.

The Toyota power rack and pinion steering gear has a removable control valve housing surrounding the control valve and pinion shaft (**Figure 13-11**). In this steering gear, claw washers are used to lock the inner tie rod ends to the rack. Apart from these minor differences, the Toyota power rack and pinion gear is similar to the Saginaw and TRW gears.

⚠ WARNING When working on any power steering system, always wear protective gloves and use caution, because the system hoses, components, and fluid can be very hot if the system has been in operation for a period of time.

Figure 13-9 Saginaw and TRW power rack and pinion steering gear.

ELECTRONIC VARIABLE ORIFICE STEERING

Input Sensors

The **electronic variable orifice (EVO) steering** system is standard on many late-model vehicles. The EVO steering system provides high-power steering assistance during low-speed cornering and parking and normal power steering assistance at higher speeds for proper road feel. High-power steering assistance during low-speed cornering and parking increases driver convenience.

The **steering wheel rotation sensor** is mounted on the steering column, and a shutter disc attached to the steering shaft rotates through the sensor when the steering wheel

Mounting grommet

Cylinder lines

Adjuster mechanism

Boot clamp

Housing, rack and pinion

Outer tie rod

Cap

Adjuster tie rod

Inner tie rod

Boot

Pivot bushing

Plate

Tie rod bolts

Figure 13-10 Power rack and pinion steering gear with center take-off tie rods.

is rotated. A row of slots is positioned near the outer edge of the shutter disc (**Figure 13-12**). When these slots rotate through the sensor, a steering wheel rotation speed signal is sent from the sensor to the control module.

The **vehicle speed sensor (VSS)** is mounted in the transaxle or transmission, and this sensor sends a signal to the control module in relation to vehicle speed (**Figure 13-13**). The VSS signal is also used for other purposes.

Control Module

On some vehicles, the control module is mounted in the trunk (**Figure 13-14**). This module continually monitors the input signals from the VSS and the steering wheel rotation sensor (**Figure 13-15**). Some models have a combined EVO steering system and rear air suspension system. On these models, the control module operates the EVO steering system and the rear air suspension system. In the combined EVO steering and rear air suspension system, the inputs and outputs from both of these systems are connected to the same control module. (The rear air suspension system is explained in Chapter 11.) If the EVO system is used alone without the air suspension system, a different control module is required.

⚠ **Caution**

Never short across, or ground, any terminals in a computer-controlled system unless you are instructed to do so in the vehicle manufacturer's service manual. Such action may damage expensive electronic components.

Figure 13-11 Toyota power rack and pinion steering gear.

Figure **13-12** Steering wheel rotation sensor.

Figure **13-13** Vehicle speed sensor (VSS).

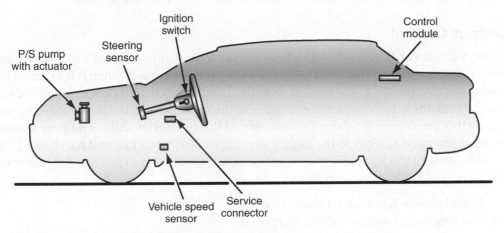

Figure **13-14** Control module and main components, electronic variable orifice steering.

Figure **13-15** Wiring diagram, electronic variable orifice steering.

Figure 13-16 Electronic variable orifice (EVO) actuator removed from the power steering pump.

Output Control

A varying current flow is sent from the control module through the EVO actuator in the power steering pump (**Figure 13-16**). The actuator swivels freely when it is installed in the power steering pump. As the control module changes the current flow through the actuator, the actuator supplies a variable pressure to the spool valve in the power steering pump. Two wires are connected from the actuator to the control module. The power steering pump is mounted directly to the engine to reduce noise, vibration, and harshness (NVH).

The control module positions the actuator and spool valve to provide full power steering assistance under these conditions:

1. Vehicle speed less than 10 mph (16 km/h)
2. Steering wheel rotation above 15 rpm

The full power-assist mode reduces driver effort required to turn the steering wheel during low-speed cornering and parking for increased convenience. In the full power-assist mode, the control module supplies 30 milliamps (mA) to the actuator.

The control module positions the actuator and spool valve to reduce power steering assistance under these conditions:

1. Vehicle speed above 25 mph (40 km/h)
2. Steering wheel rotation below 15 rpm

The reduced power-assist mode provides adequate road feel for the driver. In the reduced power-assist mode, the control module supplies 300 mA to the actuator. Above 88 mph (132 km/h), 590 mA is supplied from the control module to the actuator.

SAGINAW ELECTRONIC VARIABLE ORIFICE STEERING

Design and Operation

The term **variable effort steering (VES)** replaces the previous EVO terminology on some General Motors vehicles. In the EVO system, the vehicle speed sensor input is sent to the EVO controller. This controller supplies a pulse width modulated (PWM) voltage to the actuator solenoid in the power steering pump. The controller also provides a ground connection for the actuator solenoid (**Figure 13-17**).

Figure 13-17 Electronic variable orifice (EVO) steering system.

Figure 13-18 Actuator solenoid.

When the vehicle is operating at low speeds, the controller supplies a **pulse width modulated (PWM) voltage signal** to position the actuator solenoid plunger so the power steering pump pressure is higher (**Figure 13-18**). Under this condition, greater power assistance is provided for cornering or parking. If the vehicle is operating at higher speed, the controller changes the PWM signal to the actuator solenoid, and the solenoid plunger is positioned to reduce power steering pump pressure. This action reduces power steering assistance to provide improved road feel for the driver.

In a Magnasteer II VES system, as vehicle speed changes or lateral acceleration occurs, the amount of effort required to steer the vehicle can be varied dependent on the requirements. During low-speed operation, increased steering assist is provided for minimal steering effort to facilitate parking maneuvers and easy turning. At high speed, less assist is provided, which provides better road feel, firmer steering, and improved directional stability. If the system senses lateral acceleration, the steering effort will become firmer in an effort to reduce oversteering by the driver. The Electronic Brake Control Module

A pulse width modulated (PWM) voltage signal has a variable on time.

Figure 13-19 Magnasteer actuator solenoid.

Figure 13-20 Electronic brake and traction control module (EBTCM).

The **Magnasteer system** uses an electromagnetic actuator, a multiple pole ring permanent magnet, a pole piece, and an electromagnetic coil in the steering gear to vary steering effort. The EBCM varies steering assist by regulating the amount of current flow to the electromagnet.

(EBCM) controls a bi-directional rotary magnetic actuator solenoid mounted in the steering gear (**Figure 13-19**). The computer that operates the **Magnasteer system** is combined with the electronic brake and traction control module (EBCM), and this module is usually mounted in the left-front corner of the engine compartment (**Figure 13-20**). Two wires are connected from the steering actuator to the EBCM (**Figure 13-21**).

The steering actuator solenoid contains a pole piece with 16 magnetic segments and a coil. A matching 16-segment permanent magnet is attached to the spool valve. As the steering wheel and the spool valve rotate, the 16 segments on the spool valve move into and out of alignment with the segments on the actuator pole piece. The EBCM can reverse the current flow through the steering actuator, and this action reverses the magnetic poles on the actuator segments. At low vehicle speeds, the EBCM supplies a negative current flow, 2–3 amperes, and command through the steering actuator so the magnetic poles on the actuator repel the

Figure 13-21 Wiring connections between the Magnasteer actuator solenoid and the EBTCM.

permanent magnet segments on the spool valve. This repelling action assists the driver to turn the steering wheel and reduces steering effort. At medium speeds of about 45 mph, no current is supplied, and steering is assisted by hydraulic pressure only. At higher speeds, the EBCM provides a positive current flow of up to 2–3 amperes through the steering actuator, and this action reverses the poles on the actuator segments. Under this condition, the actuator segments are attracted to the permanent magnet segments on the spool valve. This action increases steering effort to improve driver road feel. The EBCM also has the capability to vary the current flow through the steering actuator to provide variable steering effort in relation to vehicle speed. If the EBCM detects a malfunction in the circuitry or actuator the system will supply zero amps to the actuator and the steering will be assisted by hydraulic pressure only.

Shop Manual
Chapter 13, page 580

RACK-DRIVE ELECTRONIC POWER STEERING

In electro-mechanical power steering systems, there is no need for a hydraulic pump and associated plumbing. In addition to design and control benefits, eliminating hydraulic oil in the steering system is another important contribution to environmental protection. In addition, the system saves energy by reducing fuel consumption, saving space, reducing noise, and lowering emissions. Unlike engine-driven hydraulic power steering systems, which require a constant flow of hydraulic fluid, the electro-mechanical systems only draw energy when power assist is required.

Electronic power steering systems (EPS) may be classified as rack-drive, column-drive, or pinion-drive. All three types of EPS have a rack and pinion steering gear. In a **rack-drive EPS** the electric motor that provides steering assist is coupled to the rack in the steering gear. A **column-drive EPS** has an electric-assist motor coupled to the steering shaft in the steering column, and in a **pinion-drive EPS** the electric motor is coupled to the steering gear pinion. The EPS system is light and compact compared to rack and pinion steering gear with hydraulic steering assist, because the power steering pump and hoses are not required on the EPS system. Since the EPS does not require a power steering pump driven by engine power, this system minimizes engine power loss and reduces fuel consumption. The EPS system reduces steering kickback while providing a linear steering feel.

When steering assist is required, an electric motor is actuated based on steering wheel input response. Thus, the system provides power steering assistance based on driving conditions and driver commands. The electronic power steering system has an active return function to support steering wheel return back to the center position. The active return function produces balanced, accurate straight-line stability under various driving conditions. The straight-line stability function assists the driver in maintaining straight ahead control when the vehicle is being acted upon by outside forces such as cross winds or driving up or down hills. The system will compensate for these external forces when the driver is holding the steering wheel in the straight-ahead position. Additionally, the electro-mechanical steering system can improve steering comfort over rough uneven road surfaces by limiting steering feedback. These systems require that the steering position sensors be relearned after a wheel alignment or steering component replacement. The relearn procedure is performed using a scan tool or alignment system integrated with relearn capabilities connected to the vehicle data link connector.

System Components

The rack and pinion steering gear changes rotary steering wheel motion to transverse motion of the rack. The motor that provides electric steering assist is designed into the steering gear (**Figure 13-22**). The steering sensor may be mounted in the pinion shaft or can be integrated into the steering column. This sensor sends input voltage signals to the EPS control unit in relation to the direction, amount, and torque of the steering wheel rotation. Depending on road speed, steering force, steering angle, and steering wheel rotation speed, the control unit calculates the required amount of steering assist and energizes the electric motor to provide steering assist (**Figure 13-23**).

The transmission or transaxle is equipped with a vehicle speed sensor. This sensor sends voltage input signals to the transmission control module (TCM) in relation to output shaft rotational speed. These voltage signals are transmitted from the TCM to the electronic power steering (EPS) control unit. The TCM may be located in the rear of the vehicle (**Figure 13-24**) on some platforms; it may also be integrated into the powertrain control module (PCM).

The EPS control unit may be mounted on or near the steering gear (**Figure 13-25**). This control unit receives voltage input signals from the steering sensor, VSS, or differential speed sensor. When these signals are received, the EPS control unit calculates the proper amount and direction of steering assist. The EPS control unit then commands a power module in the EPS control unit to drive the electric motor in the steering gear and provide the proper direction and amount of steering assist. The EPS control unit also contains self-diagnostic capabilities.

Figure 13-22 Steering gear with electronic assist and related system components.

Figure 13-23 Depending on road speed, steering force, steering angle, and steering wheel rotation speed, the control unit calculates the required amount of steering assist and energizes the electric motor to provide steering assist.

Figure 13-24 Electronic power steering (EPS) system component locations.

Figure 13-25 Electronic power steering (EPS) control unit and steering gear.

ELECTRONIC POWER STEERING SYSTEM OPERATION

Steering Sensor

The steering sensor contains a torque sensor and an interface. This sensor contains a slider core mounted on the pinion shaft. A spiral groove is located in each side of the slider, and two pins protruding from the pinion shaft are positioned in these grooves. The slider turns with the pinion shaft. When there is very little resistance to front wheel turning, the slider and pinion shaft rotate together and the pins remain centered in the spiral grooves in the slider. Very little resistance to front wheel turning occurs if the front tires are on a slippery road surface, or if the vehicle is raised so the front tires are off the floor. When there is resistance to front wheel turning, the torsion bar twists in the pinion shaft. This action causes the pins to move in the slider spiral grooves, and this movement causes upward or downward slider movement (**Figure 13-26**).

The slider core is surrounded by a variable differential transformer, and the slider moves upward or downward inside the transformer windings when the steering wheel is turned (**Figure 13-27**). The transformer has a primary coil and upper and lower secondary coils. When the ignition switch is turned on, an oscillation circuit in the steering sensor supplies an alternating current to the primary transformer coil. As the current alternates back and forth through the primary coil, the magnetic field is continually building up and collapsing around this coil. This rapidly expanding and collapsing magnetic field induces voltages in the upper and lower secondary coils. The position of the slider determines whether the voltage is induced in the upper or lower secondary coil.

While driving straight ahead, the slider is centered vertically between the upper and lower secondary coils. Under this condition, the voltage induced in these coils is equal. When voltages are equal in the upper and lower secondary coils, the voltage signals from these coils to the EPS control unit indicate the car is being driven straight ahead, or the steering wheel is being turned with no resistance and no electric power assist is required.

If the steering wheel is turned to the right, the slider moves upward. This slider position causes more induced voltage in the upper secondary coil and less induced voltage in

Shop Manual
Chapter 13, page 585

Figure 13-26 Pinion shaft with slider and variable differential transformer.

Figure 13-27 Slider movement when turning the steering wheel.

the lower secondary coil. When these voltage signals from the upper secondary coil are sent to the EPS control unit, this control unit supplies current to the electric motor on the rack so the motor rotates in the appropriate direction to provide the proper amount of steering assist to the right (**Figure 13-27**).

When the steering wheel is turned to the left, the slider moves downward. This slider position causes more induced voltage in the lower secondary coil and less induced voltage in the upper secondary coil. When these voltage signals from the lower secondary coil are sent to the EPS control unit, the control unit supplies current to the electric motor on the rack so the motor rotates in the appropriate direction to provide the proper amount of steering assist to the left (**Figure 13-28**).

Steering condition	Slider movement	Induction voltage on secondary coil
Steering to right (load steering)	Upward shift	Voltage on upper coil increases, and voltage on the lower decreases
Advancing straight ahead (no load steering)	Neutral	Voltage on the upper and lower coils are equal
Steering to left (load steering)	Downward shift	Voltage on lower coil increases, and voltage on the upper coil decreases

Figure 13-28 Summary of steering sensor and transformer operation.

Figure 13-29 Interface in the steering sensor.

The voltage signals from the upper and lower secondary transformer coils are sent through the interface in the steering sensor to the EPS control unit (**Figure 13-29**). The interface rectifies the alternating current (AC) voltage signals from the upper and lower transformer coils to direct current (DC) voltage signals and amplifies or increases the signal strength.

Electric Motor and Steering Gear Operation

The armature in the electric motor is hollow, and the rack extends through the center of this armature. Ball bearings are mounted between the outer diameter of the armature and the steering gear housing to support the armature. Two spring-loaded brushes are mounted on opposite sides of the commutator on one end of the armature. A gear with helical teeth is mounted on the other end of the armature. The teeth on the armature gear are in constant mesh with a matching gear on the ball screw shaft (**Figure 13-30**). A recirculating ball screw nut is mounted on the ball screw shaft. Ball bearings are mounted in grooves in the ball screw shaft and recirculating ball screw nut. The recirculating ball screw is bolted to the steering gear rack. The ball screw shaft and recirculating ball screw nut are similar in design to the worm shaft and ball nut in a recirculating ball steering gear. A permanent magnet is mounted in the steering gear housing, and this magnet surrounds the armature core. There is a small clearance between the armature core and the permanent magnet.

Figure 13-30 Steering gear and electric motor.

Recirculating ball
screw nut

Ball screw

Tube

Steel balls

Figure 13-31 Ball screw shaft and recirculating ball screw nut.

When the steering wheel is turned, electric current is supplied from the power module in the EPS control unit through the brushes and armature windings to ground. This current flow creates strong magnetic fields around the armature windings. These magnetic fields around the armature windings react with the magnetic field of the permanent magnet and cause armature rotation in the proper direction to supply steering assist. When the armature rotates, the gear on the armature shaft drives the gear on the ball screw shaft. Ball screw shaft rotation moves the recirculating ball screw nut on the shaft. Since the recirculating ball screw nut is bolted to the steering gear rack, movement of this ball screw nut provides steering assist in the proper direction. The power module can reverse the armature rotation to provide steering assist in either direction by reversing the polarity of the brushes on the commutator. When the brush polarity is reversed, the current flow through the armature windings is reversed and this changes the direction of armature rotation.

The recirculating ball screw nut is designed so the ball bearings roll between the grooves in the ball screw shaft and the grooves in the recirculating ball screw nut. Ball bearings coming out of the recirculating ball screw nut move through a tube and re enter the recirculating ball screw nut at the other end (**Figure 13-31**). The ball bearings in the grooves in the ball screw shaft and recirculating ball screw nut allow this nut to move on the shaft with very low friction.

Steering Gear Motor Current Control

The power module in the EPC control unit contains a driving circuit, current sensor, field effect transistor (FET) drive circuit, power relay, and fail-safe relay. When the ignition switch is turned on and the engine is cranked, the EPS control unit closes the power relay and fail-safe relay to make the EPS system operational. These relays actually close when the alternator begins producing voltage while the engine is cranking. Voltage signals are sent from the alternator and the ignition switch to the EPS control unit (**Figure 13-32** and **Figure 13-33**). The EPS system remains operational if the engine stalls and the ignition switch is on. If the engine stalls on a hydraulically assisted power steering system, power steering assist is lost and the steering wheel becomes very hard to turn. This action may result in a collision if the engine stalls while turning a corner. Since the EPS system is still operational if the engine stalls and the ignition switch is on, the EPS system reduces the possibility of a collision resulting from loss of power steering assist during an engine stall.

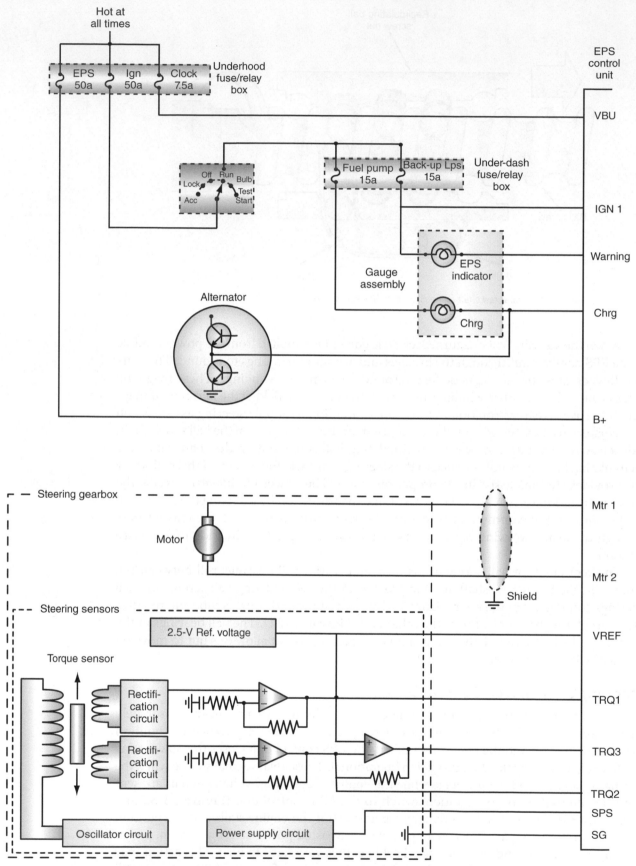

Figure 13-32 EPS system wiring diagram.

Figure 13-33 EPS system wiring diagram.

When the driver turns the steering wheel, the steering sensor input voltage signals inform the EPS control unit that steering assist is necessary. The steering sensor input voltage signals also inform the EPS control unit regarding the direction and amount of steering assist required. If the driver supplies more torque to the steering wheel, the steering sensor input voltage signal indicates to the EPS control unit that more steering assist is necessary.

When the EPS control unit receives input voltage signals from the steering sensor indicating that steering assist is required, this control unit signals the FET drive circuit in the power module. This drive circuit supplies a PWM voltage to the motor brushes with the proper polarity to provide the required direction and amount of steering assist. A PWM voltage signal is a pulsating voltage signal with a constant frequency but has a variable on time. One cycle of a PWM signal is a specific length of time that includes one Off and one On signal. If the on time lasts for 40 percent of each cycle time and the off time lasts for 60 percent of the cycle time, the motor current remains lower and this reduces motor speed and steering assist. When the on time lasts for 60 percent of each cycle time, and the off time lasts for 40 percent of the cycle time, the motor current is higher and power steering assist is increased (**Figure 13-34**). The FET drive circuit operates the four FET transistors in the FET bridge to supply the proper direction and amount of voltage and current through the armature windings to provide the necessary direction and amount of steering assist (**Figure 13-35**).

While the electric-assist motor is operating, the motor current flows through a current sensor in the power module regardless of the direction of motor rotation. The current sensor sends a feedback voltage signal to the EPS control unit. If the motor current exceeds a predetermined average motor current for the current operating condition, the EPS control unit signals the drive circuit in the power module to reduce the motor current to

Figure 13-34 Pulse width modulated (PWM) voltage signal.

Steering condition	FET (1)	FET (2)	FET (3)	FET (4)	Motor operation
Steering to right	PWM	OFF	OFF	ON	Operates in direction steering to the right
Straight ahead	OFF	OFF	OFF	OFF	Stops
Steering to left	OFF	PWM	ON	OFF	Operates in direction steering to the left

Figure 13-35 Drive circuit and field effect transistor (FET) bridge in the power module.

prevent motor overheating. If the steering wheel is turned fully in one direction and held in this position, the motor current becomes much higher. Under this condition, the current sensor signals the EPS control unit, and this unit signals the drive circuit in the power module to reduce motor current to protect the motor.

If an electrical defect occurs in the EPS system, the EPS control unit illuminates an EPS indicator light in the instrument panel to inform the driver that a fault is present in the EPS system (**Figure 13-36**). Under this condition, the EPS unit opens the fail-safe and power relays to make the EPS system inoperative. When this action occurs, manual steering without any power assist is available.

If one of the front wheels strikes a large road irregularity, force from the front wheel is transferred to the steering gear rack. Rack movement attempts to move the ball screw nut on the ball screw shaft. This action tries to rotate the ball screw shaft and armature. A specific amount of force is required to move the ball screw nut and rotate the ball screw shaft and armature because the armature windings have to rotate through the magnetic field between the permanent magnets. This resistance to movement of the ball screw nut, ball screw shaft, and armature helps prevent road shock from being supplied through the steering gear to the steering wheel. When a very high road shock is transferred to the steering wheel, this road shock moves the pinion shaft and slider in the steering sensor. When this EPS control unit receives this steering sensor input signal, the EPS control unit

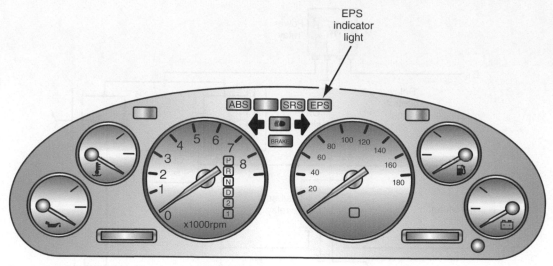

Figure 13-36 EPS indicator light.

immediately energizes the armature windings to oppose and cancel the road shock applied to the steering gear. Therefore, the EPS steering gear reduces road shock transferred from the front wheels to the steering wheel.

COLUMN-DRIVE ELECTRONIC POWER STEERING

Some vehicles are presently equipped with a column-drive EPS. In this type of EPS, a motor/module assembly is bolted to the lower end of the steering column (**Figure 13-37**). A combined steering wheel position sensor and steering shaft torque sensor is mounted

Figure 13-37 Column-drive electronic power steering system.

in the steering column at the motor/module attachment point. The module in the motor/module assembly is called the power steering control module (PSCM). Mounting the EPS motor/module assembly under the instrument panel provides underhood space for other components, subjects the assembly to less rigorous temperatures, and may provide better protection during a collision.

Input Sensors

The steering shaft torque sensor is the most important input used by the PSCM to supply the proper amount and direction of steering assist. The steering column contains a torque sensor input shaft connected from the steering wheel to the sensor, and an output shaft connected from the sensor to the steering shaft coupler. A torsion bar mounted inside the steering shaft torque sensor separates the input and output shafts. When the steering wheel is turned in either direction, the torque supplied from the steering wheel to the torsion bar causes this bar to twist. Increased torsion bar twisting results in higher-voltage signals from the steering shaft torque and position sensors. The steering shaft torque sensor is a dual sensor that provides two different voltage signals (**Figure 13-38**). During a right turn the voltage from sensor 1 increases and the voltage from sensor 2 decreases. The voltage signal range from each sensor is 0.25 V to 4.75 V. While completing a left turn, the voltage from sensor 1 decreases and the voltage from sensor 2 increases. The steering shaft torque sensor voltage signals inform the PSCM regarding the direction and amount of steering wheel torque.

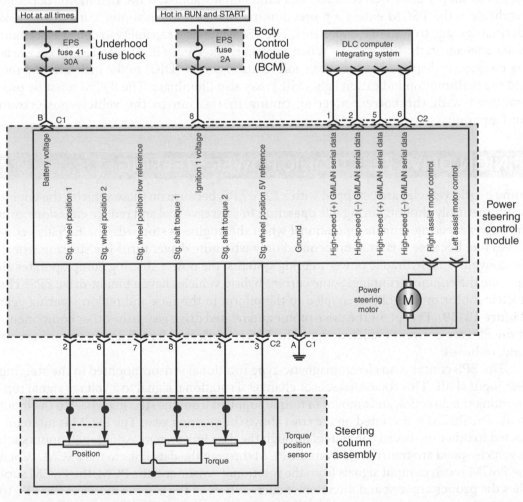

Figure 13-38 Wiring diagram for column-drive electronic power steering.

The steering wheel position sensor also contains dual voltage sensors that operate in the 0 V–5 V range. During steering wheel rotation, the voltage from these two sensors should remain within 2.5 V–2.8 V of each other. The steering wheel position sensor informs the PSCM regarding the position of the steering wheel. The combined steering shaft torque and position sensor assembly is serviced only as a unit.

The PSCM also receives a vehicle speed input signal from the PCM. This voltage signal is transmitted from the PCM to the PSCM through the interconnecting data links.

Power Steering Motor

The power steering motor is a 12-V brushless DC reversible motor rated at 65 amperes. This motor is coupled to the steering shaft through a worm gear and a reduction gear located in the steering column housing.

COLUMN-DRIVE ELECTRONIC POWER STEERING OPERATION

When the input signals indicate the driver is beginning to turn the steering wheel, the PSCM supplies the proper amount of current to the power steering motor to provide the necessary power steering assist. At low vehicle speeds, the PSCM provides more steering assist for easy steering wheel rotation during parking maneuvers. The PSCM reduces the motor current and provides reduced steering assist when the vehicle is driven at higher speeds to supply improved road feel and directional stability. If the EPS motor becomes overheated, the PSCM enters a protection mode that reduces motor current to avoid thermal damage to system components. The PSCM has the capability to detect electronic system defects in the EPS system. When a defect is detected, a POWER STEERING warning message is displayed in the driver information center (DIC) in the instrument panel and the malfunction indication light (MIL) may also illuminate. The PSCM must be programmed with the correct steering tuning in relation to the vehicle powertrain configuration.

PINION-DRIVE ELECTRONIC POWER STEERING

Some hybrid vehicles are equipped with a 12-V EPS, because on these vehicles the engine is automatically shut off during idle operation to conserve fuel and reduce emissions, and the power steering remains operational when the engine is stopped. The EPS also conserves fuel because it is used on demand only when the driver turns the steering wheel. On conventional hydraulic power steering systems, the power steering pump operates all the time the engine is running. Some current hybrid vehicles have a pinion-drive EPS. The electric motor on the EPS is coupled to the pinion in the rack and pinion steering gear (**Figure 13-39**). Except for the drive motor and related drive gears, the other components in the EPS are conventional in design. The EPS is lubricated for life and has no reservoir, fluid, or hoses.

The EPS contains an electromagnetic-type rotational sensor mounted in the steering gear input shaft. This rotational sensor changes a rotational signal to a voltage signal representing the direction and amount of torque supplied from the steering wheel to the input shaft. The PSCM is mounted on the cowl above the steering gear. The PSCM is interconnected to other on-board computers through the data link system. Additional inputs such as vehicle speed are transmitted from the PCM through the data links to the PSCM. When the PSCM receives input signals from the rotational sensor and the PCM, the PSCM supplies the proper amount and direction of current flow to the electric motor in the EPS to provide the required steering assist. A drain wire is connected from the PSCM to ground.

Figure 13-39 Pinion-drive electronic power steering (EPS) gear.

This wire prevents the EPS from emitting electromagnetic interference that could affect other electrical/electronic systems in the vehicle. If a defect occurs in the EPS electrical system, a SERVICE PWR STEERING message illuminates in the instrument panel.

Shop Manual
Chapter 13, page 587

ACTIVE STEERING SYSTEMS

Introduction

Active steering is available on a number of luxury vehicles. An active steering system contains many of the same components located in a conventional power rack and pinion steering system. The main difference in the active steering system is that the steering wheel and steering shaft are no longer connected directly to the pinion gear in the steering gear. In an active steering system, a dual planetary gear set is connected between the steering shaft and the pinion gear in the steering gear (**Figure 13-40**). The steering wheel and shaft are connected to one of the sun gears in a dual planetary gear set. The second sun gear in the planetary gear set is connected to the pinion gear in the steering gear. The two sun gears are connected by a set of planetary pinions. A brushless three-phase DC electric motor drives the ring gear on the planetary gear sets through a worm drive (Figure 13-41).

The **electronic control unit (ECU)** operates the electric motor to control the steering gear ratio and provide steering angle corrections to improve vehicle stability (**Figure 13-41**). In the active steering system, there is always a mechanical connection between the steering wheel and the pinion gear in the steering gear. The ECU may drive the electric motor and change the steering gear ratio or steering angle, but the mechanical connection between the steering wheel and the steering gear pinion always leaves the driver in control of the steering. The electric motor in the active steering system never provides turning forces to the front wheels.

ACTIVE STEERING SYSTEM COMPONENTS

DC Electric Motor and Motor Position Sensor

The electric drive motor has a wound stator and a permanent magnet rotor assembly. The ECU controls drive motor torque by a field-oriented control. The motor position sensor is mounted in the end of the electric motor, and this sensor informs the ECU regarding electric motor position.

Figure 13-40 Active steering system.

Figure 13-41 Active steering actuator.

Steering Angle Sensor

The **steering angle sensor** is mounted in the steering column switch assembly. The steering angle sensor is an optical sensor with no contacting parts, and this sensor is mounted to the circuit board near the top of the steering column (**Figure 13-43**). The main components in the steering angle sensor are an **encoded disc**, which is a slotted wheel, and an optical sensor. The encoded disc is attached to the steering shaft and rotates with this shaft. The encoded disc rotates within the optical sensor (**Figure 13-44**). The stationary part of

Figure 13-42 Active steering ECU.

Figure 13-43 Steering angle sensor mounting.

the steering angle sensor contains a light-emitting diode (LED), fiber-optical conductor, and an optical sensor. An optical steering angle sensor uses an LED on one side of a slotted ring and a photocell on the opposite side. When the sensor is operating, light from the LED is projected onto the encoded disc through the optical conductor. As the steering wheel is turned, the slotted ring rotates causing light to alternately strike the photocell as the window (slot) is opened and then closed, in effect turning the photocell on and off, producing a digital square wave output (**Figure 13-45**). The optical sensor converts these light signals to voltage signals in relation to steering wheel rotation. The steering wheel sensor is comprised of two separate sensors, and the encoding disc is segmented into at least two slotted segments, an inner and an outer (**Figure 13-46**). The inner ring is separated into five distinct segments of 72 degrees each. Each of these five segments has slotted openings that

Figure 13-44 Steering angle sensor internal design.

are equally spaced within the segment, but the spacing is different for each of the five segments. This provides a distinct digital code for each segment. The outer ring determines the angle by using six photoelectric beam pairs of LEDs and photocells. The system can detect steering angles up to 1,044 degrees of rotation. Most steering systems allow for approximately 2.75 turns of the steering wheel, or 990 degrees of rotation. The system continuously monitors the degree of rotation after each turn. Thus, the system can detect when a full steering wheel circle of 360 degrees has been exceeded. The type of sensors in an active steering system may vary depending on the model and year of vehicle, and can include non-contact optical, Hall effect, and magnetoresistive style sensors.

Pinion Angle Sensor and Summation Sensor

The **pinion angle sensor** is mounted in the steering gear at the end of the pinion housing, and this sensor operates on a magneto resistive principle. The pinion angle sensor sends a signal to the ECU regarding pinion position. The **summation sensor** is mounted in the steering gear housing, and this sensor sends a signal to the ECU in relation to total steering gear rack movement.

Electronic Control Unit (ECU)

The ECU contains two microprocessors that control the electric motor, steering pump, servotronic valve, and electric locking unit. The ECU communicates with other ECUs and the active steering system input sensors via a CAN network.

Electric Locking Unit (ELU)

The **electric locking unit (ELU)** contains an electric solenoid and a lock pin. When the ELU solenoid is not energized by the ECU, the lock pin drops into one of the slots in the worm drive gear. This action locks the worm drive and drive motor (**Figure 13-47**). This locking action is taken by the ECU if a safety-related defect occurs in the active steering system. When the worm drive is locked, the driver maintains normal steering control. Under normal conditions, the ECU operates the ELU solenoid and pulls the lock pin away from the notches in the worm drive gear, and this allows normal active steering system operation.

Figure 13-45 An optical steering angle sensor uses a light emitting diode on one side of a slotted ring and a photocell on the opposite side. As the steering wheel is turned, the slotted ring rotates causing light to alternately strike the photocell, in effect turning it on and off and producing a digital square wave output.

ACTIVE STEERING OPERATION

The actuator motor and planetary gear set have the capability to vary the steering gear ratio in relation to vehicle speed. The ECU receives voltage input signals from the system sensors, and in relation to these inputs, the ECU operates the actuator motor to vary the steering gear ratio. While cornering at low speeds, the steering gear ratio approximates 10:1. With this ratio, the driver does not have to rotate the steering wheel as much to obtain more front wheel turning action. When the vehicle is stationary, less than two steering wheel turns are required to turn the front wheels from lock-to-lock. This steering gear ratio allows the driver to apply the least amount of turning action to obtain a large amount of front wheel turning action. While parking and cornering at low speeds, this type of steering system requires less driver hand-to-hand action on the steering wheel, which increases safety and reduces driver fatigue.

Figure 13-47 Actuator motor locking mechanism.

Photoelectric beam pair

Figure 13-46 The steering wheel position sensor integrates two distinct sensors. The inner sensor has a ring separated into five distinct segments of 72 degrees each. The outer sensor ring determines the angle by using six photoelectric beam pairs of LEDs and photocells.

To reduce the steering ratio at low speeds, the actuator motor is driven in the same direction as the steering input, which decreases the steering ratio, and this action over-drives the steering input. This steering gear action may be compared to walking on an escalator. If you walk in the same direction as the escalator is moving, you multiply the total walking result.

As the vehicle speed increases, the ECU operates the actuator motor and planetary gear set to increase the steering ratio. At higher vehicle speeds, the steering gear ratio may be 20:1 and the increase in steering gear ratio dampens any sudden or excessive steering input by the driver. The increase in steering gear ratio under-drives the steering input. This action may be compared to walking against the movement of an escalator. The walking action of the person is canceled to some extent by the escalator movement, and the person has to walk more to cover the same distance. This steering gear action reduces the possibility of the driver causing the vehicle to go into a sideways swerve (**yaw motion**) by excessive steering wheel rotation. Any excessive steering wheel rotation by the driver is immediately counteracted by an increase in steering ratio by the active steering system. This active steering system action reduces the possibility of yaw vehicle motion. Many vehicles with active steering systems are also equipped with dynamic stability control (DSC), which greatly reduces the possibility of the vehicle swerving sideways out of control. On a vehicle equipped with active steering, the DSC system will not have to operate as frequently. If an electronic defect occurs in the active steering system, the system enters a fail silent mode in which the active steering system is inoperative and the driver has normal control of the steering.

Yaw motion is an erratic deviation from an intended course such as a sideways swerve when driving a vehicle.

A BIT OF HISTORY

Electronically controlled systems on vehicles have decreased emissions levels, increased fuel economy, and improved driver comfort and convenience. In the near future, electronic systems on vehicles may be interconnected with intelligent transportation systems (ITS) to speed up traffic flow and reduce accidents. Some traffic experts indicate that Americans lose 2 billion working hours each year because of traffic congestion and related delays. It is estimated these traffic delays cost American employers $40 billion per year. Therefore, we can appreciate that a system to increase traffic movement could save us billions of dollars each year.

Oakland County, Michigan, claims to be the world's largest operational test of an ITS system. This system uses 400 video cameras at 200 intersections to transmit information to a central computer that switches traffic light operation to match traffic flow. A similar system called Faster And Safer Travel Through Traffic Routing and Advanced Controls (FAST-TRAC) is installed in Oakland County, California. The ITS or FAST-TRAC systems may contain Advanced Traveler Information Systems (ATIS).

In these systems, on-board computers on vehicles receive information regarding driver information and route guidance from roadside beacons. The FAST-TRAC system in Oakland County, California, has decreased left-turn accidents by 69 percent and increased rush hour vehicle speeds by 19 percent. The U.S. Secretary of Transportation has indicated a commitment to installing ITS systems in 75 metropolitan areas in the United States with a goal of reducing travel time by 15 percent in these areas.

POWER STEERING SYSTEM

The electric-drive power steering pump and the active steering system are controlled by the same ECU, which contains two microprocessors. The power steering pump is a high-capacity vane-cell pump that is mounted immediately in front of the steering gear (**Figure 13-48**). The active steering system provides faster front wheel steering angles, and this action requires a higher-capacity power steering pump compared with conventional hydraulic-assisted power steering systems. The active steering ECU operates the **electronically controlled orifice (ECO) valve** in the power steering pump to control the fluid flow from the pump (**Figure 13-49**). The power steering pump supplies fluid

Figure 13-48 Electric-drive power steering system.

Figure 13-49 Electric-drive power steering pump.

pressure from the ECO valve in the pump to the **servotronic valve** in the steering gear. The active steering ECU operates the servotronic valve to increase power steering pump pressure and reduce steering effort and driver fatigue at low speeds (**Figure 13-50**). At higher vehicle speeds, the ECU operates the servotronic valve to reduce power steering pump pressure and increase steering effort to provide improved road feel and steering control. On some models, the active steering ECU directly operates the servotronic valve.

On other models, the software for the servotronic valve operation is stored in the active steering ECU, and this ECU sends signals through one of the CAN networks to the **safety and gateway module (SGM)**. On the basis of these input signals, the SGM operates the servotronic valve and the ECO valve.

The input signals required for servotronic valve operation are these:

1. Road speed signals from the wheel speed sensors to the DSC unit via one of the CAN networks to the active steering ECU.

Servotronic valve

Electric servomotor lock

Electric servomotor with motor position sensor

Planetary gearbox with overdrive function

Figure 13-50 At low speeds the ECU operates the servotronic valve to increase power steering pump pressure.

2. Engine status signal from the **digital motor electronics (DME)** control unit via one of the CAN networks to the active steering ECU.
3. Terminal status from the **car access system (CAS)** via one of the CAN networks to the active steering ECU.

The servotronic valve is operated only when the engine is running. The servotronic valve enters a default mode if any of the input signals are incorrect, or if there is an electrical defect in the servotronic valve winding or connecting wires. In the default mode, the servotronic valve does not affect steering assist.

STEER-BY-WIRE SYSTEMS

The main difference between conventional steering systems and steer-by-wire systems is that steer-by-wire systems do not have any mechanical linkage between the steering wheel and the front wheels. In a steer-by-wire system, steering wheel input is supplied to torque sensor, gear, DC motor, and motor angle sensor (**Figure 13-51**). Input signals are sent to a controller from the motor angle sensor, torque sensor, and motor current sensor. On the basis of these input signals, the controller supplies output voltage signals to the motor drivers. The motor driver connected to the DC motor in the steering gear operates this motor to supply the desired steering angle.

Steer-by-wire systems are not available in automotive showrooms at present, but they may be available in the future. Steer-by-wire systems could have many possible advantages such as active steering control, improved vehicle maneuverability and stability, and steering system tuning to specific types of driving conditions. However, before steer-by-wire steering systems are commercialized, concerns regarding reliability and confirmation of advantages must be completely satisfied.

Figure 13-51 Steer-by-wire system.

ELECTRONICALLY CONTROLLED FOUR-WHEEL STEERING

Four-wheel (4WS) or all-wheel steering was first introduced by Honda as an option on the 1987 Prelude. Since then, 4WS has come and gone on several manufacturers' platforms, with one of the more notable and publicized being Quadra-steer offered on some General Motors trucks and full-size SUVs between 2002 and 2005. More recent entrants into 4WS were the 2016 Cadillac CT6 and the 2017 Audi Q7 platform, with other manufacturers—including Porsche, Infiniti, Lexus, Volvo, Renault, Ferrari, and Lamborghini—offering four-wheel steering on some models. The Audi Q7 offers a five-link independent suspension system with a rear-mounted electronic rear steering actuator assembly (**Figure 13-52**). Dynamic 4WS is designed to improve vehicle stability, comfort, and agility. While all the systems are somewhat similar with the utilization of a rear-mounted steering actuator assembly that resembles a rack and pinion assembly. Newer systems being introduced use the latest technology in electronic power assist, along with sophisticated computer programs that make full use of the CAN network system, communicating with other nodes on the network, including but not limited to active suspension system, ABS, stability control, and anti-rollover technology. While there are many benefits to 4WS, cost and system integration complexity are still major limiting factors that at least currently limit auto manufacturers desire to integrate 4WS. At least for now, 4WS will continue to be a novelty and not a mainstream feature.

In a **four-wheel steering (4WS)** system, all four wheels are turned to steer the vehicle.

If a car with conventional front wheel steering is parallel parked at a curb between two vehicles, this car may be driven from the parking space without hitting the car in front if the front wheels are turned all the way to the left (**Figure 13-53**, view A). When the same car is equipped with 4WS and the rear wheels steer in the same direction as the front wheels at low speed, the car will not steer out of the parking space without striking the vehicle in front (**Figure 13-53**, view B).

Steering angle is the number of degrees that the wheels turn when steered.

When the car in the same parking space has a 4WS system that steers the rear wheels in the opposite direction to the front wheels at low speed, the car steers out of the parking space with plenty of distance between the vehicle parked in front (**Figure 13-53**, view C). When the rear wheels steer in the opposite direction to the front wheels, the rear wheels steer toward the curb. This action causes the right rear tire to strike the curb if the car is parked close to the curb. Therefore, the maximum rear **steering angle** must be considerably less than the maximum front wheel steering angle to help prevent this problem. A car with 4WS has a smaller **turning circle**, or turning radius, compared to a vehicle with

Turning circle is the diameter of a circle completed by a vehicle when the front wheels are turned fully to the right or left.

Figure 13-52 A five-link independent rear suspension assembly with rear steering actuator assembly that resembles a rack and pinion steering gear.

Figure 13-53 Parallel parking with conventional front wheel steering and four-wheel steering.

Figure 13-54 On all-wheel steering systems, rear wheel turning angle is generally limited to a maximum of 5 degrees or less.

conventional front wheel steering. On all-wheel steering systems, rear wheel turning angle is generally limited to a maximum of 5 degrees or less (**Figure 13-54**). This improves maneuverability while parking.

 A BIT OF HISTORY

The **four-wheel steering (4WS)** concept has been researched by automotive engineers for several decades.

In 1962, an engineer representing one of the Japanese manufacturers stated at the Japanese Automotive Engineering Association's technical meeting, "A major improvement in control and stability may be anticipated by means of an automatic rear wheel steering system." In the late 1970s, Honda and Mazda were actively engaged in 4WS development. However, it was not until the late 1980s that 4WS was introduced on a significant number of Honda and Mazda cars.

The rear wheel steering in a 4WS system may be controlled in relation to vehicle speed or the amount of steering wheel rotation. On all-wheel steering systems, the rear wheel turning angle is counter (opposite) to the direction the front wheels are turned at low speeds below 30 mph (48 km/h), resulting in tighter turning radius (**Figure 13-55**).

Counter steering
Counter steering the rear wheels results in a tighter turning radius when driving under 31 mph.

Figure 13-55 On all-wheel steering systems, the rear wheel turning angle is counter (opposite) to the direction the front wheels are turned at speeds below 30 mph, resulting in tighter all-wheel radius.

Parallel steering
When all four wheels turn the same direction, this results in precise lateral tracking at highway speeds.

Figure 13-56 On all wheel steering systems, the rear wheel turning angle is counter (opposite) to the direction the front wheels are turned at speeds below 30 mph resulting in tighter turning radius.

Allowing the rear wheels to steer in the opposite direction to the front wheels at low vehicle speeds has the following benefits:

■ Smaller turning radius
■ Improved agility
■ Improved handling in low-speed traffic and urban driving
■ Improved parking maneuverability

When the vehicle is operating at higher speeds or with a small amount of steering wheel rotation, the rear wheels are steered in the same direction as the front wheels (**Figure 13-56**). The advantage of steering the rear wheels in the same direction as the front wheels at highway speeds is precise lateral tracking; this is referred to as parallel steering. When a vehicle is cornering at higher speeds, centrifugal force tends to move the rear of the vehicle sideways. This action causes the rear tires to slip sideways on the road surface. This process may be called **sideslip**. The vehicle speed and the severity of the turn determine the amount of sideslip. If sideslip is excessive, the car may spin around, causing the driver to lose control. When the 4WS system steers the rear wheels in the same direction as the front wheels at higher speeds, sideslip is reduced and vehicle stability is improved. Allowing the rear wheels to steer in the same direction to the front wheels at higher vehicle speeds has these additional benefits:

■ Improved vehicle stability
■ Increased driving safety
■ Improved driving dynamics

The higher-speed same-direction steering angle is considerably less than the low-speed opposite-direction steering angle. As was stated earlier, the maximum rear wheel steering angle is 5 degrees on most platforms during any steering maneuver, and even less on some platforms.

Shop Manual
Chapter 13, page 594

Figure 13-57 Rear steering actuator and adjustable lower links.

Entering corner at mid-speed; rear toe angle initially opposite phase for improving response

Figure 13-58 Rear wheel steering during negative phase operation.

Exiting corner at mid-speed or lane change; rear toe angle goes to same phase for improved yaw damping

Figure 13-59 Rear wheel steering during positive phase operation.

The four-wheel active steering (4WAS) improves steering and handling by adjusting the geometry of the rear suspension in relation to steering input and vehicle speed. The 4WAS control unit calculates the ultimate vehicle dynamics from the input signals from a group of sensors. In response to these inputs, the control unit operates the electric actuator to lengthen or shorten each lower rear suspension link (**Figure 13-57**).

When the vehicle begins turning a corner, driving around a curve, or making a lane change at mid-speed, the control unit operates the 4WAS actuator to initially turn the rear wheels in the negative phase mode in relation to the front wheels (**Figure 13-58**). This action provides improved turn-in response.

When the steering wheel rotation indicates the vehicle is exiting a corner or making a lane change at mid-speed, the control unit operates the 4WAS actuator to turn the rear wheels in the positive phase mode in relation to the front wheels (**Figure 13-59**). Under this condition yaw forces are reduced on the rear of the vehicle to provide improved vehicle stability.

Figure 13-60 Electronically controlled 4WS with control unit mounted in the trunk.

On the electronically controlled 4WS system, there is no mechanical connection between the front steering gear and the **rear steering actuator**. The rear steering actuator is controlled by a 4WS control unit (**Figure 13-60**). The 4WS control unit in the electronically controlled system uses steering wheel rotational speed, vehicle speed, and front steering angle information to calculate and control the rear steering angle.

Input sensors information accessed by the electronically controlled 4WS system include, but are not limited to, the following:

1. Main rear wheel angle sensor.
2. Sub rear wheel angle sensor in the rear steering actuator.
3. Main steering wheel angle sensor in the steering column under the combination switch.
4. Sub front wheel steering angle sensor.
5. ABS wheel speed sensors.
6. Vehicle speed sensors.

The rear steering actuator may be compared to an electric steering gear. This actuator contains an electric motor that drives a steering rack through a ball screw mechanism (**Figure 13-61**).Tie rods are connected from the steering actuator to the rear steering arms and spindles. A return spring inside the actuator moves the rear wheels to the straight-ahead position when the ignition switch is turned off or when a defect occurs in the 4WS system. Some manufacturers, such as Porsche, use individual rear steering actuators for both the left and right rear wheels (**Figure 13-62**).

When the engine is running, the 4WS control unit continually receives information from all the input sensors and associated control modules on the network. If the steering wheel is turned, the 4WS control unit analyzes information from the vehicle speed sensor, main steering wheel angle sensor, sub front wheel angle sensor, main rear wheel angle sensor, sub rear wheel angle sensor, and the rear wheel speed sensors, among others. The 4WS control unit calculates the proper rear wheel steering angle and then sends voltage command to the rear steering actuator(s) motor to provide this rear steering angle (**Figure 13-63**).

In an **electronically controlled 4WS system**, rear wheel steering is controlled electronically.

 Caution

Some electronic sensor wires have a special shield surrounding them to prevent electromagnetic interference (EMI) from other voltage sources. If this shield is damaged or removed, computer system operation may be adversely affected. Do not reroute sensor wires close to other voltage sources such as spark plug wires.

Figure 13-61 Rear steering actuator, electronically controlled 4WS system.

Figure 13-62 Some manufacturers use individual rear wheel steering actuators in unison to control rear wheel turning angle.

Transistors may be defined as automatic electronic relays that have no moving parts and are made from semiconductor materials.

Battery voltage is sent to the rear steering actuator motor through two heavy-duty output **transistors**. One of these transistors conducts current during a right turn, whereas the other transistor is activated during a left turn. The main rear wheel angle sensor and the sub rear wheel angle sensor send feedback signals to the 4WS control unit, indicating the proper rear steering angle has been supplied.

Figure 13-63 The 4WS control unit analyzes input sensor information, calculates the required rear steering angle, and operates the rear steering actuator motor to provide the proper rear steering angle.

Figure 13-64 Rear steering angle in relation to vehicle speed and steering wheel rotation.

When the vehicle is operating at speeds less than 18 mph (29 km/h) to 30 mph (48 km/h), depending on the manufacturer, the rear wheels immediately begin to steer in the opposite direction to the front wheels if the steering wheel is turned (**Figure 13-64**). Maximum rear steering angle is generally 5° at 0 mph, but this maximum angle does vary depending on the manufacturer, from as little as 2.5° to as much as 12° (always refer to specific vehicle manufacturer service information for exact specifications). The rate of rear steering angle decreases in relation to vehicle speed.

When the vehicle speed increases above 30 mph (48 km/h), this speed varies depending on model year and manufacturer, the rear wheels steer in the same direction as the front wheels through the first 200° of steering wheel rotation. The rear steering angle reverts to the opposite phase if the steering wheel is rotated more than 200° in this vehicle speed range. When the vehicle speed is 60 mph (96 km/h) and the steering wheel rotation is 100°, the rear wheels steer about 1° in the same direction as the front wheels. If the steering wheel is rotated 500° slowly at this speed, the rear wheels are steered about 1° in the opposite direction to the front wheels.

4WAS adjusts the rear steering angle and ratio in relation to vehicle speed to create a more fluid sense of control. The system improves low-speed feel, provides more responsive mid-speed turn-in, and enhanced high-speed stability. The following description will contain details on the major components that are often found in a 4WS system.

Steering Wheel Position Sensor

The steering wheel speed/position sensor (SWPS) is mounted at the lower end of the steering column and this sensor is controlled by steering shaft rotation (**Figure 13-65**). The steering wheel position sensor provides an analog signal and three digital signals to the control module. The body control module (BCM) supplies a 5-V reference signal to the SWPS, and a low reference or ground wire is also connected from the SWPS to the BCM. The SWPS contains a potentiometer, which sends an **analog voltage signal** to the BCM in relation to steering wheel rotation (**Figure 13-66**). The analog voltage signal from the SWPS to the BCM ranges from 0.25 V with the steering wheel positioned one turn to the left of the center position to 4.75 V when the steering wheel is positioned one turn to the right of the center position. With the steering wheel in the center position, the SWPS analog voltage signal is 2.5 V. When the steering wheel is turned more than one turn to the right or left of the center position, the SWPS signal does not change. The BCM relays the SWPS analog voltage signal through class 2 data links to the rear wheel steering control module.

The SWPS sends digital signals through the phase A, phase B, and marker pulse circuits directly to the control module. The marker pulse digital signal is displayed on a scan tool as High if the steering wheel is positioned between 10° to the left or 10° to the right of the center position. If the steering wheel is positioned more than 10° to the right or left of the center position, the pulse marker signal is displayed as Low. The phase A and phase B signals are displayed on a scan tool as High or Low as the steering wheel is rotated. These signals change from High to Low every one degree of steering wheel rotation.

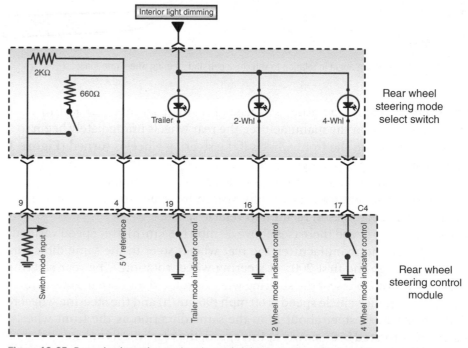

Figure 13-65 Rear wheel steering mode select switch inputs to the control module.

Figure 13-66 Steering wheel speed/position sensor.

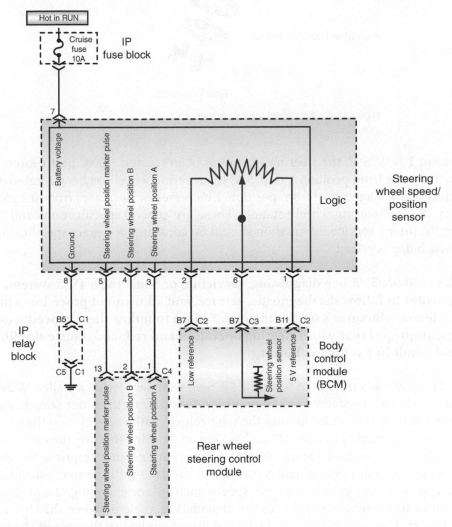

Figure 13-67 SWPS inputs to the control module.

Rear Wheel Position Sensor

The rear wheel position sensor is mounted on the lower side of the rear wheel steering gear (**Figure 13-67**). The rear wheel position sensor has two signal circuits connected to the rear wheel steering control module (**Figure 13-68**). The position 1 signal is a linear measurement of voltage per degree of rear steering position sensor rotation. For the position 1 input, the measurement in degrees is from −620° to the left to +620° to the right. The voltage signal from the position 1 input is 0.25 V to 4.75 V. If the signal voltage from

Sensor O-ring

Rear wheel position sensor

Sensor cover

Retaining bolts

Figure 13-68 Rear wheel position sensor.

position 1 is 0.25 V, the steering wheel has been rotated −600° past center. When the signal voltage from position 1 is 4.75 V, the steering wheel has been rotated +600° past center. The voltage signal from position 2 increases or decreases from 0.25 V to 4.75 V every 180° of steering wheel rotation. These are general specifications and the vehicle manufacturer's service information should be consulted for exact specifications for the vehicle being serviced.

⚡ **WARNING When diagnosing, servicing, or adjusting a 4WS system, it is very important to follow the diagnostic, service, and adjustment procedures in the vehicle manufacturer's service manual. Failure to follow these procedures may cause improper rear wheel steering operation and reduced vehicle stability that could result in a vehicle collision.**

There are advantages to having a 4WS on vehicles towing a trailer. When the rear vehicle wheels are steered in the opposite direction to the front wheels during low-speed turning, the trailer follows the true vehicle path more closely than it does with a two-wheel steering system. When backing up a trailer, steering the rear wheels in the opposite direction to the front wheels provides better trailer response to vehicle steering inputs, and this action makes it easier to back the trailer into the desired position. When steering the vehicle at higher speeds such as lane changing, the positive steering action of the rear wheels reduces the articulation angle between the tow vehicle and the trailer. This action reduces the lateral forces applied to the rear of the tow vehicle by the trailer, which in turn reduces yaw velocity gain and improves trailer stability. A 4WS system reduces turning circle diameter from 45 ft. to 33.9 ft. (**Figure 13-69**) on some SUVs, and a similar proportional reduction in turning circles is experienced on small performance car platforms as well.

Full size SUV (45.0 ft.)

Ford Excursion (43.7 ft.)

Toyota Land Cruiser (39.7 ft.)

Land Rover Range Rover (39.0 ft.)

Jeep Grand Cherokee Laredo (37.6 ft.)

Mercedes M-class (37.0 ft.)

Chevrolet Metro (31.5 ft.)

Honda Civic coupe (32.8 ft.)

Figure 13-69 Turning circle diameter, four-wheel steering system.

FOUR-WHEEL ACTIVE STEERING (4WAS)

4WAS System Design

The 4WAS has a front ECU and a main (rear) ECU. The 4WAS system has a front steering actuator mounted coaxially in the front steering shaft (**Figure 13-70**). The front actuator contains a front wheel steering angle sensor, a front lock solenoid valve, a front motor, and a gear shaft (**Figure 13-71**). The front ECU operates the motor and gear shaft in the front actuator to change the steering ratio. The lock solenoid valve in the front steering actuator locks this actuator so the steering ratio cannot change if a defect occurs in the system.

The rear steering actuator body is attached to a chassis member, and the outer ends of the actuator shaft are linked to the rear wheels. The rear steering actuator contains a rod attached to the rear steering arms (item 1), a motor (item 4), motor shaft and drive gear (item 5), a driven gear (item 7), a housing assembly (item 3), and a rear wheel steering angle sensor (item 6) (**Figure 13-72**).

As an automotive technician, you must keep up to date on the changes in automotive electronics, and these changes are occurring at a very rapid pace. It has been my experience that one of the best ways to keep up to date is to join professional technicians' organizations such as the National Automotive Service Task Force (NASTF), Automotive Service Association (ASA), or the International Automotive Technicians' Network (IATN). As a member of these organizations, you will be able to obtain information on the latest automotive electronics technology and the solutions to diagnostic problems related to automotive electronics.

Figure 13-70 4WAS system.

Four-wheel active steering (4WAS) may be defined as front and rear wheel steering in which a computer(s) can electronically change the front and rear steering angles.

Figure 13-71 4WAS front actuator.

Shop Manual
Chapter 13, page 533

The main ECU is mounted in the trunk, and controls the rear steering actuator, and the front ECU is mounted under the dash and controls the front steering actuator. The software in the ECUs contains a reference model that contains the desired dynamic steering characteristics. The ECUs operate the front and rear steering actuators to conform to the reference model in the software.

Figure 13-72 4WAS rear actuator.

Figure 13-73 Steering angles while cornering at low speed.

The main control unit calculates the front and rear steering angles that will provide the best steering performance and vehicle stability based on the front and rear steering angle sensor inputs and the vehicle speed input. The vehicle speed input is sent from the antilock brake system (ABS) control unit through the CAN communication network to the main control unit. Engine speed signals are transmitted from the engine electronic control unit (ECU) via the CAN network to the main control unit. The 4WAS warning light is mounted in the unified meter in the instrument panel, and operated by the main control unit.

4WAS System Operation

When cornering at low speed the front control unit operates the front steering actuator to reduce the steering gear ratio. This action increases the front steering angle with less steering wheel movement and reduced driver steering effort (**Figure 13-73**).

When changing lanes in the medium-speed range, the front steering actuator can increase the front steering angle in relation to steering wheel rotation, and simultaneously the rear steering actuator steers the rear wheels a small amount in the same direction. When this action is taken, yaw motion, lateral force, and vehicle slide slip are reduced (**Figure 13-74**).

Figure 13-74 Steering angles while changing lanes at medium speed.

Figure 13-75 Steering angles while changing lanes at high speed.

When changing lanes at high speed, the front steering actuator increases the front steering ratio so the driver steering effort is increased, steering wheel movement is also increased, and steering angle is decreased in relation to steering wheel rotation. This action provides more road feel to the driver. Simultaneously, the rear steering actuator steers the rear wheels a few degrees in the same direction (**Figure 13-75**). This action provides improved steering response and vehicle stability.

On some Infiniti luxury car model packages, the 4WAS feature is an available option. This technology may or may not stay part of the luxury car segment. Often it depends more on whether the feature sells cars or not than it does on the viability of the technology. The 4WAS makes the steering and handling characteristics on this long-wheelbase car more nimble and agile compared to models without this technology.

SUMMARY

- Manual or power rack and pinion steering systems are lighter and more compact than recirculating ball steering gears and parallelogram steering linkages.
- The rack and pinion steering gear must be mounted so the rack maintains the tie rods in a parallel position in relation to the lower control arms.
- The rack takes the place of the idler arm and pitman arm in a parallelogram steering linkage.
- The pinion teeth may be spur or helical. The inner tie rods are connected to the rack through spring-loaded inner ball sockets. The rack bearing and adjuster plug maintains proper preload between the rack and pinion teeth.
- A rack and pinion steering gear may be mounted on the firewall or front crossmember. A rack and pinion steering system has fewer friction points than a recirculating ball steering gear and parallelogram steering linkage.
- Reducing the number of friction points in a steering system provides more road feel and reduces the steering system's ability to isolate road vibration and noise.
- Steering ratio is the relationship between steering wheel movement and front wheel movement to the right or left.
- During a left turn with a power rack and pinion steering gear, fluid pressure is directed to the left side of the rack piston, and fluid is released from the right side of this piston. This action moves the rack to the right to complete the left turn.
- During a right turn with a power rack and pinion steering gear, fluid pressure is directed to the right side of the rack piston, and fluid is released from the left side of this piston. Under this condition, the rack is forced to the left to help the driver complete the right turn.
- The fluid pressure in a power rack and pinion steering gear is directed to the appropriate side of the rack piston by the spool valve and rotary valve position.
- The spool valve position is controlled by torsion bar twisting during a right or left turn. If hydraulic pressure is not available in a power rack and pinion steering gear, stop tangs on the stub shaft and pinion provide steering action, but steering effort is much higher.

- The differences between Saginaw and TRW power rack and pinion steering gears are mainly in the method of tie rod attachment, bulkhead oil seal and retainer, and upper and lower pinion bearing hardware.
- The Toyota power rack and pinion steering gear has removable control valve housing and claw washers to lock the inner tie rods to the rack.
- The electronic variable orifice (EVO) steering system provides high-power steering assistance during low-speed cornering and parking and normal power assistance at higher speeds.
- The two inputs in the EVO system are the steering wheel rotation sensor and the vehicle speed sensor.
- The rear wheel steering system in a 4WS system is usually electronically controlled.
- Steering the rear wheels in the opposite direction to the front wheels at low speed provides a shorter turning circle, or radius, for easier maneuvering.
- Steering the rear wheels in the same direction as the front wheels at higher speeds reduces rear side-slip and improves vehicle stability while cornering or changing lanes.
- In an electronically controlled 4WS system, there is no mechanical connection between the front steering gear and the rear steering gear.
- In an electronically controlled 4WS system, the rear steering actuator contains an electric motor that is controlled by the 4WS control unit in response to various input sensor signals.
- In a 4WS system, the control unit uses inputs from the vehicle speed sensor (VSS) and the front steering position sensor to control the rear steering actuator.
- A higher output alternator is required on a 4WS system because of the higher current draw of the rear actuator.
- Negative phase steering occurs at lower speeds when the rear wheels are steered in the opposite direction as the front wheels.
- Positive phase steering occurs at higher speeds when the rear wheels are steered in the same direction as the front wheels.
- Steering the rear wheels in the same direction as the front wheels at higher speeds reduces the lateral forces applied to the rear of a vehicle, and this action reduces yaw velocity.
- When towing a trailer, steering the rear wheels in the same direction as the front wheels at higher speeds improves trailer stability.

REVIEW QUESTIONS

Short Answer Essays

1. Explain why the rack and pinion steering gear is ideally suited for unibody front-wheel-drive vehicles.

2. Explain the advantage of a 4WS system while parking a vehicle.

3. Explain the purpose of the front main steering wheel angle sensor in an electronically controlled 4WS system.

4. Explain the purpose of the bellows boots in a rack and pinion steering gear.

5. Describe two possible mounting positions for rack and pinion steering gears.

6. Explain why a rack and pinion steering gear provides improved feel of the road compared with a recirculating ball steering gear and parallelogram steering linkage.

7. Describe the fluid movement in the rack piston chamber during a right turn.

8. Explain how the spool valve is moved inside the rotary valve in a power rack and pinion steering gear.

9. Explain the difference between spur and helical pinion gear teeth and explain the advantage of the helical design.

10. Describe the driving conditions required for an electronic variable orifice (EVO) steering system to provide high-power steering assistance.

Fill-in-the-Blanks

1. The preload on the rack and pinion teeth affects steering harshness, noise, and_____.

2. Compared to a recirculating ball steering gear and parallelogram steering linkage, a rack and pinion steering system has a reduced ability to isolate road noise and vibration because the rack and pinion system has fewer_____.

3. During a left turn, the power steering pump forces fluid into the _____ side of the rack chamber, and fluid is removed from the _____ side of this chamber if the rack is mounted behind the front wheel spindle.

4. During a turn, fluid is directed to the appropriate side of the rack piston chamber by the _____ valve and _____ valve position.

5. When the torsion bar twists, it changes the position of the _____ valve.

6. In an electronically controlled 4WS system, the electronic module operates a(n) _____ _____to steer the rear wheels.

7. The front main steering angle sensor in an electronically controlled 4WS system is mounted in the_____ _____.

8. In an EVO steering system, the power steering assistance is decreased at higher speeds to provide improved_____ _____.

9. The input sensors in an EVO steering system are the vehicle speed sensor and the_____ _____ _____.

10. During negative phase steering, the rear wheels are steered in the _____ _____ to the front wheels.

Multiple Choice

1. Compared to a recirculating ball steering gear, a rack and pinion steering gear provides:
   A. Improved capability to isolate road noise.
   B. Better capability to reduce steering harshness on road irregularities.
   C. Increased capability to absorb steering kickback.
   D. Increased road feel.

2. All of these statements about a 4WS system are true EXCEPT:
   A. A 4WS system reduces yaw forces on the vehicle at higher speeds.
   B. A 4WS system provides a smaller turning circle.
   C. A 4WS system reduces tire wear.
   D. A 4WS system improves vehicle maneuverability at low speeds.

3. All of these statements about a power rack and pinion steering gear mounted behind *the front wheel spindle are* true EXCEPT:

   A. During a left turn fluid pressure is directed to the left side of the rack piston.

   B. During a right turn fluid pressure is directed to the right side of the rack piston.

   C. Fluid is returned through the vent tube positioned in the bellows boot.

   D. The spool valve is moved by torsion bar twisting during a turn.

4. In a column-drive electric power steering system all of the following are true EXCEPT:

   A. A motor module assembly is bolted to the lower end of the steering column.

   B. The steering shaft torque sensor is the most important input used by the power steering control module.

   C. At low speeds, more steering assist is provided.

   D. The steering system may have a 20:1 steering ratio during low-speed cornering.

5. In a power rack and pinion steering gear the breather tube:

   A. Vents the right rack piston chamber during a left turn.

   B. Moves air from one boot to the other during a turn.

   C. Allows air into the rotary valve area while driving straight ahead.

   D. Prevents pressure buildup in both rack piston chambers when driving straight ahead.

6. The power steering pump belt breaks on a rack and pinion power steering system. Under this condition:

   A. The torsion bar may be broken when the front wheels are turned to the right or left.

   B. The steering operates like a manual steering system, and steering effort is higher.

   C. The spool valve no longer moves inside the rotary valve when the steering wheel is turned.

   D. Power steering fluid may be forced from the rack seals when the steering wheel is turned.

7. In electronic variable orifice (EVO) steering systems:

   A. Steering assistance is increased if the vehicle is cornering at high speeds.

   B. Steering assistance is increased if the vehicle is driven straight ahead at highway speeds.

   C. Full power steering assistance is provided if the vehicle speed is below 10 mph (16 km/h) and steering wheel rotation is above 15 rpm.

   D. The control unit increases steering assistance at higher speed to provide improved road feel.

8. If the lock-to-lock steering wheel rotation is 3.25 turns and the total front wheel movement is 78 degrees, the steering ratio is:

   A. 15.5:1   C. 16:1

   B. 15:1   D. 19.5:1

9. If an electrical defect occurs in a 4WS system, the:

   A. Rear wheels remain in the centered position.

   B. Positive phase rear wheel steering is canceled.

   C. Negative phase rear wheel steering is increased.

   D. Positive phase rear wheel steering angle is increased, and negative phase rear wheel steering is decreased.

10. All of the following are true about a Magnasteer power steering system EXCEPT:

    A. As vehicle speed changes or lateral acceleration occurs the amount of effort required to steer the vehicle can be varied depending on the requirements.

    B. At high speeds more assist is provided, which provides better road feel, firmer steering, and improved directional stability.

    C. If the system senses lateral acceleration the steering effort will become firmer in an effort to reduce oversteering by the driver.

    D. During low-speed operation increased steering assist is provided for minimal steering effort to facilitate parking maneuvers and easy turning.

CHAPTER 14

RECIRCULATING BALL STEERING GEARS

Upon completion and review of this chapter, you should be able to understand and describe:

- The purpose of a steering gear.
- The advantage of a recirculating ball steering gear compared with earlier worm and roller or cam and lever steering gears.
- The purpose of the worm shaft preload adjustment.
- How the sector shaft is rotated in a manual recirculating ball steering gear.
- The purpose of the interference fit between the sector shaft teeth and the recirculating ball teeth, and explain how this fit is obtained.
- The difference between constant ratio sector teeth and variable ratio sector teeth in a recirculating ball steering

gear, and explain the advantage of the variable ratio sector teeth.
- Gear ratio.
- The term "faster" steering as it relates to steering gears.
- The power steering fluid movement in a power recirculating ball steering gear with the engine running and the front wheels straight ahead.
- The power steering fluid movement in a power recirculating ball steering gear during a left turn.
- The power steering fluid movement during a right turn.
- How kickback action is prevented in a power recirculating ball steering gear.

Terms To Know

Faster steering	Rotary valve	Worm and roller steering gear
Gear ratio	Sector	Worm and sector design
Gear tooth backlash	Spool valve	Worm shaft
Interference fit	Steering gear free play	Worm shaft bearing preload
Kickback	Torsion bar	
Pitman shaft sector	Vehicle wander	Worm shaft end play
Ross cam and lever steering gear	Worm and gear	

INTRODUCTION

In a **worm and gear** or **worm and sector design**, the sector teeth are in direct contact with the worm.

The purpose of the steering gear box is to provide a mechanical advantage that allows the driver to turn the front wheels with a reasonable amount of effort. In the early 1900s, steering gears were a **worm and gear** or **worm and sector design**. These steering gears gave the driver a mechanical advantage to turn the front wheels, but they created a lot of friction.

A BIT OF HISTORY

Steering gear design progressed from the crude, high-friction worm and gear of the early 1900s to the Ross cam and lever gear and Saginaw worm and roller gear of the 1920s. The Saginaw worm and roller steering gear was the forerunner of the modern, low-friction recirculating ball steering gear.

The **Ross cam and lever steering gear** was introduced in 1923. The cam in this gear was a spiral groove machined into the end of the steering shaft. A pin on the pitman shaft was mounted in the spiral groove in the steering shaft. When the steering wheel and shaft were turned, the pin was forced to move, and this action rotated the pitman shaft. When a front wheel struck a road irregularity, this steering gear design prevented serious **kickback** on the steering wheel. However, this steering gear design still created a considerable amount of friction and required higher steering effort.

In the mid-1920s, Saginaw Steering Division of General Motors Corporation developed the **worm and roller steering gear**. In this steering gear, the **sector** became a roller, which greatly reduced friction and steering effort. The Saginaw worm and roller steering gear was the forerunner of the recirculating ball steering gear that has been widely used on rear-wheel-drive cars for many years.

MANUAL RECIRCULATING BALL STEERING GEARS

Design and Operation

In a recirculating ball steering gear, the steering wheel and shaft are connected to the **worm shaft**. Ball bearings support both ends of the worm shaft in the steering gear housing. A seal above the upper worm shaft bearing prevents oil leaks, and an adjusting plug is provided on the upper worm shaft bearing to adjust **worm shaft bearing preload**. Proper preloading of the worm shaft bearing is necessary to eliminate **worm shaft end play** and to prevent **steering gear free play** and **vehicle wander**. A ball nut is mounted over the worm shaft, and internal threads or grooves on the ball nut match the grooves on the worm shaft. Ball bearings run in ball nut and worm shaft grooves (**Figure 14-1**).

When the worm shaft is rotated by the steering wheel, the ball nut is moved up or down on the worm shaft. The gear teeth on the ball nut are meshed with matching gear teeth on the **pitman shaft sector**. Therefore, ball nut movement causes pitman shaft sector rotation. Since the pitman shaft sector is connected through the pitman arm and steering linkage to the front wheels, the front wheels are turned by the pitman shaft sector. The lower end of the pitman shaft sector is usually supported by a bushing or a needle bearing in the steering gear housing. A bushing in the side cover supports the upper end of this shaft.

When the front wheels are straight ahead, an **interference fit** exists between the sector shaft teeth and ball nut teeth. This interference fit eliminates **gear tooth backlash** when the front wheels are straight ahead and provides the driver with a positive feel of the road. Proper axial adjustment of the sector shaft is necessary to obtain the necessary interference fit between the sector shaft and worm shaft teeth. A sector shaft adjuster screw is threaded into the side cover to provide axial sector shaft adjustment (**Figure 14-2**).

Manual recirculating ball steering gears have sector gear teeth designed to provide a constant ratio, whereas power recirculating ball steering gears usually have sector gear teeth with a variable ratio (**Figure 14-3**). The sector gear teeth have equal lengths in a constant ratio steering gear, but the center sector gear tooth is longer compared with the other teeth in a variable ratio gear. The variable ratio steering gear varies the amount of

Figure 14-1 Manual recirculating ball steering gear design.

mechanical advantage provided by the steering gear in relation to steering wheel position. This variable ratio provides **faster steering**. The steering **gear ratio** in a constant ratio manual steering gear is usually 15:1 or 16:1, whereas the average variable ratio steering gear ratio may be 13:1. When the same types of steering gears are compared, a higher numerical ratio provides reduced steering effort and increased steering wheel movement in relation to the amount of front wheel movement.

Many recirculating ball steering gears are bolted to the frame with hard steel bolts (**Figure 14-4**). These bolts must be tightened to the vehicle manufacturer's specified torque.

POWER RECIRCULATING BALL STEERING GEARS

Design and Operation

The ball nut and pitman shaft sector are similar in manual and power recirculating ball steering gears. In the power steering gear, a **torsion bar** is connected between the steering shaft and the worm shaft. Since the front wheels are resting on the road surface, they resist turning, and the parts attached to the worm shaft also resist turning. This turning resistance causes torsion bar twist when the wheels are turned, and this twist is limited to a predetermined amount. The worm shaft is connected to the **rotary valve** body, and the torsion bar pin also connects the torsion bar to the worm shaft. The upper end of the torsion bar is attached to the steering shaft and wheel. A stub shaft is mounted inside the

A steering gear with a lower numerical ratio may be called a **faster steering** gear compared with a steering gear with a higher numerical ratio.

Gear ratio refers to the relationship between the rotation of the drive and driven gears. If 13 turns of the drive gear are necessary to obtain one turn of the driven gear, the gear ratio is 13:1.

The **rotary valve** is mounted over the top of the spool valve in a steering gear.

Figure 14-2 Sector shaft adjusting nut.

Figure 14-3 Constant and variable ratio steering gears.

The **spool valve** is mounted inside the rotary valve. The twisting action of the torsion bar moves the spool valve in relation to the rotary valve.

Shop Manual
Chapter 14, page 611

rotary valve and a pin connects the outer end of this shaft to the torsion bar. The pin on the inner end of the stub shaft is connected to the **spool valve** in the center of the rotary valve (**Figure 14-5**).

⚡ **WARNING** **Many recirculating ball steering gears are mounted near the exhaust manifold, which can be extremely hot. Use caution and wear protective gloves when inspecting or servicing the steering gear to avoid burns to hands and arms.**

When the car is driven with the front wheels straight ahead, oil flows from the power steering pump through the spool valve, rotary valve, and low-pressure return line to the pump inlet (**Figure 14-6**). In the straight-ahead steering gear position, oil pressure is equal on both sides of the recirculating ball piston, and the oil acts as a cushion that prevents road shocks from reaching the steering wheel.

If the driver makes a left turn, torsion bar twist moves the valve spool inside the rotary valve body so that oil flow is directed through the rotary valve to the left-turn holes in the spool valve (**Figure 14-7**). Since power steering fluid is directed from these left-turn holes to the upper side of the recirculating ball piston (**Figure 14-8**), this hydraulic pressure on the piston assists the driver in turning the wheels to the left.

When the driver makes a right turn, torsion bar twist moves the spool valve so that oil flows through the spool valve, rotary valve, and a passage in the housing to the pressure

⚠ **Caution**

Hard steel bolts can be used for steering gear mounting. Bolt hardness is indicated by the number of ribs on the bolt head. Harder bolts have five, six, or seven ribs on the bolt heads. Never substitute softer steel bolts in place of the original harder bolts, because these softer bolts may break, allowing the steering gear box to detach from the frame. This action results in a loss of steering control.

Figure 14-4 Steering gear mounting on the vehicle frame.

Figure 14-5 Torsion bar and stub shaft.

chamber at the lower end of the ball nut piston (**Figure 14-9**). During a right turn, hydraulic pressure applied to the lower end of the recirculating ball piston helps the driver turn the wheels.

During a turn, if a front wheel strikes a bump and the front wheels are driven in the direction opposite the turning direction, the recirculating ball piston tends to move against the hydraulic pressure and force oil back out the pressure inlet port. This action would create a kickback on the steering wheel, but a poppet valve in the pressure inlet fitting closes and prevents kickback action.

AUTHOR'S NOTE After road testing many vehicles to diagnose steering problems, it has been my experience that the rack and pinion steering gear and related linkage tends to transfer more road shock from the front wheels to the steering wheel compared with a recirculating ball steering gear. The design of the recirculating ball steering gear and the related steering linkage resists the transfer of road shock from the front wheels to the steering wheel. The reason for this resistance is the linkage design in which the steering arms are connected through the tie rods to the center link, and this link is connected through the pitman arm to the steering gear. In a rack and pinion steering gear, the tie rods are connected directly between the steering arms and the rack, and the rack is connected through the pinion to the steering wheel. Compared with a recirculating ball steering gear, a rack and pinion steering gear with its related linkage has fewer friction points and a more direct connection between the front wheels and the steering wheel. However, the rack and pinion steering gear does provide the driver with a greater feel of the road.

You need to be aware of the characteristics of these steering systems when diagnosing steering problems. When you understand the basic characteristics of these steering systems, you are able to immediately recognize normal and abnormal conditions, and this allows you to provide a fast, accurate diagnosis.

Figure 14-6 Power steering gear fluid flow with the wheels straight ahead.

Figure 14-7 Spool valve position during a left turn.

Figure 14-8 Power steering gear fluid flow during a left turn.

Figure 14-9 Power steering gear fluid flow during a right turn.

SUMMARY

- Early model steering gears such as the worm and gear or Ross cam and lever created a lot of friction and required higher steering effort.
- The modern recirculating ball steering gear reduces friction and steering effort. In a manual recirculating ball steering gear, a worm shaft adjusting plug provides a worm shaft preload adjustment.
- In a recirculating ball steering gear, rotation of the steering wheel causes the ball nut to move up and down on the worm shaft.

- With the front wheels straight ahead in a recirculating ball steering gear, an interference fit exists between the ball nut teeth and the sector shaft teeth.
- Axial sector shaft adjustment provides the proper interference fit between the ball nut teeth and the sector shaft teeth.
- Sector shaft teeth may have a constant ratio design or a variable ratio design. In a power recirculating ball steering gear with the front wheels straight ahead, equal pressure is applied to both sides of the recirculating ball piston, and

the fluid from the pump is directed through the spool valve and rotary valve to the return hose and pump inlet.

■ In a power recirculating ball steering gear, the spool valve movement inside the rotary valve is controlled by torsion bar twist.

■ When the front wheels are turned in a power recirculating ball steering gear, torsion bar twist moves the spool valve inside the rotary valve, and this valve movement directs the power steering fluid to the appropriate side of the recirculating ball piston to assist steering.

REVIEW QUESTIONS

Short Answer Essays

1. Describe the advantage of the manual recirculating ball steering gear compared with the previous cam and lever steering gear, and explain how this advantage is obtained.

2. Explain three purposes of the worm shaft preload adjustment.

3. Explain the purpose of the interference fit between the ball nut teeth and the sector shaft teeth in a manual recirculating ball steering gear, and describe how this interference fit is obtained.

4. Explain the difference in design between constant ratio and variable ratio sector shaft teeth.

5. Define the term faster steering.

6. Define steering gear ratio, and explain the effect of this ratio on steering effort with a manual recirculating ball steering gear.

7. Describe the flow of power steering fluid with the engine running and the front wheels straight ahead in a power recirculating ball steering gear.

8. Describe how torsion bar twist occurs in a recirculating ball steering gear.

9. Explain the purpose of torsion bar twist in a power recirculating ball steering gear.

10. Explain how kickbacks are prevented in a power recirculating ball steering gear.

Fill-in-the-Blanks

1. A manual recirculating ball steering gear provides reduced _____ and _____ compared with the earlier cam and lever manual steering gear.

2. If a bearing has preload, there is a slight _____ on the bearing.

3. When the steering has free play, there is some _____ movement before the front wheels start to turn.

4. The worm shaft end plug provides a worm shaft bearing _____ adjustment.

5. In a manual recirculating ball steering gear, the interference fit between the ball nut and sector shaft teeth is obtained by proper _____ sector shaft adjustment.

6. Compared with a constant ratio steering gear with a ratio of 13:1, a constant ratio steering gear with a ratio of 16:1 requires _____ steering effort.

7. When a faster steering system is compared with a slower steering system, the faster system provides more front wheel movement with _____ steering wheel rotation.

8. A variable ratio steering gear varies the amount of _____ in relation to steering wheel position.

9. In a power recirculating ball steering gear with the front wheels straight ahead and the engine running, the fluid on each side of the recirculating ball piston helps prevent _____ from reaching the steering wheel.

10. In a power recirculating ball steering gear during a turn, the _____ movement directs the power steering fluid pressure to the appropriate side of the recirculating ball piston.

Multiple Choice

1. In a power recirculating ball steering gear, kickback is prevented by:
   A. The preload between the ball nut teeth and sector gear teeth.
   B. The bending of the torsion bar.
   C. A poppet valve in the pressure inlet fitting.
   D. The flexible coupling in the steering shaft.

2. When the worm shaft and bearings are properly installed and adjusted in a power recirculating ball steering gear:

   A. The worm shaft bearings should be preloaded.

   B. The worm shaft should have the specified endplay.

   C. The worm shaft endplay should be eliminated without any preload on the bearings.

   D. There must be an interference fit between the ball nut teeth and the worm shaft teeth.

3. All of these statements about recirculating ball steering gears are true EXCEPT:

   A. The steering gear provides a mechanical advantage.

   B. The steering gear reduces steering effort.

   C. Some early model steering gears had a worm and sector design.

   D. Ross cam and lever steering gears provided a very low internal friction.

4. A typical steering gear ratio in a manual steering gear is:

   A. 16:1.   C. 19:1.

   B. 18:1.   D. 22:1.

5. When comparing manual recirculating ball steering gears a higher numerical ratio provides:

   A. Increased steering effort.

   B. Increased steering wheel movement in relation to front wheel movement.

   C. Reduced kickback force on the steering wheel.

   D. Improved steering control.

6. In a manual recirculating ball steering gear:

   A. The interference fit between the sector shaft teeth and ball nut teeth is present when the front wheels are straight ahead.

   B. The proper interference fit between the sector shaft teeth and ball nut teeth is obtained by radial sector shaft movement.

   C. The interference fit between the sector shaft teeth and ball nut teeth becomes tighter when the front wheels are turned to the right or left.

   D. Tightening the worm shaft adjusting plug tightens the interference fit between the sector shaft teeth and ball nut teeth.

7. All of these statements about steering gear ratios are true EXCEPT:

   A. When a steering gear with a 15:1 ratio is compared with a steering gear with a 13:1 ratio, the gear with the 15:1 ratio requires more steering wheel rotation to turn the front wheels.

   B. When a steering gear with a 15:1 ratio is compared with a steering gear with a 13:1 ratio, the gear with the 15:1 ratio provides reduced steering effort.

   C. When a steering gear with a 15:1 ratio is compared with a steering gear with a 13:1 ratio, the gear with the 15:1 ratio may be called a faster steering gear.

   D. The 13:1 steering gear ratio may be used on a smaller, lighter car and the 15:1 steering gear ratio may be used on a larger, heavier vehicle.

8. In power recirculating ball steering gears:

   A. A variable ratio steering gear has sector shaft teeth with equal lengths.

   B. A variable ratio steering gear has a constant mechanical advantage regardless of steering wheel position.

   C. The torsion bar is connected between the steering shaft and the sector shaft.

   D. During a turn, power steering fluid is directed to the proper end of the ball nut piston by spool valve movement.

9. While driving with the front wheels straight ahead with a power recirculating ball steering gear:

   A. The torsion bar is twisted to move the spool valve inside the valve body.

   B. The fluid pressure is higher on the upper side of the ball nut piston compared with the lower side.

   C. The fluid pressure on each side of the ball nut piston cushions road shocks from reaching the steering wheel.

   D. The return passage from the steering gear valve body to the power steering pump is nearly closed.

10. When a vehicle with a power recirculating ball steering gear is making a right turn:

A. Torsion bar twisting and spool valve movement has moved the spool valve and aligned the pressure ports with the right-turn ports.

B. Fluid pressure is supplied to the upper side of the ball nut piston.

C. Fluid pressure is exhausted from the lower side of the ball nut piston.

D. Fluid pressure gradually leaks past the ball nut piston to reduce steering assist during long, gradual turns.

GLOSSARY

Note: **Terms are highlighted in color**, followed by **Spanish translation in bold**.

Accelerometer An input sensor that senses vehicle acceleration in a computer-controlled suspension system.

Acelerómetro Sensor de entrada que advierte la aceleración del vehículo en un sistema de suspensión controlado por computadora.

Active coils Located in the center area of a coil spring.

Muelles activos Ubicado en el centro de un muelle espiral.

Active suspension system A computer-controlled suspension system with double-acting solenoids at each wheel.

Sistema activo de suspensión Sistema de suspensión controlado por computadora con solenoides de doble acción montados en cada una de las ruedas.

Actuator An electronically operated solenoid in a strut or shock absorber that controls firmness.

Actuador Un solenoide operado electrónicamente en un tirante o un amortiguador que controla la firmeza.

Adaptive cruise control system Maintains a specific distance between the vehicle being driven and the vehicle being followed.

Sistema adaptable de marcha a velocidad de crucero Mantiene una distancia específica entre el vehículo que se conduce y el vehículo que se sigue.

Adjustable strut A strut with a manual-operated adjustment for strut firmness.

Tirante ajustable Un tirante de ajuste operado por mano que determina la firmeza del tirante.

Air bag deployment module The air bag and deployment canister assembly that is mounted in the steering wheel for the driver's side air bag or in the dash panel for the passenger's side air bag.

Unidad de despliegue del Airbag El conjunto del Airbag y elemento de despliegue montado en el volante de dirección para proteger al conductor, o en el tablero de instrumentos para proteger al pasajero.

Air spring An air-filled membrane that replaces the conventional coil springs in an air suspension system.

Muelle de aire Membrana llena de aire que reemplaza los muelles helicoidales convencionales en un sistema de suspensión de aire.

Air spring solenoid valve An electrically operated solenoid that allows air to flow in and out of an air spring.

Muelle de aire con válvula activada por solenoide Solenoide activado eléctricamente que permite el flujo libre de aire en un muelle de aire.

All-season tires Tires with special tread designed to improve traction on snow or ice, while providing acceptable noise levels on smooth road surfaces.

Neumáticos para toda época Neumáticos con una huella especial diseñada para mejorar la tracción en la nieve o en el hielo, a la misma vez que provee niveles aceptables de ruido cuando se conduce el vehículo en un camino cuya superficie es lisa.

American Petroleum Institute (API) An organization in charge of engine oil classifications and many other areas in the petroleum industry.

Instituto Americano del Petróleo (API) Organización que tiene a su cargo la clasificación del aceite de motor y varias otras áreas en la industria del petróleo.

Ampere (A) A unit of measurement of electrical current flow.

Amperio (A) El amperio es una unidad de medida del flujo de la corriente eléctrica.

Amplitude The extent of a vibratory movement.

Amplitud El magnitud de un movimiento de vibración.

Analog voltage signal A voltage signal that varies continuously within a specific range.

Señal de voltaje análogo Un señal que varía continuamente dentro de una banda específica.

Angular bearing load A load applied to a bearing in a direction between horizontal and vertical.

Carga del cojinete angular Una carga aplicada a un cojinete en una dirección entre horizontal y vertical.

Antilock brake system (ABS) A computer-controlled brake system that prevents wheel lockup during brake applications.

Sistema antiblocante de frenos Un sistema controlado por computadora que previene el bloqueo en las ruedas durante la aplicación de los frenos.

Asymmetrical An arrangement marked by irregularity such as different shapes, sizes, and positions.

Asimétrico Un arreglo caracterizado por la irregularidad tal como las diferencias de forma, tamaño y posición.

Asymmetrical leaf spring Has the same distance from the spring center bolt to each end of the spring.

Ballesta de hojas asimétricas Tiene la misma distancia desde el perno del centro del resorte a cada extremo del resorte.

Atmospheric pressure The pressure exerted on the earth by the atmosphere, which is 14.7 psi at sea level.

Presión atmosférica Presión que la atmósfera ejerce sobre la tierra. Dicha presión es 14.7 psi [peso sobre unidad de superficie] al nivel del mar.

Automatic level control (ALC) System maintains the rear suspension trim height regardless of the rear suspension load. If a heavy object is placed in the trunk, the rear wheel position sensors send trim height signals to the ESC module.

Sistema de Control de nivel automático (CNA) Mantiene la altura del centrado de la suspensión trasera sin tomar en cuenta la carga de la suspensión trasera. Si un objeto se coloca en la cajuela, los sensores de posición de la rueda envían señales de altura por debajo del centrado al módulo de CSE.

Axle offset A condition in which the complete rear axle assembly has turned so one rear wheel has moved forward and the opposite rear wheel has moved rearward.

Descentralización del eje Condición que ocurre cuando todo el conjunto del eje trasero ha girado de manera que una de las ruedas traseras se ha movido hacia adelante y la opuesta se ha movido hacia atrás.

Axle sideset A condition in which the rear axle assembly has moved sideways from its original position.

Resbalamiento lateral del eje Condición que ocurre cuando el conjunto del eje trasero se ha movido lateralmente desde su posición original.

Axle thrustline Is a line extending forward from the center of the rear axle at a 90° angle.

Eje de directriz de presiones Es una línea que se extiende hacia enfrente desde el centro del eje trasero a un ángulo de 90°.

Axle tramp The repeated lifting of the differential housing during extremely hard acceleration.

Barreta del eje Levantamiento repetido del alojamiento del diferencial durante una aceleración sumamente rápida.

Axle windup The tendency of the rear axle housing to rotate in the opposite direction to the wheel-and-tire rotation during hard vehicle acceleration.

Bobinado del eje La tendencia de la carcasa del eje trasero de rotar en la dirección opuesta a la rotación de la rueda y el neumático durante la aceleración fuerte del vehículo.

Ball bearings Have round steel balls between the inner and outer races.

Cojinete de bolas Tiene bolas de acero entre las carreras interiores y exteriores.

Bead filler A piece of rubber positioned above the bead that reinforces the sidewall and acts as a rim extender.

Relleno de la pestaña de la llanta Pieza de caucho ubicada sobre la pestaña que sirve para reforzar la pared lateral y actúa como una extensión de la llanta.

Bead wire A group of circular wire strands molded into the inner circumference of the tire that anchors the tire on the rim.

Cable de pestaña Grupo de cordones circulares de cables moldeados en la circunferencia interior del neumático que sujetan el neumático a la llanta.

Bearing hub unit A complete wheel bearing and hub unit to which the wheel rim is bolted.

Conjunto de cubo y cojinete Conjunto total de cojinete y cubo de rueda al que se emperna la llanta de la rueda.

Bearing seals Circular metal rings with a sealing lip on the inner edge. Bearing seals are usually mounted on both sides of a bearing to prevent dirt and moisture from entering the bearing while sealing the lubricant in the bearing.

Juntas de estanqueidad del cojinete Anillos metálicos circulares con un reborde de estanqueidad en el borde interior. Normalmente las juntas de estanqueidad del cojinete se montan en ambos lados de un cojinete para evitar la entrada de suciedad y humedad mientras sellan el lubricante dentro del mismo.

Bearing shields Circular metal rings attached to the outer race that prevent dirt from entering the bearing but allow excess lubricant to flow out of the bearing.

Protectores para cojinetes Anillos metálicos circulares fijados a la arandela exterior que evitan la entrada de suciedad al cojinete, pero permiten la salida de exceso de lubricante del mismo.

Belt cover A nylon cover positioned over the belts in a tire that helps to hold the tire together at high speed, and provides longer tire life.

Cubierta de correa Cubierta de nilón ubicada sobre las correas de un neumático. Esta cubierta ayuda a conservar la solidez del neumático cuando se conduce el vehículo a gran velocidad, además de proporcionarle mayor durabilidad.

Belted bias-ply tire A type of tire construction with the cord plies wound at an angle to the center of the tire, and fiberglass or steel belts mounted under the tread area of the tire.

Neumático de correas con estrías diagonales Tipo de neumático fabricado con estrias arregladas a un ángulo con respecto al centro del neumático, y correas de fibra de vidrio o de acero montadas debajo de la huella del mismo.

Bias-ply tire A type of tire construction with the cord plies wound at an angle to the center of the tire and no belts surrounding the cords.

Neumático de estrías diagonales Un tipo de construcción de neumático con estrías arregladas en un ángulo con respecto al centro del neumático y sín correas envolviendo las estrías.

Brake-by-wire A computer-controlled brake system with no direct mechanical connection between the brake pedal and the master cylinder.

Frenado alámbrico Un sistema de frenos controlado por computadora sin una conexión mecánica directa entre el pedal de freno y el cilíndro maestro.

Brake pressure modulator valve (BPMV) A group of solenoid valves mounted in a single valve body that controls fluid pressure in the brake system during antilock brake, traction control, and vehicle stability control functions.

Válvula moderador de presión de frenado (BPMV) Un grupo de válvulas de solenoide montado en una sóla caja de válvula que controla la presión de los fluidos en el sistema de frenos durante las operaciones de freno antibloqueante, control de tracción, y control de la estabilidad del vehículo.

Brake pressure switch An electrical switch operated by brake fluid pressure.

Interruptor de presión del freno Interruptor eléctrico activado por la presión del disco de freno.

Braking and deceleration torque The torque applied to the rear axle assembly during deceleration and braking in a rear-wheel-drive vehicle.

Torsión de deceleración y frenaje La torsión impuesta en el ensamblado del eje trasero durante la deceleración y el frenado en un vehículo de tracción trasera.

Breather tube A small metal tube that allows air to flow between the bellows boots in a rack and pinion steering gear during a turn.

Tubo respiradero Pequeño tubo metálico que permite el flujo libre de aire entre las botas de fuelles en un mecanismo de dirección de cremallera y piñón durante un viraje.

Bump steer The tendency of the steering to veer suddenly in one direction when one or both front wheels strike a bump.

Cambio de dirección ocacionado por promontorios en el terreno Tendencia de la dirección a cambiar repentinamente de sentido cuando una o ambas ruedas delanteras golpea un promontorio.

Cam ring A metal ring with an elliptical-shaped inner surface on which the vanes make contact in a power steering pump.

Anillo de levas Un anillo metálico con una superficie interior de forma elíptica en la cual las aletas hacen contacto en una bomba de dirección hidraulica.

Car access system (CAS) A system that controls specific vehicle functions and operates in conjunction with the safety and gateway module on a vehicle with an active steering system.

Sistema de acceso al automóvil (CAS, en inglés) Un sistema que controla las funciones específicas del vehículo y que opera en conjunción con el módulo de la puerta de enlace y de seguridad de un vehículo con un sistema de dirección activo.

Carbon dioxide (CO_2) A greenhouse gas that is a by-product of gasoline or diesel fuel combustion.

Dióxido de carbono (CO2) Un gas de efecto invernadero que es un subproducto de la combustión de la gasolina o del combustible diesel.

Caster angle The angle between an imaginary line through the center of the tire and wheel and another imaginary line through the centers of the lower ball joint and upper strut mount viewed from the side.

ángulo de inclinación El ángulo entre una línea imaginaria que pasa por el centro de un neumático y la rueda y otra línea que pasa por los centros de la articulación esférica inferior y la montadura superior del apoyadero visto de un lado.

Center link A long rod connected from the pitman arm to the idler arm in a steering linkage.

Biela motriz central Varilla larga conectada del brazo pitman al brazo auxiliar en un cuadrilátero de la dirección.

Center take-off (CTO) The tie rods are attached to a movable sleeve in the center of the gear.

Conexión central (Center take-off, CTO) Los brazos de dirección se conectan a una manga móvil en el centro del engranaje.

Channel frame A type of frame construction in which a steel channel is positioned on the top and bottom of the vertical frame web.

Armazón fabricado con ranuras Tipo de fabricación del armazón en la que una ranura de acero es colocada en las partes superior e inferior de la malla del armazón vertical.

City safety system On some vehicles, the collision mitigation system is called a city safety system.

Sistema de seguridad de la ciudad En algunos vehículos, el sistema de mitigación de choques se conoce como sistema de seguridad de la ciudad.

Clock spring electrical connector A conductive ribbon in a plastic case mounted on top of the steering column that maintains electrical contact between the air bag inflator module and the air bag electrical system.

Conector eléctrico de cuerda de reloj Cinta conductiva envuelta en una cubierta plástica montada en la parte superior de la columna de dirección, que mantiene contacto eléctrico entre la unidad infladora y el sistema eléctrico del Airbag.

Collision mitigation system Operates specific vehicle functions to help prevent a collision.

Sistema de alivio de colisiones Opera funciones especificas del vehiculo para ayudar a evitar una colisión.

Column-drive EPS Has an electric assist motor coupled to the steering shaft in the steering column.

Transmisión de columna de GPE (guiado de propulsión electrónica) Tiene un motor de corriente eléctrica que acompaña al eje de dirección en la columna de la dirección.

Compact spare tire A spare tire and wheel that is much smaller than the other tires on the vehicle, and is designed for short-distance driving at low speed.

Neumático compacto de repuesto Conjunto de neumático y rueda de repuesto mucho más pequeño que los demás neumáticos del vehículo, diseñado para viajes de corta distancia a poca velocidad.

Complete box frame A type of vehicle frame shaped like a rectangular steel box.

Fabricación completa Tipo de armazón en forma de caja rectangular de acero.

Compressible A substance is compressible when an increase in pressure causes a decrease in volume.

Compresibile Una sustancia es compresible cuando un aumento en la presión produce una disminución en el volumen.

Compression loaded A suspension ball joint mounted so the vehicle weight is forcing the ball into the joint.

Cargada de presión Junta esférica de la suspensión montada de manera que el peso del vehículo presiona la bola hacia el interior de la junta.

Conicity Tire conicity refers to a condition where the plies and/or belts are not level across the tire tread. When a tire has conicity, the plies and/or belts are somewhat cone shaped and may cause the vehicle to pull in one direction when driven.

Conicidad La conicidad de un neumático se refiere al estado en el que las capas y/o las fajas no están niveladas en la trocha del neumático. Cuando un neumático presenta conicidad, las capas y/o las fajas tienen de alguna manera forma cónica y pueden hacer que el vehículo tire hacia una dirección cuando se conduce.

Controller Area Network (CAN) A high-speed serial bus communication network. The CAN protocol has been standardized by the International Standards Organization (ISO) as ISO 11898 standard for high-speed and ISO 11519 for low-speed data transfer.

Red de área de controlador (CAN) Una red de comunicaciones de bus serial de alta velocidad. El protocolo CAN ha sido estandarizado por la Organización Internacional de Estandarización (ISO) como la norma ISO 11898 para la transferencia de datos a alta velocidad y la norma ISO 11519 para la transferencia de datos a baja velocidad.

Cord plies Surround both tire beads and extend around the inner surface of the tire to enable the tire to carry its load.

Pliegues de cuerda Rodean ambas cejas de los neumáticos y se extienden alrededor de la superficie interior del neumático para permitir que el nuemático soporta su carga.

Cross-axis ball joints Contain large insulating bushings in place of typical ball joints.

Articulaciones de rótula a través del eje Contienen pasa tapas grandes en lugar de las rótulas esféricas comunes.

Cycle A series of events that repeat themselves regularly.

Ciclo Una serie de eventos que se repitan regularmente.

Damper solenoid valve An electronically operated hydraulic valve that controls shock absorber or strut damping.

Válvula amortiguador del solenoide Una válvula hidráulica operada electronicamente que controla el amortiguador o el amortiguamiento del apoyadero.

Data bus network Circuit over which serial data are transmitted.

Red de datos de bus Circuito por el cual se transmiten datos seriales.

Diamond-frame condition A frame that is diamond-shaped from collision damage.

Condición forma de diamante del armazón Armazón que ha adquirido la forma de un diamante a causa del impacto recibido durante una colisión.

Digital motor electronics (DME) A computer that controls specific vehicle functions and operates in conjunction with the safety and gateway module on a vehicle with an active steering system.

Sistema electrónico digital del motor (DME, en inglés) Una computadora que controla funciones específicas del vehículo y opera en conjunción con el módulo de la puerta de enlace y de seguridad de un vehículo con un sistema de dirección activo.

Digital voltage signal A voltage signal that is either high or low.

Señal digital de tensión Señal de tensión que es alta o baja.

Directional stability The tendency of the vehicle steering to remain in the straight-ahead position when driven straight ahead on a smooth, level road surface.

Estabilidad direccional Tendencia de la dirección del vehículo a permanecer en línea recta al ser así conducido en un camino cuya superficie es lisa y nivelada.

Double-row ball bearing A bearing with two adjacent rows of circular steel balls.

Cojinete de bolas de doble hilera Cojinete que contiene dos hileras adyacentes de bolas circulares de acero.

Double wishbone rear suspension system A multilink suspension system in which the links may be shaped like a wishbone.

Suspensión trasera de doble horquilla Sistema de suspensión de empalme múltiple en el que a las bielas motrices se les puede dar forma de espoleta.

Dynamic balance The balance of a wheel in motion.

Equilibrio dinámico Equilibrio de una rueda en movi-miento.

Electric locking unit (ELU) Locks the worm drive and drive motor in an active steering system if a safety-related defect occurs.

Unidad de bloqueo eléctrico (ELU, en inglés) Bloquea la transmisión por tornillo sinfín y el motor de impulsión de un sistema de dirección activo si se produce un defecto relacionado con la seguridad.

Electronic brake and traction control module (EBTCM) A module that controls antilock brake, traction control, and vehicle stability control functions.

Módulo electrónico de control de freno y tracción (EBTCM) Un módulo que controla las funciones de control del freno antibloqueo, control de tracción, y la estabilidad del vehículo.

Electronic control unit (ECU) A computer that receives input signals and controls output functions.

Unidad de control electrónico (ECU, en inglés) Una computadora que recibe señales de entrada y controla las funciones de salida.

Electronic rotary height sensor Has an internal rotating element and this sensor sends voltage signals to the control module in relation to the curb riding height.

Sensores electrónicos rotativos de altura Tienen un elemento rotativo interno, y estos sensores envían señales de voltaje al módulo de control con relación a la curva de la altura de rodaje.

Electronic suspension control (ESC) system Controls damping forces in the front struts and rear shock absorbers in response to various road and driving conditions.

Sistema de control de suspensión electrónica (CSE) Controla las fuerzas de amortiguamiento en las barras transversales frontales y los amortiguadores neumáticos traseros como respuesta a las variadas condiciones de manejo y del camino.

Electronic variable orifice (EVO) steering A computer-controlled power steering system in which the computer operates a solenoid to control power steering pump pressure in relation to vehicle speed.

Dirección de orificio variable electrónico Sistema de dirección hidráulica controlado por computadora en el que la computadora activa un solenoide para controlar la presión de la bomba de la dirección hidráulica de acuerdo a la velocidad del vehículo.

Electronic vibration analyzer (EVA) A tester that measures vibration amplitude.

Analizador electrónico de vibraciones (EVA) Un comprobador que mide el amplitud de las vibraciones.

Electronically controlled orifice (ECO) valve A computer-controlled solenoid valve in the power steering pump that controls fluid flow from the pump to the servotronic valve.

Válvula de orificio controlado electrónicamente (ECO, en inglés) Una válvula solenoide controlada por computadora que se encuentra en la bomba de dirección de potencia que controla el flujo de fluidos desde la bomba hasta la válvula servotrónica.

Electronically controlled 4WS system A four-wheel steering system in which the rear wheel steering is controlled electronically.

Dirección en las cuatro ruedas controlada electrónicamente Sistema de dirección en las cuatro ruedas en el que la dirección de la rueda trasera se controla electrónicamente.

Encoded disc Works with an optical steering angle sensor to supply a voltage signal in relation to steering wheel rotation.

Disco codificado Funciona con un sensor de ángulo de la dirección óptico para proporcionar una señal de voltaje en relación con la rotación del volante.

End take-off (ETO) The tie rods are connected to the ends of the rack.

Conexión en los extremos (End take-off, ETO) Los brazos de dirección se conectan a los extremos de la cremallera.

Energy The ability to do work.

Energía Capacidad para realizar un trabajo.

Energy-absorbing lower bracket A steering column bracket that protects the driver by allowing the column to move away from the driver when impacted by the driver in a collision.

Soporte absorbente de energía Un soporte de la columna de dirección que proteja al conductor permitiendo que la columna se aleja del conductor cuando éste la golpea en una colisión.

Faster steering A steering gear that turns the front wheels from lock-to-lock with less steering wheel rotation.

Dirección más rápida Mecanismo de dirección que gira las ruedas delanteras de un extremo al otro con menos movimiento del volante de dirección.

Road feel A feeling experienced by a driver during a turn when the driver has a positive feeling that the front wheels are turning in the intended direction.

Sensación del camino Sensación experimentada por un conductor durante un viraje cuando está completamente seguro de que las ruedas delanteras están girando en la dirección correcta.

Firm relay A relay in a computer-controlled suspension system that supplies voltage to the strut actuators and moves these actuators to the firm position.

Relé de fijación Relé en un sistema de suspensión controlado por computadora que le provee tensión a los accionadores de los montantes y los lleva a una posición fija.

Five-link suspension A five-link suspension has five attachment locations between each side of the suspension and the frame.

Suspensión de cinco enlaces Una suspensión de cinco enlaces tiene cinco lugares de conexión entre cada lado de la suspensión y el chasis.

Flow control valve A special valve that controls fluid movement in relation to system demands.

Válvula de control de flujo Válvula especial que controla el movimiento del fluido de acuerdo a las exigencias del sistema.

Fluted lip seal A seal lip with fine grooves that direct the lubricant back into the reservoir rather than leaking past the seal lip.

Junta de estanqueidad de reborde estriado Reborde de una junta de estanqueidad con ranuras finas que dirigen el lubricante hacia el tanque en vez de permitir que el mismo se escape del reborde.

Force Implies the exertion of strength, which may be physical or mechanical.

Fuerza Implica el esfuerzo excesivo físico o mecánico.

Four-wheel active steering (4WAS) A steering system in which the rear wheels are steered electronically in relation to the front wheel steering angle and vehicle speed.

Dirección activa con tracción en las cuatro ruedas (4WAS, en inglés) Un sistema de dirección en el que las ruedas traseras se direccionan electrónicamente en relación con el ángulo de dirección de las ruedas delanteras y la velocidad del vehículo.

Four-wheel alignment Measurement and adjustment of wheel alignment angles at the front and rear wheels.

Alineación de cuatro ruedas La medida y el ajuste de los ángulos de alineación de las ruedas en las ruedas delanteras y traseras.

Four-wheel steering (4WS) A steering system in which both the front and rear wheels turn right or left to steer the vehicle.

Dirección en las cuatro ruedas Sistema de dirección en el que tanto las ruedas traseras como las delanteras giran hacia la derecha o la izquierda para dirigir el vehículo.

Frame buckle A frame that is accordion-shaped to some extent from collision damage.

Encorvamiento del armazón Armazón que ha adquirido la forma de un acordeón a causa del impacto recibido durante una colisión.

Frame sag A frame that is bent downward in the center from collision damage.

Hundimiento del armazón Armazón torcido hacia abajo en el centro a causa del impacto recibido durante una colisión.

Frame twist A frame that is bent from collision damage so one corner is higher in relation to the opposite corner.

Torcedura del armazón Armazón torcido a causa del impacto recibido durante una colisión; como resultado de esta torcedura un ángulo será más alto con relación al ángulo opuesto.

Friction The resistance to motion when the surface of one object is moved over the surface of another object.

Fricción Resistencia al movimiento cuando la superficie de un objeto se mueve sobre la superficie de otro.

Full-wire open-end springs Coil spring ends that are cut straight off and sometimes flattened, squared, or ground to a D-shape.

Muelles de extremo abierto con cable cerrado Extremos de muelles helicoidales cortados en línea que a veces son aplanados, cuadrados o afilados en forma de D.

Garter spring A circular spring behind a seal lip.

Muelle jarretera Muelle circular ubicado detrás de un reborde de una junta de estanqueidad.

Gas-filled shock absorber Contains a gas charge and hydraulic fluid.

Amortiguadore lleno de gas Contienens una carga de gas y fluido hidráulico.

Gear ratio The relationship between the drive gear and the driven gear in a gear set.

Relación de engranaje La relación entre el engranaje de impulso y el engranaje mandado en un juego de engranajes.

Gear tooth backlash Movement between gear teeth that are meshed together.

Juego entre los dientes del engranaje Movimiento entre los dientes del engranaje que están endentados entre sí.

Geometric centerline The exact centerline of the vehicle chassis.

Línea central geométrica La línea central exacta del chasis del vehículo.

Geometric centerline alignment A wheel alignment procedure in which the front wheel toe is adjusted to specifications using the geometric centerline as a reference.

Alineación de la línea central geométrica Procedimiento de alineación de las ruedas en el que se ajusta el tope de la rueda delantera según las especificaciones utilizando la línea central geométrica como punto de referencia.

Geometric vehicle centerline An imaginary line through the exact center of the front and rear wheels.

Linea de eje central geométrica Línea imaginaria a través del centro exacto de las ruedas delanteras y traseras.

Greenhouse gas A gas that helps to form a blanket above the earth's atmosphere.

Gas de efecto invernadero Un gas que ayuda a formar un manto por encima de la atmósfera de la tierra.

Gussets Pieces of metal welded into a corner where two pieces of metal meet to increase component strength.

Esquinero Las piezas de metal soldadas en una esquina en donde se juntan dos piezas de metal para aumenter la fuerza del componente.

Hall element An electronic device that produces a voltage signal when the magnetic field approaches or moves away from the element.

Elemento Hall Es un dispositivo electrónico que produce una señal de voltaje cuando el campo magnético se acerca o se aleja del elemento.

Heavy-duty coil spring Compared with conventional coil springs, heavy-duty springs have larger wire diameter and 3 percent to 5 percent greater load-carrying capability.

Muelle helicoidal para servicio pesado Comparado con los muelles helicoidales convencionales, los muelles para servicio pesado tienen cables de diámetro más grande y una capacidad de carga de un 3% a un 5% mayor.

Heavy-duty shock absorbers Compared with conventional shock absorbers, heavy-duty shock absorbers have improved seals, a single tube to reduce heat, and a rising rate valve for precise spring control.

Amortiguadores para servicio pesado Comparados con los amortiguadores convencionales, los amortiguadores para servicio pesado tienen juntas de estanqueidad mejoradas, un tubo único para reducir el calor, y una válvula de vástago ascendente para un control preciso del muelle.

Height sensor Sends a signal to the suspension computer in relation to chassis height.

Sensor de altura Envía una señal a la computadora de suspensión referente a la altura del chasis.

Helical gear teeth Are positioned at an angle in relation to the gear centerline.

Dientes de engranaje helicoidales Ubicados a un ángulo con relación a la línea central del engranaje.

Hertz (Hz) A measurement for the speed at which an electronic signal or vibration cycles from high to low.

Hercio Una medición de la velocidad a la cual una señal electrónica o una vibración pasa de alta a baja.

High-frequency transmitter An electronic device that sends high-frequency voltage signals to a receiver. Some wheel sensors on computer-controlled wheel aligners send high-frequency signals to a receiver in the wheel aligner.

Transmisor de alta frecuencia Un dispositivo electróncio que manda las señales de alta frecuencia a un receptor. Algunos sensores de ruedas en los alineadores de ruedas controlados por computadora mandan los señales de alta frecuencia a un receptor en el alineador de ruedas.

High-strength steels (HSS) Steels that have above-average strength compared with other body components. High-strength steels are used in some unitized body parts.

Aceros de alta resistencia (HSS) Los aceros que tienen una resistencia superior a lo normal comparado a los otros componentes de la carrocería. Los aceros de alta resistencia se usan en algunos partes de un monocasco.

Hold valve A valve that prevents fluid return to the master cylinder during traction control or vehicle stability control functions.

Válvula de retención Una válvula que previene que los fluidos regresan al cilíndro maestro duante las funciones de control de tracción o control de estabilidad del vehículo.

Horsepower A measurement for the amount of power delivered by an internal combustion engine or an electric motor.

Caballo de fuerza Una medida de la cantidad de potencia entregada por un motor de combustión interna o por un motor eléctrico.

Hybrid vehicles Have two power sources that can supply torque to the drive wheels. In most applications the two power sources are a gasoline engine and an electric motor(s).

Vehículos híbridos Tienen dos fuentes de potencia que pueden proporcionar par motor a las ruedas de arranque. En la mayoría de las aplicaciones las dos fuentes de potencia son los motores de gasolina y eléctricos.

Hydroboost power-assisted steering system Uses fluid pressure from the power steering pump to supply brake power assistance.

Sistema de dirección asistido hidráulicamente Utiliza la presión del fluido de la bomba de la dirección hidráulica para reforzar la potencia del freno.

Hydro-forming Is the process of using extreme fluid pressure to shape metal.

Hidroformación Es el proceso de usar presión extrema de líquidos para darle forma a un metal.

Hydroplaning Occurs when water on the pavement is allowed to remain between the pavement and the tire tread contact area. This action reduces friction between the tire tread and the road surface, and can contribute to a loss of steering control.

Hidrodeslizamiento Sucede cuando se deja agua entre el pavimento y la superficie de rodadura. Esta acción reduce la fricción entre la superficie de rodadura y la superficie del camino, y puede contribuir a una pérdida de control en la dirección.

Idler arm A short, pivoted steering arm bolted to the vehicle frame and connected to the steering center link.

Brazo auxiliar Brazo de dirección corto y articulado, empernado al armazón del vehículo y conectado a la biela motriz central de dirección.

Inactive coils Inactive coils located at the top and bottom ends of a coil spring introduce force into the spring when a wheel strikes a road irregularity.

Bobinas inactivas Las bobinas inactivas localizadas en las extremidades superiores e inferiores de un resorte helicoidal introducen la fuerza en el resorte cuando una rueda choque contra una irregularidad en el camino.

Included angle The sum of the camber and steering axis inclination (SAI) angles.

Ángulo incluído La suma de los ángulos de la inclinación y la inclinación del eje de dirección (SAI).

Independent rear suspension A rear suspension system in which one rear wheel moves upward or downward without affecting the opposite rear wheel.

Suspensión trasera independiente Sistema de suspensión trasera en el que una rueda trasera se mueve de manera ascendente o descendente sin afectar la rueda trasera opuesta.

Inertia The tendency of an object to remain at rest, or the tendency of an object to stay in motion.

Inercia La tendencia de un objeto a permanecer inmóvil, o la tendencia de un objeto a continuar en movimiento.

Inner race The race in the center of a ball or tapered roller bearing that supports the balls or rollers.

Anillo interior El anillo en el centro de una bola o de un cojinete de rodillos cónicos que apoya las bolas o los rodillos.

Integral fluid reservoir A reservoir that is part of the power steering pump.

Tanque de líquido integral Tanque que forma parte de la bomba de la dirección hidráulica.

Integral power-assisted steering system A power steering system in which the power-assisted components are integral with the steering gear.

Sistema de dirección de astencia hidráulica integral Sistema de dirección hidráulica en el que los componentes asistidos forman parte integral del mecanismo de dirección.

Integral reservoir A reservoir that is part of the power steering pump.

Tanque de la bomba Tanque que forma parte de la bomba de la dirección hidráulica.

Interference fit A precision fit between two components that provides a specific amount of friction between the components.

Ajuste a interferencia Ajuste de precisión entre dos componentes que provee una cantidad específica de fricción entre los mismos.

Jounce Jounce is the action of the spring deflecting or compressing, storing energy.

Rebote Rebote es la acción del resorte al desviar o comprimir la energía de almacenamiento.

Jounce travel Upward wheel and suspension movement.

Sacudida Movimiento ascendente de la rueda y de la suspensión.

Kickback A force supplied to the steering wheel when one of the front wheels strikes a road irregularity.

Contragolpe Fuerza que se le suminista al volante de dirección cuando una de las ruedas delanteras golpea una irregularidad en el camino.

King pin inclination (KPI) The angle of a line through the center of the king pin in relation to the true vertical centerline of the tire viewed from the front of the vehicle.

Inclinación de la clavija maestra Ángulo de una línea a través del centro de la clavija maestra con relación a la línea central vertical real del neumático vista desde la parte frontal del vehículo.

Ladder frame design A frame with crossmembers between the side rails. These crossmembers resemble steps on a ladder.

Diseño de armazón de tipo escalera Armazón con travesaños entre las vigas laterales. Dichos travesaños se parecen a los peldaños de una escalera.

Lane departure warning (LDW) system Warns the driver if the vehicle drifts out of the lane in which it is driven.

Sistema de advertencia de salida del carril (LDW, en inglés) Advierte al conductor si el vehículo se aparta del carril en el que se está desplazando.

Lateral accelerometer Sends a voltage signal to the vehicle stability control computer in relation to lateral rear chassis movement.

Acelómetro lateral Manda una señal de voltaje a la computador de control de estabilidad del vehículo en relación con el movimiento lateral del trasero del chasis.

Lift/dive input A computer-controlled suspension system input signal regarding front end lift during acceleration and front end dive while braking.

Señal de entrada de ascenso y descenso Señal de entrada del sistema de suspensión controlado por computadora referente al ascenso del extremo delantero durante la aceleración y al descenso del extremo delantero durante el frenado.

Light-emitting diode (LED) A diode that emits light when current flows through the diode.

Diodo emisor de luz (LED) Diodo que emite luz cuando una corriente fluye a través del mismo.

Linear-rate coil spring A coil spring with equal spacing between the coils, one basic shape, and constant wire diameter. These springs have a constant deflection rate regardless of load. If 200 pounds compress the spring 1 inch, 400 pounds compress the spring 2 inches.

Muelle helicoidal de capacidad lineal Muelle helicoidal con separación igual entre los espirales, una forma básica, y un diámetro de cable constante. Estos muelles tienen una capacidad de desviación constante sin importar la carga. Si 200 libras comprimen el muelle una pulgada, 400 libras comprimen el muelle 2 pulgadas.

Liner A synthetic gum rubber layer molded to the inner surface of a tire for sealing purposes.

Revestimiento Capa de caucho sintético moldeada en la superficie interior de un neumático para sellarlo herméticamente.

Lithium-based grease A special lubricant that is used on the rack bearing in manual rack and pinion steering gears.

Grasa con base de litio Lubricante especial utilizado en el cojinete de la cremallera en mecanismos de dirección de cremallera y piñón manuales.

Live axle rear suspension system A live axle rear suspension system is one in which the differential housing, wheel bearings, and brakes act as a unit.

Sistemas de suspensión trasera del eje motriz La suspensión trasera del eje motriz es aquella en la cual la caja del diferencial, los rodamientos y los frenos actúan como una unidad.

Load-carrying ball joint A ball joint that supports the weight of the chassis.

Junta esférica con capacidad de carga Junta esférica que apoya el peso del chasis.

Load-leveling shock absorbers Shock absorbers to which air pressure is supplied to increase their load-carrying capability.

Amortiguadores con nivelación de carga Amortiguadores a los que se suministra presión de aire para aumentar su capacidad de carga.

Load rating A rating that indicates the load-carrying capability of a tire.

Clasificación de carga Clasificación que indica la capacidad de carga de un neumático.

Lock-to-lock A complete turn of the front wheels from full right to full left, or vice versa.

Vuelta completa de las ruedas Viraje completo de las ruedas delanteras desde la extrema derecha hasta la extrema izquierda, o viceversa.

Lower vehicle command A command, or signal, sent from a height sensor to the suspension computer that indicates the suspension height must be lowered.

Orden para bajar el vehículo Orden o señal enviada desde un sensor de altura a la computadora del sistema de suspensión que indica que debe bajarse la altura de la suspensión.

MacPherson strut front suspension system A suspension system in which the strut is connected from the steering knuckle to an upper strut mount, and the strut replaces the shock absorber.

Sistema de suspensión delantera de montante MacPherson Sistema de suspensión en el que el montante se conecta del muñón de dirección a un montaje del montante superior; el montante reemplaza el amortiguador.

Magnasteer system A power steering system that varies the steering effort in relation to vehicle speed.

Sistema Magnasteer Un sistema de dirección de potencia que varía el esfuerzo de dirección en relación con la velocidad del vehículo.

MagneRide system A suspension system with computer-controlled shock absorbers containing magneto-rheological fluid.

Sistema MagneRide Un sistema de suspensión con amortiguadores controlados por computadora que contienen fluido magnetoreológico.

Magnesium alloy wheels Wheels manufactured from magnesium mixed with other metals.

Ruedas de aleación de magnesio Ruedas fabricadas con magnesio mezclado con otros metales.

Magneto-rheological fluid (MR fluid) A synthetic fluid that contains numerous small, suspended metal particles. It can be found in some electronically controlled shock absorbers and struts.

Líquido magnetorreológico Líquido sintético que contiene numerosas partículas metálicas suspendidas y pequeñas. Puede encontrarse en algunos amortiguadores y barras transversales electrónicamente controlados.

Mass The measurement of an object's inertia.

Masa La medida de la inercia de un objeto.

Memory steer Occurs when the steering does not return to the straight-ahead position after a turn, and the steering attempts to continue turning in the original turn direction.

Dirección de memoria Condición que ocurre cuando la dirección no regresa a la posición de línea recta después de un viraje, y la dirección intenta continuar girando en el sentido original.

Momentum An object gains momentum when a force overcomes static inertia and moves the object.

Impulso Un objeto cobra impulso cuando una fuerza supera la inercia estática y mueve el objeto.

Mono leaf spring A leaf spring with a single leaf.

Muelle de lámina singular Muelle de lámina con una sola hoja.

Motorist Assurance Program (MAP) A program that establishes uniform parts inspection guidelines to improve customer satisfaction with the automotive industry.

Programa de aseguranza del conductor (MAP) Una programa que establece los requerimientos uniformes de la inspección de partes para mejorar la satisfacción del cliente con la industria automotríz.

Mud and snow tires Tires with a special tread that provides improved traction when driving in mud or snow.

Neumáticos para condiciones de lodo y nieve Los neumáticos que tienen una huella especial que provee tracción mejorada para conducir en el lodo o la nieve.

Multilink front suspension A suspension system in which the top of the knuckle is supported by two links connected to the chassis.

Suspensión delantera de bielas múltiples Un sistema de suspensión en el cual la parte superior del muñon se soporta por las dos bielas conectadas al chasis.

Multilink independent rear suspension system It has upper and lower links, and a third link connects the upper link to the top of the knuckle through a bearing.

Sistema de enlaces múltiples de suspensión trasera independiente Posee un enlace superior e inferior, y un tercer enlace que conecta el enlace superior a la parte superior del nudillo a través de un cojinete.

Multiple-leaf spring A leaf spring with more than one leaf.

Muelle de láminas múltiples Muelle de lámina con más de una hoja.

Natural vibration frequencies The frequency at which an object tends to vibrate.

Frecuencias de vibración naturales La frecuencia a la cual tiende a vibrar un objeto.

Needle roller bearing A bearing containing small circular rollers.

Cojinete de rodillos con agujas Cojinete que contiene rodillos circulares pequeños.

Negative camber Occurs when the camber line through the center of the tire is tilted inward in relation to the true vertical centerline of the tire viewed from the front of the vehicle.

Combadura negativa Condición que ocurre cuando la combadura a través del centro de un neumático se inclina hacia adentro con relación a la línea central vertical real del neumático vista desde la parte frontal del vehículo.

Negative caster The angle of a line through the center of the tire that is tilted forward in relation to the true vertical centerline of the tire viewed from the side.

Avance negativo del pivote de la rueda Es el ángulo de una línea a través del centro de la llanta que está inclinado hacia enfrente en relación con la verdadera línea central vertical de la rueda vista por un lado.

Negative offset A rim with negative offset has the centerline outboard of the mounting face.

Desviación negativa La llanta con desviación negativa tiene la línea central fuera de la borda de la superficie de montaje.

Negative-phase steering Occurs when the rear wheels are steered in the opposite direction to the front wheels.

Dirección de fase negativa Se aplica al modo cuando las ruedas traseras se mueven en dirección opuesta a las ruedas frontales.

Negative scrub radius Is present when the steering axis inclination line meets the road surface outside the true vertical centerline of the tire at the road surface.

Radio matorral negativo Condición que ocurre cuando la inclinación del pivote de dirección entra en contacto con la superficie del camino fuera de la línea central vertical real del neumático en la superficie del camino.

Network Data-transmitting wires that interconnect various computers on a vehicle.

Red Cables que transmiten datos que interconectan diversas computadoras de un vehículo.

Neutral phase steering Occurs when the rear wheels are centered in the straight-ahead position on a four-wheel steering system.

Direccional de fase neutral Sucede cuando las ruedas traseras se centran en una posición hacia enfrente.

Nitrogen tire inflation Tires inflated with nitrogen rather than air.

Inflado de neumáticos con nitrógeno Neumáticos inflados con nitrógeno en lugar de aire.

Non-load-carrying ball joint A ball joint that maintains suspension component location, but does not support the chassis weight.

Junta esférica sin capacidad de carga Junta esférica que mantiene la ubicación del componente de la suspensión, pero no apoya el peso del chasis.

Occupational Safety and Health Administration (OSHA) Regulates working conditions in the United States.

Dirección para Seguridad y Salud Industrial Rige las condiciones de trabajo en los Estados Unidos.

Ohms (Ω) The ohm is a unit of measurement of electrical resistance.

Ohmios (Ω) Ohmio es una unidad de medida para la resistencia eléctrica.

Ohm's law Ohm's law defines the relationship between voltage, resistance, and current.

Ley de Ohmio La Ley de Ohmio define la relación entre la tensión (o voltaje), la resistencia y la corriente.

Outer race The outer part of a bearing that supports the balls or tapered rollers.

Anillo exterior La parte exterior de un cojinete que apoya las bolas o los cojinetes de rodillo cónico.

Outlet fitting venturi A narrowing of the passage through the outlet fitting in a power steering pump.

Conexiones de salida venturi Estrechamiento del pasaje a través de la conexión de salida en la bomba de la dirección hidráulica.

Parallel HEVs A hybrid electric vehicle in which the drive wheels may be driven by the internal combustion engine, electric motor, or both.

HEV paralelos Un vehículo híbrido eléctrico en el cual las ruedas motrices se pueden impulsar mediante el motor de combustión interno, el motor eléctrico o ambos.

Parallelogram steering linkage A steering linkage in which the tie rods are parallel to the lower control arms.

Cuadrilátero de la dirección en paralelograma Cuadrilátero de la dirección en el que las barras de acoplamiento son paralelas a los brazos de mando inferiores.

Perimeter-type frame A vehicle frame in which the side rails are bent outward so these rails are near the side of the body.

Armazón de tipo perímetro Armazón del vehículo en el que las vigas laterales están dobladas hacia afuera para que queden cerca del lado de la carrocería.

Photo diode A diode that provides a voltage signal when light shines on the diode. The light source may be a light-emitting diode (LED).

Fotodiodo Diodo que provee una señal de tensión cuando es iluminado por la luz. La fuente de luz puede ser un diodo emisor de luz (LED).

Pigtail spring ends An end of a coil spring that is wound to a smaller diameter.

Extremos de muelle enrollados en forma de espiral Extremo de un muelle helicoidal devanado a un diámetro más pequeño.

Pinion The drive gear in rack and pinion steering gear.

Piñón Engranaje de mando en el mecanismo de dirección de cremallera y piñón.

Pinion angle sensor Provides a voltage signal in relation to steering pinion position.

Sensor de ángulo del piñón Proporciona una señal de voltaje de acuerdo con la posición del piñón de dirección.

Pinion-drive EPS The electric motor is coupled to the steering gear pinion.

GPE(guiado de propulsión electrónica) por tracción del piñón El motor eléctrico acompaña al piñón de la dirección.

Pitman arm A short steel arm connected from the steering gear sector shaft to the steering center link.

Brazo pitman Brazo corto de acero conectado del árbol del mecanismo de dirección a la biela motriz central de dirección.

Pitman shaft sector A shaft in a recirculating ball steering gear that is connected to the pitman shaft, and the gear teeth on the sector shaft are meshed with the worm gear.

Sector de eje pitman Un eje de un engranaje de bola recirculatoria que se conecta al eje pitman, y los dientes del engranaje del eje del sector se endentan con el engranaje sinfín.

Positive camber Occurs when the camber line through the centerline of the tire is tilted outward in relation to the true vertical tire centerline viewed from the front.

Combadura positiva Condición que ocurre cuando la combadura a través de la línea central del neumático se inclina hacia afuera con relación a la línea central vertical real vista desde la parte frontal.Is the angle of a line through the center of the tire that is tilted rearward in relation to the true vertical centerline of the tire viewed from the side.

Positive caster Is the angle of a line through the center of the tire that is tilted rearward in relation to the true vertical centerline of the tire viewed from the side.

Avance positivo del pivote de la rueda Es el ángulo de una línea a través del centro de la llanta que está inclinada hacia atrás en relación con la verdadera línea central vertical de la llanta vista por el lado.

Positive-phase steering Is applied to the mode when the rear wheels are steered in the same direction as the front wheels.

Guiado de fase positiva Se aplica cuando las ruedas traseras se guían en la misma dirección de las delanteras.

Positive offset Occurs when the wheel centerline is located inboard from the vertical wheel-mounting surface.

Desfasaje positivo Se produce cuando la línea central de las ruedas se encuentra dentro de la superficie vertical de montaje de las ruedas.

Positive scrub radius Occurs when the SAI line through the center of the lower ball joint and upper strut mount meets the road surface inside the true vertical centerline of the tire at the road surface as viewed from the front.

Radio matorral positivo Condición que ocurre cuando la línea SAI a través del centro de la junta esférica inferior y el montaje del montante superior entra en contacto con la superficie del camino dentro de la línea central vertical real del neumático en la superficie del camino vista desde la parte frontal.

Pre-safe systems React during the few milliseconds before a collision occurs to increase driver and passenger safety.

Presistema de protección Reacción durante los pocos milisegundos antes de que ocurra un choque para aumentar la seguridad del conductor y del pasajero.

Pressure relief ball A spring-loaded ball that opens at a specific pressure and limits pressure in a hydraulic system.

Bola de alivio de presión Bola con cierre automático que se abre a una presión específica y limita la presión en un sistema hidráulico.

Programmed ride control (PRC) system A computer-controlled suspension system in which the computer operates an actuator in each strut to control strut firmness.

Sistema de control programado del viaje Sistema de suspensión controlado por computadora en el que la computadora opera un accionador en cada uno de los montantes para controlar la firmeza de los mismos.

Pulse width modulated (PWM) voltage signal Is a signal with a variable on time that may be used by a computer to control an output.

Señal modulada de la anchura entre impulsos Es una señal con una variable de tiempo que puede usar una computadora para controlar una potencia de salida.

Pulse width modulation (PWM) A method of computer control in which the computer cycles an output on and off with a variable on time.

Modulación de duración de impulsos Método de control de computadoras en el que la computadora produce un ciclo de rendimiento a intérvalos; lo que produce un trabajo efectivo variable en cada ciclo.

Puncture sealing tires A tire with a special compound on the inner tire surface that seals punctures when the puncturing object is removed from the tire.

Neumáticos autoselladores Neumático con un compuesto especial en la superficie de la parte interior que sella pinchazos cuando se le remueve el objecto punzante al neumático.

Pyrotechnic Device that contains an explosive and an ignition source.

Pirotecnia Dispositivo que contiene un explosivo y una fuente de ignición.

Quadrasteer® A type of 4WS system used on some SUVs.

Quadrasteer® Un tipo de suspensión de cuatro ruedas utilizado en algunos modelos de SUV.

Rack A horizontal shaft in a rack and pinion steering gear containing teeth that are meshed with the pinion teeth.

Cremallera Eje horizontal en un mecanismo de dirección de cremallera y piñón que contiene los dientes que se engranan con los del piñón.

Rack and pinion power steering system A type of power steering in which the steering gear contains a pinion gear that is meshed with teeth on the horizontal rack.

Sistema de dirección hidráulica de cremallera y piñón Tipo de dirección hidráulica en la que el mecanismo de dirección contiene un piñón que se engrana con los dientes de la cremallera horizontal.

Rack and pinion steering linkage A rack and pinion steering gear with the tie rods connected from the rack ends or rack center to the steering arms.

Cuadrilátero de la dirección de cremallera y piñón Mecanismo de dirección de cremallera y piñón con las barras de acoplamiento conectadas de los extremos de la cremallera o del centro de la misma a los brazos de dirección.

Rack-drive EPS The electric motor that provides steering assist is coupled to the rack in the steering gear.

Cremallera de dirección GPE(guiado de propulsión electrónica) El motor eléctrico que proporciona ayuda de dirección acompaña a la cremallera en el mecanismo de dirección.

Rack piston A piston mounted on the rack in a power rack and pinion steering gear. Hydraulic pressure is supplied to this piston for power steering assistance.

Pistón de la cremallera Pistón montado en la cremallera en un mecanismo de dirección hidráulica de cremallera y piñón. A este pistón se le suministra presión hidráulica para reforzar la dirección hidráulica.

Rack seals Are positioned between the rack and the housing in a rack and pinion steering gear.

Juntas de estanqueidad de la cremallera Ubicadas entre la cremallera y el alojamiento en un mecanismo de dirección de cremallera y piñón.

Radial bearing load A load applied in a vertical direction.

Carga de marcación radial Carga aplicada en dirección vertical.

Radial-ply tires A tire in which the carcass plies are positioned at 90° in relation to the center of the tire, and steel or fiberglass belts are mounted under the tread.

Neumáticos de cordón radial Neumático fabricado con las estrías de armazón colocadas a un ángulo de 90° con relación al centro del neumático, y correas de acero o de fibra de vidrio montadas debajo de la huella.

Raise vehicle command A command, or signal, sent from a height sensor to the computer in a computer-controlled suspension system indicating the chassis height must be raised.

Orden para levantar el vehículo Orden o señal enviada desde un sensor de altura a la computadora en un sistema de suspensión controlado por computadora que indica que debe levantarse la altura del chasis.

Reactive Having the capability to cause a chemical reaction with another substance.

Reactivo Que tiene la capacidad de producir una reacción química con otra sustancia.

Real-time damping (RTD) system A term used to designate the road-sensing suspension system in on-board diagnostics.

Sistema de amortiguamiento en tiempo real Término utilizado para designar el sistema de suspensión con equipo sensor en pruebas de diagnóstico realizadas en el vehículo mismo.

Rear active steering (RAS) Has recently been introduced as standard equipment on the 2006 Infinity M series luxury car. The RAS makes the steering and handling characteristics on this long-wheelbase car more nimble and agile compared with models without this technology.

Dirección activa trasera (DAT) Últimamente se presentó como un equipo estándar en la serie del automóvil de lujo de la serie Infinity M del 2006. La DAT hace que la dirección y las características de manipulación de este automóvil de paso largo sean más ligeras y ágiles comparadas con los modelos que no poseen esta tecnología.

Rear axle bearing A bearing that supports the rear axle in the housing on a rear-wheel-drive vehicle.

Cojinete del eje trasero Cojinete que apoya el eje trasero en el alojamiento en un vehículo de tracción trasera.

Rear axle offset A condition in which the complete rear axle assembly has turned so one rear wheel has moved forward, and the opposite rear wheel has moved rearward.

Desviación del eje trasero Condición que ocurre cuando todo el conjunto del eje trasero ha girado de manera que una de las ruedas traseras se ha movido hacia adelante y la opuesta se ha movido hacia atrás.

Rear axle sideset A condition in which the rear axle assembly has moved sideways from its original position.

Resbalamiento lateral del eje trasero Condición que ocurre cuando el conjunto del eje trasero se ha movido lateralmente desde su posición original.

Rear steering actuator An assembly that controls rear wheel steering in some four-wheel steering systems.

Accionador de la dirección trasera Conjunto que controla la dirección de la rueda trasera en algunos sistemas de dirección en las cuatro ruedas.

Rear suspension system The rear suspension system plays a very important part in ride quality and in the control of suspension and differential noise, vibration, and shock.

Sistema de suspensión trasera El sistema de suspensión trasera desempeña un papel muy importante en la calidad de la conducción y en el control de la suspensión y el ruido, la vibración y el impacto en el diferencial.

Rear wheel camber The tilt of a line through the rear tire and wheel centerline in relation to the true vertical centerline of the rear tire and wheel.

Comba de las ruedas traseras El ángulo de una línea que atraviesa el eje mediano del neumático trasero y de la rueda en relación al eje mediano verdadero del neumático trasero y la rueda.

Rear wheel toe The distance between the front edges of the rear wheels in relation to the distance between the rear edges of the rear wheels.

Tope de la rueda trasera Distancia entre los bordes frontales de las ruedas traseras con relación a la distancia entre los bordes traseros de las ruedas traseras.

Rebound Rebound is the action of the spring expanding and releasing stored energy.

Rebote Rebote es la acción del resorte al expandir y liberar la energía de almacenamiento.

Rebound travel The downward spring and wheel action resulting from rebound. This downward spring and wheel action is called rebound travel.

Viaje del rebote Rebote es la acción del resorte al expandir y liberar la energía de almacenamiento. Esta acción descendente del resorte y la rueda se conoce como viaje del rebote.

Receiver An electronic device in some computer-controlled wheel aligners that receives high-frequency voltage signals from the wheel sensors.

Receptor Un dispositivo electrónico en algunos alineadores controlados por computadoras que recibe los señales de alta frecuencia de los sensores de las ruedas.

Regenerative braking A system on a hybrid or electric vehicle that allows the drive wheels to turn the electric drive motor so it recharges the battery during vehicle deceleration.

Frenado regenerativo Un sistema en un vehículo híbrido o eléctrico que permite que las ruedas motrices giren el motor de impulsión eléctrica para que recargue la batería durante la desaceleración del vehículo.

Regular-duty coil spring A coil spring supplied to handle average loads to which the vehicle is subjected. This type of spring has smaller wire diameter compared with a heavy-duty spring.

Muelle helicoidal para servicio normal Muelle helicoidal provisto para sostener la carga normal a la que el vehículo está expuesto. Este tipo de muelle tiene un cable de un diámetro más pequeño comparado con un muelle para servicio pesado.

Release valve A valve that releases fluid pressure from a wheel caliper during antilock brake operation.

Válvula descargador Una válvula que descarga la presión de fluido del calíbre de la rueda durante la operación de frenado antibloqueante.

Remote fluid reservoir A reservoir mounted separately from the power steering pump.

Tanque de líquido remoto Un tanque montado por separado de la bomba de dirección hidráulica.

Remote reservoir A reservoir mounted separately from the power steering pump.

Depósito remoto Un depósito montado separadamente de la bomba de dirección hidráulica.

Replacement tires Tires purchased to replace the original tires that were supplied by the vehicle manufacturer.

Neumáticos de repuesto Neumáticos que se compran para reemplazar los neumáticos originales provistos por el fabricante del vehículo.

Resonance A reinforcement of sound in a vibrating body caused by sound waves from another body.

Resonancia Una fortificación del sonido en un cuerpo vibrante causada por las ondas sónicas de otro cuerpo.

Resource Conservation and Recovery Act (RCRA) States that hazardous material users are responsible for hazardous materials from the time they become a waste until the proper waste disposal is completed.

Ley de Conservación y Recuperación de Recursos Establece que los usuarios de materiales peligrosos se encarguen de estos materiales desde el momento en que se convierten en desperdicios hasta que se lleve a cabo la eliminación adecuada de los mismos.

Road crown The higher center of a road surface in relation to the edges of this surface.

Corona de camino Centro más alto de la superficie de un camino con relación a los bordes de la misma.

Road force variation Is a condition that occurs when tire sidewalls do not have equal stiffness around the complete area of the sidewalls.

Variación de la fuerza en carretera Es una condición que sucede cuando las paredes laterales de las llantas no tienen la rigidez equitativa alrededor del área completa de las paredes laterales.

Road variables Variables such as weight in the vehicle or road surface that affect wheel alignment while driving.

Variables del camino Condiciones variables, como por ejemplo el peso en el vehículo o la superficie del camino que afectan la alineación de la rueda durante un viaje.

Roller bearing A roller bearing contains precision machined rollers that have the same diameter at both ends. Designed primarily to carry radial loads, but they can withstand some thrust load.

Cilindro del cojinete El cilindro del cojinete contiene cilindros con torneado de precisión con el mismo diámetro en ambos extremos. Se diseñaron inicialmente para transportar cargas radiales, pero pueden soportar algunas cargas de empuje.

Rolling elements The balls or rollers and the separator in a bearing.

Elementos rodantes Las bolas o rodillos y el separador en un cojinete.

Ross cam and lever steering gear A type of steering gear used in the early 1900s that had a spiral groove in the lower end of the steering shaft meshed with a pin on the pitman shaft.

Mecanismo de dirección de leva y palanca Ross Tipo de mecanismo de dirección utilizado a principios de siglo que tenía una ranura espiral en el extremo inferior del eje de dirección engranado con un pasador en el árbol pitman.

Rotary valve A valve mounted with the spool valve in a power steering gear. The position of these two valves directs power steering fluid to the appropriate side of the rack or power piston.

Válvula rotativa Válvula montada con la válvula de carrete en un mecansimo de dirección hidráulica. La posición de estas dos válvulas dirige el fluido de la dirección hidráulica al lado apropiado de la cremallera o del pistón impulsor.

Run-flat tires Tires designed to operate safely without any air pressure for a specific distance.

Neumáticos de no presión Los neumáticos diseñados a operar sin peligro sín presión por distancias específicas.

Safety and gateway module (SGM) Operates the servotronic valve and the electronically controlled orifice valve in some power steering systems.

Módulo de puerta de enlace y seguridad (SGM, en inglés) Opera la válvula servotrónica y la válvula de orifi cio controlado electrónicamente en algunos sistemas de dirección de potencia.

Scrub radius The distance between the SAI line and the true vertical centerline of the tire at the road surface.

Radio matorral Distancia entre la línea SAI y la línea central vertical real del neumático en la superficie del camino.

Sector A shaft in a recirculating ball steering gear that is connected to the pitman shaft. The gear teeth on the sector shaft are meshed with the worm gear.

Sector Árbol en un mecanismo de dirección con bola recirculante que se conecta al árbol pitman. Los dientes del engranaje en el eje sector se endentan con el engranaje sinfín.

Semi-elliptical springs Semi-elliptical springs have individual leaves stacked with the shortest leaf at the bottom and the longest leaf at the top.

Resortes semielípticos Los resortes semielípticos poseen láminas individuales apiladas con la lámina más corta en el fondo y la lámina más larga en la parte superior.

Semi-independent rear suspension A rear suspension system in which one rear wheel has a limited amount of movement without affecting the opposite rear wheel.

Suspensión trasera semi-independiente Sistema de suspensión trasera en el que una de las ruedas traseras tiene una cantidad limitada de movimiento sin afectar la rueda trasera opuesta.

Separator A component that prevents contact between two other parts.

Separador Componente que evita que otras dos piezas entren en contacto.

Series HEVs A hybrid electric vehicle in which the internal combustion engine and the electric drive motor both supply torque to the drive wheels.

HEV de serie Un vehículo híbrido eléctrico en el cual tanto el motor de combustión interna como el motor de impulsión eléctrica proporcionan potencia a las ruedas motrices.

Serpentine belt A ribbed V-belt drive system in which all the belt-driven components are on the same vertical plane.

Correa serpentina Sistema de transmisión con correa nervada en V en el que todos los componentes accionados por una correa se encuentran sobre el mismo plano vertical.

Servotronic valve A computer-controlled valve in an active steering system that controls power steering pump pressure in relation to the vehicle speed.

Válvula servotrónica Una válvula controlada por computadora en un sistema de dirección activo que controla la presión de la bomba de dirección de potencia de acuerdo con la velocidad del vehículo.

Shock absorber A hydraulic mechanism connected between the chassis and the suspension to control spring action.

Amortiguador Un mecanismo hidráulico conectado entre el chasis y la suspensión para controlar la acción del muelle.

Shock absorber ratio The amount of extension control compared with the amount of compression control.

Relación del amortiguador Cantidad de control de extensión comparado con la cantidad de control de compresión.

Short-and-long arm front suspension systems Suspension systems in which the upper control arm is shorter than the lower control arm.

Sistemas de suspensión de brazos largos y cortos Un sistema suspensión en el cual el brazo de control superior es más corto que el brazo de control inferior.

Sideslip The tendency of the rear wheels to slip sideways because of centrifugal force while a vehicle is cornering at high speed.

Patinaje Tendencia de las ruedas traseras a deslizarse lateralmente a causa de la fuerza centrífuga mientras un vehículo hace un viraje a gran velocidad.

Side sway Occurs when one side of a vehicle frame is bent inward.

Desviación Ocurre cuando un lado del bastidor del vehículo esta torcido hacia adentro.

Sidewall The area between the tread and the bead of a tire.

Pared lateral Área entre la huella y la pestaña de un neumático.

Silencer A component designed to reduce noise.

Silenciador Componente diseñado para disminuir el ruido.

Single-row ball bearing A bearing with a single row of balls.

Cojinete de bola de una sola fila Cojinete que contiene una sola fila de bolas.

Slip angle The actual angle of the front wheels during a turn compared with the turning angle of the front wheels with the vehicle at rest.

Ángulo de patinaje Ángulo real de las ruedas delanteras durante un viraje comparado con el ángulo de giro de las ruedas delanteras cuando se ha detenido la marcha del vehículo.

Snapring A circular steel ring with some tension designed to snap into a groove and retain a component.

Anillo de resorte Anillo circular de acero con un poco de tensión diseñado para ajustarse dentro de una ranura y sujetar un componente en su posición.

Society of Automotive Engineers (SAE) A society of professional engineers that provides many member and industry services, such as the development of industry standards and the communication of engineering information through publications and conferences.

Sociedad de Ingenieros de Automóviles (SAE) Sociedad de ingenieros profesionales que les provee muchos servicios a sus miembros y a la industria, como por ejemplo, el desarrollo de normas para la industria, y la comunicación de información sobre ingeniería mediante publicaciones y conferencias.

Sodium-based grease A special lubricant with a sodium base that may be required on some steering components.

Grasa a base de sodio Lubricante especial con una base de sodio que podrían necesitar algunos componentes de la dirección.

Soft relay A relay in a computer-controlled suspension system that supplies voltage to the strut actuators in the soft mode.

Relé blando Un relé en un sistema de suspensión controlado por computadora que suministra el voltaje a los accionadores de los montantes en un modo blando.

Solid axle beam A solid axle beam is usually a transverse inverted U-section channel connected between the rear wheels in a semi-independent rear suspension system.

Viga del eje sólido Una viga del eje sólido suele ser un canal transversal de sección U invertida, conectado entre las ruedas traseras en un sistema de suspensión trasera semi-independiente.

Speed rating A tire rating that indicates the maximum safe vehicle speed that a tire will withstand.

Clasificación de la velocidad Clasificación de un neumático que indica la velocidad máxima que podrá resistir un neumático.

Speed-sensitive steering (SSS) system A computer-controlled steering system that varies the steering effort in relation to vehicle speed.

Dirección sensible a la velocidad Sistema de dirección controlado por computadora que varía el esfuerzo necesario de la dirección de acuerdo a la velocidad del vehículo.

Spherical bearing A bearing shaped like a sphere and used in some tilt steering wheels.

Cojinete esférico Cojinete en forma de esfera utilizado en algunos volantes de dirección inclinables.

Spiral cable A conductive ribbon that is mounted in a plastic container on top of the steering column and maintains electrical contact between the air bag inflator module and the air bag electrical system. A spiral cable may be called a clock spring electrical connector.

Cable espiral Cinta conductiva montada en un recipiente plástico sobre la columna de dirección que mantiene el contacto eléctrico entre la unidad infladora y el sistema eléctrico del Airbag. El cable espiral se conoce también como conector eléctrico de cuerda de reloj.

Spool valve Positioned with the rotary valve in a power steering gear. These two valves direct power steering fluid to the appropriate side of the rack or power piston to provide steering assistance.

Válvula de carrete Ubicada con la válvula rotativa en un mecanismo de dirección hidráulica. Estas dos válvulas dirigen el fluido de la dirección hidráulica al lado apropiado de la cremallera o del pistón impulsor para reforzar la dirección.

Spring insulator Positioned between the ends of a coil spring and the spring mounting surfaces to reduce the transfer of noise and vibration from the spring to the chassis.

Aisladore de muelle Ubicados entre los extremos de un muelle helicoidal y las superficies para el montaje de muelles con el propósito de reducir la transferencia de ruido y la vibración del muelle al chasis.

Springless seal A seal lip with no spring behind the lip.

Junta de estanqueidad sin muelle Reborde de una junta de estanqueidad que no tiene un muelle detrás del mismo.

Spring-loaded seal A seal lip with a garter spring behind the lip to increase lip tension.

Junta de estanqueidad con cierre automático Reborde de una junta de estanqueidad con un muelle jarretera detrás del reborde para aumentar la tensión del mismo.

Sprung weight Is the vehicle weight that is supported by the coil springs.

Peso suspendido Es el peso del vehículo que soportan los muelles en espiral cilíndrica.

Spur gear teeth Gear teeth that are parallel to the centerline of the gear.

Dientes de engranaje rectos Dientes del engranaje que son paralelos a la línea central del engranaje.

Stabilitrak® A computer-controlled system that provides vehicle stability control by reducing sideways swerving.

Stabilitrak® Un sistema controlado por computadora que provee el control de estabilidad del vehículo asi disminuyendo viraje lateral.

Stabilizer bar A round steel bar, connected between the front or rear lower control arms, that reduces body sway.

Barra estabilizadora Una barra de acero circular conectada entre los brazos de control inferiores delanteros o traseros, que reducen el balanceo del cuerpo.

Static balance Refers to the balance of a tire and wheel at rest.

Equilibrio estático Se refiere al equilirio de un neumático y una rueda cuando se ha detenido la marcha del vehículo.

Steering angle The angle of the front wheel on the inside of a turn compared to the angle of the front wheel on the outside of the turn.

Ángulo de la dirección Ángulo de la rueda delantera en el interior de un viraje comparado con el ángulo de la rueda delantera en el exterior del viraje.

Steering angle sensor An optical-type sensor that supplies a voltage signal in relation to steering wheel rotation.

Sensor del ángulo de la dirección Un sensor de tipo óptico que proporciona una señal de voltaje de acuerdo con la rotación del volante

Steering axis inclination (SAI) The angle of a line through the center of the upper strut mount and lower ball joint in relation to the true vertical centerline of the tire viewed from the front of the vehicle.

Inclinación del pivote de dirección Ángulo de una línea a través del centro del montaje del montante superior y la junta esférica inferior con relación a la línea central vertical real del neumático vista desde la parte frontal del vehículo.

Steer-by-wire A computer-controlled steering system with no direct mechanical connection between the steering wheel and the steering gear.

Dirección alámbrica Un sistema de dirección controlado por computadora sín una conexión mecánica directa entre el volante de dirección y el aparato de dirección.

Steering drift The tendency of the steering to drift slowly to one side when the vehicle is driven straight ahead on a smooth, straight road surface.

Desviación de la dirección La tendencia de la dirección a desviarse poco a poco hacia un lado mientras que el vehículo se conduce en línea recta sobre un camino liso y nivelado.

Steering gear free play The amount of steering wheel movement before the front wheels begin to turn.

Juego en el aparato de dirección Cantidad de movimiento del volante de dirección antes de que las ruedas delanteras comiencen a girar.

Steering kickback Road force supplied back to the steering wheel.

Tensión de retroceso de la dirección Fuerza del camino que se vuelve a suministrar al volante. Sensor de posició.

Steering pull The tendency of the steering to pull constantly to one side or the other when driving the vehicle straight ahead on a smooth, level road surface.

Tiro en la dirección La tendencia de la dirección a tirar en una manera constante hacia un lado u otro mientras que el vehículo se conduce en línea recta sobre un camino liso y nivelado.

Steering ratio The number of degrees of steering wheel rotation in relation to the number of degrees of front wheel movement.

Relación de la dirección Número de grados de rotación del volante de dirección con relación al número de grados de movimiento de la rueda delantera.

Steering sensor A sensor that sends a voltage signal to the computer in relation to the amount and speed of steering wheel rotation in a programmed ride control (PRC) system.

Sensor de dirección Un sensor que manda una señal de voltaje a la computadora en relación a la cantidad y la velocidad de la rotación del volante de dirección en un sistema de viaje controlado (PRC).

Steering wander The tendency of the steering to pull to the right or left when the vehicle is driven straight ahead on a smooth road surface.

Desviación de la dirección Tendencia de la dirección a desviarse hacia la derecha o hacia la izquierda cuando se conduce el vehículo en línea recta en un camino cuya superficie es lisa.

Steering wheel angle sensor (SWAS) Provides a voltage signal in relation to steering wheel rotation.

Sensor del ángulo del volante (SWAS, en inglés) Proporciona una señal de voltaje de acuerdo con la rotación del volante.

Steering wheel position sensor (SWPS) A sensor that produces a voltage signal in relation to the amount and velocity of steering wheel rotation.

Sensor de posición de la dirección (SWPS) Un sensor que produce una señal de voltaje en relación a la cantidad y la velocidad de rotación del volante de dirección.

Steering wheel rotation sensor An input sensor in a computer-controlled suspension or steering system that sends a signal to the computer in relation to the amount and speed of steering wheel rotation.

Sensor de la rotación del volante de dirección Sensor de entrada en un sistema de suspensión controlado por computadora o en un sistema de dirección que le envía una señal a la computadora referente a la cantidad y a la velocidad de la rotación del volante de dirección.

Stop-to-stop Steering wheel rotation from extreme left to extreme right.

Parada a parada Rotación del volante de dirección desde la extrema izquierda hasta la extrema derecha.

Struts Components connected from the top of the steering knuckle to the upper strut mount that maintain the knuckle position and act as shock absorbers to control spring action.

Montantes Componentes conectados de la parte superior del muñón de dirección al montaje del montante superior que mantienen la posición del muñón y actúan como amortiguadores para controlar el movimiento de ascenso y descenso del muelle.

Summation sensor Provides a voltage signal in relation to steering rack position.

Sensor de suma Proporciona una señal de voltaje de acuerdo con la posición del soporte de la dirección.

Symmetrical leaf spring A symmetrical leaf spring has the same distance from the center bolt to the front and rear of the spring.

Hoja de muelle simétrica La hoja de muelle simétrica posee la misma distancia desde el centro del perno hasta la parte frontal y trasera del resorte.

Synthetic rubber A type of rubber developed in a laboratory.

Caucho sintético Tipo de caucho fabricado en un laboratorio.

Tapered roller bearing A bearing containing tapered roller bearings mounted between the inner and outer races.

Cojinete de rodillos cónicos Cojinete que contiene cojinetes de rodillos cónicos montados entre los anillos interiores y exteriores.

Taper-wire closed-end springs Coil spring ends that are tapered to a smaller diameter.

Muelles helicoidales de extremos cónicos Extremos de muelles helicoidales a los que se les da forma de cono para reducir su diámetro.

Telematics Using an in-vehicle phone for communication with the vehicle driver regarding vehicle diagnostics, emissions, or traffic- and safety-related concerns.

Telemática Uso de un teléfono incorporado al vehículo para comunicarse con el conductor del vehículo acerca de problemas de diagnóstico, emisiones, tránsito y seguridad.

Temperature rating A tire rating that indicates the ability of the tire to withstand heat.

Clasificación de la temperatura Clasificación de un neumático que indica la capacidad del neumático de resistir el calor.

Tension loaded A ball joint mounted so the vehicle weight tends to pull the ball out of the joint.

Cargada de tensión Junta esférica montada de manera que el peso del vehículo tiende a remover la bola de la junta.

Throttle position sensor (TPS) Sends an analog voltage signal to the engine computer in relation to throttle opening.

Sensor de posición del acelerador (TPS, en inglés) Envía una señal de voltaje analógica a la computadora del motor en relación con la apertura del acelerador.

Thrust angle The angle between the thrust line and the vertical centerline of the vehicle.

Ángulo de empuje El ángulo entre la línea de empuje y el eje mediano vertical del vehículo.

Thrust bearing load This type of load may be called an axial load.

Carga del cojinete de empuje Este tipo de carga puede llamarse una carga.

Thrust line An imaginary line positioned at a 90° angle to the center of the rear axle and extending forward.

Directriz de presiones Una línea imaginaria en posición de un ángulo de 90° con relación al centro del eje trasero y que se extiende hacia enfrente.

Thrust line alignment A wheel alignment in which the thrust line is used as a reference for front wheel toe adjustment.

Alineación de la línea de empuje Alineación de la rueda en la que se utiliza la línea de empuje como punto de referencia para el ajuste del tope de la rueda delantera.

Tie rod A rod connected from the steering arm to the rack or center link, depending on the type of steering linkage.

Barra de acoplamiento Varilla conectada del brazo de dirección a la cremallera o a la biela motriz central, dependiendo del tipo de cuadrilátero de la dirección.

Tire belts Belts that are placed under the tire tread to provide longer tread wear. Belts are usually made from steel or polyester.

Banda de refuerzo del neumático Las bandas que se colocan debajo de la banda de rodamiento para aumentar la durabilidad de la banda de rodamiento. Las bandas suelen ser fabricadas del acero o del poliester.

Tire chains May be placed over the tires to improve traction when driving on ice or snow.

Cadenas antideslizantes Pueden colocarse sobre los neumáticos para mejorar la tracción cuando se conduce el vehículo en el hielo o la nieve.

Tire contact area The part of the tire in contact with the road surface when the tire is supporting the vehicle weight.

Área de contacto del neumático Parte del neumático en contacto con la superficie del camino cuando el neumático apoya el peso del vehículo.

Tire free diameter The distance between the outer edges of the tread measured on a horizontal line through the center of the wheel.

Diámetro libre del neumático Distancia entre los bordes exteriores de la huella que se mide sobre una línea horizontal a través del centro de la rueda.

Tire performance criteria (TPC) Information molded on the tire sidewall indicating that the tire meets the manufacturer's performance standards for traction, endurance, dimensions, noise, handling, and rolling resistance.

Criterio sobre el rendimiento del neumático Información moldeada en la pared lateral del neumático que indica que el neumático cumple con las normas de rendimiento establecidas por el fabricante sobre la tracción, acción, dimensiones, ruido, movilización, y resistencia al rodaje.

Tire placard Often attached to the rear face of the driver's door, the tire placard provides information regarding maximum vehicle load, tire size, and cold inflation pressure.

Cartel del neumático Comúnmente fijado a la cara posterior de la puerta del conductor, el cartel del neumático provee información referente a la carga máxima del vehículo, el tamaño del neumático, y la presión fría de inflación.

Tire pressure The amount of air pressure contained in the tire to allow the tire to carry a load.

Presión del neumático La cantidad del presión de aire contenido en el neumático para permitir que el neumático soporta una carga.

Tire rolling diameter The distance between the outer edges of the tread measured on a vertical line through the center of the wheel when the tire is supporting the vehicle weight.

Diámetro del rodaje del neumático Distancia entre los bordes exteriores de la huella que se mide sobre una línea vertical a través del centro de la rueda cuando el neumático apoya el peso del vehículo.

Tire treads The part of the tire in contact with road surface.

Bandas de rodamiento del neumático La parte del neumático que se pone en contacto con la superficie del camino.

Tire valves Mounted in the wheel rim, the tire valves allow air to enter or be exhausted from the tire.

Válvulas del neumático Estas válvulas están montadas en la llanta de la rueda y permiten la entrada o la salida de aire desde el neumático.

Toe-in A condition that is present when the distance between the front edges of the front or rear wheels is less than the distance between the rear edges of the wheels.

Convergencia Condición que ocurre cuando la distancia entre los bordes frontales de las ruedas delanteras o traseras es menor que la distancia entre los bordes traseros de las mismas.

Toe-out A condition that is present when the distance between the front edges of the front or rear wheels is more than the distance between the rear edges of the wheels.

Divergencia Condición que ocurre cuando la distancia entre los bordes delanteros de las ruedas delanteras o traseras es mayor que la distancia entre los bordes traseros de las mismas.

Toe-out on turns The steering angle of the wheel on the inside of a turn compared with the steering angle of the wheel on the outside of the turn.

Divergencia durante un viraje El ángulo de la dirección de la rueda en el interior de un viraje comparado con el ángulo de la dirección de la rueda en el exterior del viraje.

Toe plate A metal plate surrounding the steering column and attached to the vehicle floor.

Placa de pie Una placa de metal que rodea la columna de dirección y conectada al piso del vehículo.

Torque A twisting force.

Par de torsión Fuerza de torcimiento.

Torque steer The tendency of the steering on a front-wheel-drive vehicle with unequal drive axles to pull to one side during hard acceleration.

Dirección de torsión Tendencia de la dirección en un vehículo de tracción delantera con ejes de mando desiguales a desviarse hacia un lado durante una aceleración rápida.

Torsion bar A steel bar connected from the chassis to the lower control arm. As the vehicle weight pushes the chassis downward, the torsion bar twists to support this weight. Torsion bars are used in place of coil springs.

Barra de torsión Barra de acero conectada del chasis al brazo de mando inferior. Mientras el peso del vehículo presiona el chasis hacia abajo, la barra de torsión se tuerce para apoyar este peso. Las barras de torsión se utilizan en lugar de los muelles helicoidales.

Track bar Some rear suspension systems have a track bar connected from one side of the differential housing to the chassis to prevent lateral chassis movement.

Barra transversal Algunos sistemas de suspensión trasera poseen una barra transversal conectada de un lado de la caja del diferencial al chasis, para evitar el movimiento lateral del chasis.

Tracking Refers to the position of the rear wheels in relation to the front wheels.

Encarrilamiento Se refiere a la posición de las ruedas traseras con relación a las ruedas delanteras.

Traction control system (TCS) A computer-controlled system that prevents drive wheel spinning.

Sistema de control de tracción (TCS) Un sistema controlado por computador que previene que giran las ruedas de propulsión.

Traction rating A tire rating indicating the traction capabilities of the tire.

Clasificación de tracción Clasificación de un neumático que indica las capacidades de tracción del mismo.

Transistors Fast-acting electronic switches with no moving parts.

Transistores Interuptores electrónicas de acción rápida sín partes móviles.

Transitional coils Coils located between the inactive and active coils in variable-rate coil springs.

Bobinas de transición Las bobinas localizadas entre las bobinas inactivas y activas en los resortes helicoidales de relación variable.

Travel-sensitive strut A strut with the capability to adjust its firmness in relation to the amount of piston travel inside the strut.

Montante sensible al movimiento Montante con la capacidad de ajustar su firmeza de acuerdo a la cantidad de movimiento del pistón dentro del montante.

Tread wear rating A tire rating indicating the wear capabilities of the tread that allow customers to compare tire life expectancy.

Clasificación del desgaste de un neumático Clasificación de un neumático que indica las capacidades de desgaste de la huella y le permite a los clientes comparar el índice de durabilidad del neumático.

Trim height The specified, or normal, suspension height in a computer-controlled suspension system.

Altura de la suspensión Altura especificada, o normal, de la suspensión en un sistema de suspensión controlado por computadora.

Tubular frame A frame member with a circular, or tubular, design.

Fabricación del armazón en forma tubular Pieza del armazón diseñada en forma circular o tubular.

Turning circle The turning angle of one front wheel in relation to the opposite front wheel during a turn.

Círculo de giro Ángulo de giro de una de las ruedas delanteras con relación a la rueda delantera opuesta durante un viraje.

Turning radius The turning angles of the front wheels around a common center point.

Diámetro del giro Los ángulos de giro de las ruedas frontales alrededor de un punto central común.

Uniform Tire Quality Grading (UTQG) System Information including tread wear, traction, and temperature ratings that is molded into the tire sidewall, and is required by the Department of Transportation (DOT).

Sistema de sobre la clasificación de la calidad uniforme de neumáticos Información moldeada en la pared lateral del neumático que incluye el desgaste de la huella, la tracción, y las clasificaciones de temperatura y que es requerida por el Ministerio de Transporte.

Unitized body A body design that does not have a frame because each body component is a supporting member.

Carrocería unificada Diseño de la carrocería que no tiene armazón porque cada uno de sus componentes es una pieza suplementaria.

Unsprung weight The vehicle weight that is not supported by the coil springs.

Peso no suspendido Es el peso del vehículo que no soportan los muelles en espiral cilíndrica.

Upper strut mount A mount connected between the strut and the strut tower. Front upper strut mounts contain a bearing that allows strut rotation.

Montaje de los montantes superiores Montaje conectado entre el montante y la torre del montante. Los montajes frontales y superiores contienen un cojinete que permite la rotación del montante.

Vacuum A pressure that is less than the atmospheric pressure.

Vacío Presión menor que la de la atmósfera.

Vane-type power steering pumps Pumps that have rotors with metal vanes that provide a seal between the rotor and the pump cam.

Bombas de dirección hidráulica tipo aletas Las bombas que tienen rotores con aletas de metal que proveen un sello entre el rotor y la leva de la bomba.

Variable effort steering (VES) A computer-controlled steering system that provides reduced steering effort at low vehicle speeds and increased steering effort at higher vehicle speeds.

Dirección de esfuerzo variable Sistema de dirección controlado por computadora que provee un menor esfuerzo de dirección cuando el vehículo viaja a velocidad baja y un mayor esfuerzo de dirección cuando el vehículo viaja a gran velocidad.

Variable-rate coil springs Rather than having a standard spring deflection rate, these springs have an average spring rate based on load at a predetermined deflection.

Muelles helicodal de capacidad variable En vez de tener una capacidad de desviación de muelle estándar, estos muelles tienen un valor promedio de elasticidad basado en la carga a una desviación predeterminada.

V-belt A drive belt with a V-shape.

Correa en V Correa de transmisión en forma de V.

Vehicle dynamic suspension (VDS) System found in some SUVs that is similar to air suspension systems.

Suspensión dinámica del vehículo (SDV) Sistema que se encuentra en algunas camionetas SUV que es similar a los sistemas de suspensión de aire.

Vehicle speed sensor (VSS) An input sensor that sends a voltage signal to the engine computer in relation to vehicle speed.

Sensor de la velocidad del vehículo Sensor de entrada que le envía una señal de tensión a la computadora del motor referente a la velocidad del vehículo.

Vehicle stability control system A computer-controlled system that prevents vehicle swerving, especially during hard acceleration on slippery road surfaces.

Sistema de control de estabilidad del vehículo Un sistema controlado por computadora que previene que se desvía el vehículo, especialmente durante una aceleración fuerte sobre un camino cuyo superficie es resbalosa.

Vehicle wander The tendency of the steering to pull to the left or the right when the vehicle is driven straight ahead on a smooth road surface.

Desviación de la marcha del vehículo Tendencia de la dirección a desviarse hacia la izquierda o hacia la derecha cuando se conduce el vehículo en línea recta en un camino cuya superficie es lisa.

Vent valve An electrically operated valve that vents air from an air spring in a computer-controlled suspension system.

Válvula de respiración Válvula activada eléctricamente que da salida al aire desde un muelle de aire en un sistema de suspensión controlado por computadora.

Venturi A narrow portion of an air passage.

Venturi Una porción estrecha de un pasaje de aire.

Vibration A rapid motion of particles, or a component that produces sound.

Vibración Un movimiento rápido de los partículos, o de un componente que produce un sonido.

Voltage (V) Voltage is a unit of measurement of electrical pressure.

Voltaje (V) El voltaje es una unidad de medida de la presión eléctrica.

Volume Volume is the length, width, and height of the space occupied by an object.

Volumen El volumen es la longitud, la anchura y la altura del espacio ocupado por un objeto.

Watts rod A rod connected from the chassis to the rear suspension to reduce body side sway, usually referred to as a tracking bar.

Barra wats Una barra conectada del chasis a la suspensión trasera para disminuir la oscilación lateral de la carrocería, suele referirse como una barra de tracción.

Wear indicators Rubber bars located near the bottom of tire treads. When the tread is worn to a specific depth, these bars become visible.

Indicadores del desgaste Las barras del caucho localizadas cerca la parte inferior de las bandas de rodamiento de los neumáticos. Cuando las bandas de rodamiento se han gastado a una profundidad específica, estas barras son visibles.

Weight The measurement of the earth's gravitational pull on an object.

Peso La medida de la atracción gravitacional de la tierra en un objeto.

Wheel alignment An adjustment and refitting of suspension parts to original specifications that ensures design performance.

Alineación de una rueda Puede definirse como un ajuste y una reparación de las piezas de la suspensión según las especificaciones originales, lo que asegura el rendimiento del diseño.

Wheelbase The distance between the center of the front and rear wheels.

Distancia entre ejes Distancia entre el centro de las ruedas delanteras y traseras.

Wheel offset The distance between the vertical wheel-mounting surface and the centerline of the wheel.

Desfasaje de rueda La distancia entre la superficie vertical de montaje de ruedas y la línea central de la rueda.

Wheel position sensor Is connected between each front and rear lower control arm and the chassis in a road-sensing suspension system. This sensor provides a computer input signal in relation to the amount and velocity of wheel movement.

Sensor para la posición de las ruedas Se conecta entre cada uno de los brazos de mando delantero y trasero y el chasis en un sistema de suspensión con equipo sensor. Este sensor envía una señal de entrada a la computadora referente a la cantidad y la velocidad del movimiento de la rueda.

Wheel rims Circular steel, aluminum, or magnesium components on which the tires are mounted. The wheel rim and tire assembly is bolted to the wheel hub.

Llantas de la rueda Componentes circulares de acero, aluminio o magnesio sobre los que se montan los neumáticos. El conjunto de la llanta de la rueda y el neumático se emperna al cubo de la rueda.

Wheel sensor An electronic unit attached to each wheel and connected to a four-wheel aligner.

Sensor de la rueda Una unedad electrónica prendida a cada rueda y conectada a un alineador de cuatro ruedas.

Wheel setback Occurs when one front or rear wheel is moved rearward in relation to the opposite front or rear wheel.

Retroceso de la rueda Ocurre cuando una rueda delantera o trasera se mueve hacia atrás con relación a la rueda delantera o trasera opuesta.

Wheel shimmy Rapid inward and outward oscillations of the front wheels.

Bailoteo de la rueda Oscilaciones rápidas hacia adentro y hacia afuera de las ruedas delanteras.

Wheel speed sensor Sends an AC voltage signal to the antilock brake system computer in relation to wheel speed.

Sensor de velocidad de la rueda Manda una señal de voltaje de corriente alterna a la computadora del sistema de freno antibloqueo en relación a la velocidad de la rueda.

Wheel tramp Rapid upward and downward wheel and tire movement.

Recorrido de la rueda Movimiento rápido ascendente y descendente de la rueda y el neumático.

Work The result of applying a force.

Trabajo Resultado de la aplicación de una fuerza.

Worm and gear or worm and sector design A steering gear developed in the early 1900s that required high steering effort because of internal friction.

Diseño del sinfín y engranaje o del sinfín y sector Un engranaje de dirección desarrollado en los principios de los años 1900 que requerían un gran esfuerzo en maniobro de dirección debido a la fricción interna.

Worm and roller steering gear A gear that contains a wormshaft and a roller-type sector.

Engranaje de dirección con sinfín y rodillo Un engranaje que contiene un eje de sinfín y un sector tipo rodillo.

Worm shaft The gear meshed with the pitman shaft sector in a recirculating ball steering gear.

Árbol del sinfín Engranaje endentado con el sector del árbol pitman en un mecanismo de dirección de bola recirculante.

Worm shaft bearing preload The amount of tension placed on the bearing by the adjustment procedure.

Carga previa del cojinete del árbol del sinfin Cantidad de tensión aplicada al cojinete por el procedimiento de ajuste.

Worm shaft end play The distance between the fully upward and fully downward worm shaft positions in a recirculating ball steering gear.

Holgadura del árbol del sinfín Distancia entre las posiciones completamente ascendente y descendente del árbol del sinfín en un mecanismo de dirección de bola recirculante.

Yaw forces Tend to cause the rear of the vehicle to swerve sideways.

Fuerzas de dirección tienden a causar que la parte posterior del vehículo viren bruscamente hacia los lados.

Yaw motion The tendency of the rear of a vehicle to swerve sideways during a turn.

Movimiento de derrape La tendencia de la parte posterior de un vehículo de desplazarse lateralmente durante un giro.

Yaw rate sensor Sends a voltage signal to the vehicle stability control computer in relation to sideways chassis movement.

Sensor de cantidad de desviación Manda una señal de voltaje a la computadora de control de estabilidad del vehículo en relación del movimiento lateral del chasis.

INDEX